THE MODERN KITCHEN COMBINES BEAUTY AND UTILITY

Courtesy of Crane Company, Chicago

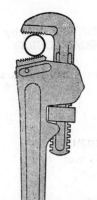

HOW TO DESIGN AND INSTALL
Plumbing

FOURTH EDITION

BY
A. J. MATTHIAS, JR.

Supervisor of Training, Nash-Kelvinator Corporation, Kenosha, Wisconsin. Formerly Plumbing Instructor, Wisconsin Vocational Schools. Licensed Journeyman Plumber.

REVISED BY
ESLES SMITH, SR.

Instructor, Palomar College, Vista, California. Former Instructor, Plumbing Departments, San Diego, California, Vocational High School and Junior College. Member, American Society of Sanitary Engineering, and Western Plumbing Officials Assoc.

CONTRIBUTOR TO FOURTH EDITION
ROBERT J. VOLLAND

Supervising Director, Trade and Industrial Education, Public Schools of the District of Columbia. Former Principal, Alexander Graham Bell Vocational Evening High School, Washington, D. C. Licensed Master Plumber, District of Columbia.

AMERICAN TECHNICAL SOCIETY·CHICAGO·U·S·A

FIRST EDITION
1st Printing 1940

SECOND EDITION (REVISED)
2d Printing 1941
3d Printing 1942
4th Printing 1943
5th Printing 1944
6th Printing 1945
7th Printing 1946
8th Printing 1947
9th Printing 1948
10th Printing 1950

THIRD EDITION (REVISED)
11th Printing 1952
12th Printing 1953
13th Printing 1953
14th Printing 1956
15th Printing 1958
16th Printing 1959
17th Printing 1959

FOURTH EDITION (REVISED)
16th Printing 1959
17th Printing 1959
18th Printing 1960
19th Printing 1961
20th Printing 1962
21st Printing 1963
22nd Printing 1964
23rd Printing 1965
24th Printing 1966
25th Printing 1966
26th Printing 1967
27th Printing 1968
28th Printing 1969
29th Printing 1970
30th Printing 1971
31st Printing 1972
32nd Printing 1973
33rd Printing 1973
34th Printing 1975

35th Printing 1977

Library of Congress Catalog Number 60–1968
ISBN 0–8269–0595–1

PREFACE

Since its publication in 1940 *How to Design and Install Plumbing* has undergone two major revisions. The need for these revisions was brought about by the advances in the design of plumbing installations, more efficient work practices, and new materials. Further advances in these areas have now made necessary the fourth edition of this popular book.

In addition to achieving these goals, the fourth edition of *How to Design and Install Plumbing* explains and illustrates standard practices so that students, journeymen, architects, and other tradesmen in building and construction can have a source of information for study and reference or a means of familiarizing themselves with plumbing problems.

This book also satisfies the demands of teachers, students, and journeymen for additional material showing how plumbing principles are employed in designing and installing typical plumbing systems in modern buildings. In order to achieve this, the design and installation of a complete plumbing system for a modern six-room, two-bath residence is explained and illustrated. A complete set of blueprints is included with this material.

In addition to the acknowledgments given in the book, appreciation is expressed for the assistance and cooperation given by the following companies:

Crane Co., Chicago

Kohler Co., Kohler, Wisc.

Ruud Manufacturing Co., Pittsburgh, Pa.

The Humphryes Manufacturing Co., Mansfield, Ohio

The Duro Company, Dayton, Ohio

Chicago Pump Co., Chicago

Sloan Valve Co., Chicago

General Electric Co., Bridgeport, Conn.

The Strauss Electric Appliance Co., Waukesha, Wisc.

Appreciation is also due Prof. F. M. Dawson, College of Engineering, The State University of Iowa; The University of Wisconsin; the Wisconsin State Board of Health; The Metropolitan Sewage Commission of Milwaukee; and Mr. O. Matthias.

CONTENTS

* NOTE—A complete set of blueprints is bound at the rear of the book, following page 374.

INTRODUCTION

The history of plumbing is an interesting one. Prehistoric man of a hundred thousand years ago left indications of sanitation and plumbing skill. Crude as these devices were, they offered proof that even these primitive people realized the consequences of poor plumbing. The rulers of Egypt, Greece, and Rome, thousands of years before Christ, advocated sanitary facilities of one kind or another. Bathtubs which were mere holes in the ground lined with tile, and water-conveying aqueducts, constructed of terra cotta and brick and terminating in a reservoir, were some of these historic accomplishments.

The individual who worked in the sanitary field of ancient Rome was called a *plumbarius*, taken from the Latin word *plumbum*, meaning lead. Because his work consisted of shaping lead, this name seemed to fit him well. It is interesting to note that much lead is still used for waste and water supply, and after two thousand years the sanitarian is still called a plumber. Evidence of the skill of these artisans can be seen in the aqueducts they built, some of which are still in use today.

During the period known as the Dark Ages (A.D. 400 to 1400), the culture of the early Romans deteriorated. Disease was rampant, and unsanitary conditions were responsible for destroying at least one quarter of the population of ancient Europe. In the fight for supremacy during this period, the Goths, Christians, and other invaders destroyed what remained of Roman culture. Europe was dormant for almost ten centuries.

During the Renaissance a gradual upbuilding of plumbing again began. Early in the seventeenth century the first plumbing apprentice laws were passed in England. France began building public water service installations in the eighteenth century. In general, Europe was in a period of building, including in this progress the art of sanitary science.

In the United States, which was largely devoted to agricultural pursuits, very little progress in plumbing was made up to the year 1800. Some of the notables of that time had built crude plumbing installations in their homes, the efficacy of which was somewhat doubtful. These installations consisted of a sink and a portable bathtub. The outside privy was a common means of disposing of waste matter. Water closets, imported from England, were in use in a few instances, but it is doubtful whether scientific principles were applied in installations of that day.

After the Civil War, plumbing improvements came slowly but steadily. Patents were issued on traps and methods of ventilation. Public water supply and sewage disposal systems became more evident, and plumbing came to be regarded as a necessity rather than a luxury, as it was considered twenty years before. Up to 1900 very few homes in urban localities provided more than a slop hopper and hydrant for the disposal of waste. After the turn of the century, plumbing progressed more rapidly. Water closets of the hopper and washout varieties as well as sinks, wash basins, and bathtubs were provided within the walls of a building. Scientific methods were used in constructing plumbing installations.

Fixture traps were ventilated, and hot and cold running water were introduced. The siphon wash-down closet appeared during this period and states developed legislation for the control of sanitation. The greatest progress in plumbing took place after the year of 1910, which is rather recent for a trade that has a background of thousands of years. Modern manufacturing methods provided materials and equipment which could be scientifically incorporated into a plumbing system. Buildings became larger, and the people who occupied them demanded more sanitary facilities. Although there are still many homes which do not have complete plumbing systems, the trend will ultimately correct this unsatisfactory condition.

SEWAGE DISPOSAL PLANT OF THE METROPOLITAN SEWAGE COMMISSION, MILWAUKEE COUNTY, WISCONSIN

(1) Administration Building; (2) Grit Chambers; (3) Fine Screen House; (4) Coarse Screen House; (5) Boiler Room, Power House;

HOW TO DESIGN AND INSTALL PLUMBING

CHAPTER I

MUNICIPAL SEWAGE DISPOSAL

Practically all municipalities of more than 2,500 population maintain systems of sewage disposal consisting of a sewage treatment unit of either activated sludge or filter type and a system of sewers installed below the streets by which the raw sewage is conveyed to the disposal plant or unit. In recent years health authorities have realized more than ever before, that the source of drinking water must be protected against contamination caused by the discharge of large volumes of domestic and industrial sewage into it. With the aid, in some instances, of federal grants, this phase of sanitation has progressed rapidly within the last ten years.

The installation of public sewage systems does not come under the jurisdiction of the plumber. It offers him, however, a new field into which he may easily be fitted, but as far as the installation of public sewers and disposal units is concerned he does not play a very active part. The work, especially that of constructing concrete and vitrified clay sewers, usually is done by semi-skilled labor or by workers trained in this field. The plumber does, however, use the public installation as a terminal for private plumbing systems, and it is quite necessary that he should be familiar with the public system, at least to the extent of being well-informed. It is quite essential, too, that he know the various types of sewers and the principles involved in construction and operation of a modern disposal plant.

CLASSIFICATION OF PUBLIC SEWERS

There are three types of public sewers, each classified according to the kind of wastes it is required to handle.

Combination public sewers are the oldest variety of the three types of sewers and they are required to carry storm and sanitary wastes to some safe terminal. These sewers are obsolete, but still make up a large part of public sewage systems. Rain water should be carried, through separate storm sewers, to some terminal not associated with the disposal plant, as large quantities of water affect operation and necessitate an exceptionally large installation.

Sanitary sewers are those which are required to carry regular sanitary wastes only. All rain water must be excluded from them. The terminal of these sewers is a modern sewage disposal plant. Combination and sanitary sewers generally are placed about ten feet under the street grade and usually are found below the center line of the street.

Storm sewers are a comparatively new installation, made neces-

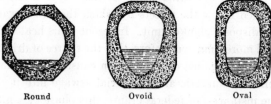

Round　　　　　　Ovoid　　　　　　Oval
Fig. 1. Cross Section of Three Types of Intercepting Sewers

sary because of sewage treatment. These sewers are made to carry only storm water and may terminate at any natural drainage area.

Sanitary sewer systems are of two kinds: the intercepting or trunk-line sewer and the tributary or contributing sewer.

Intercepting Sewers. Intercepting sewers may rightly be called collecting sewers. They are constructed of poured concrete and are placed at depths in the ground varying from 15 to 100 feet, depending entirely on the natural contour of the soil. The intercepting sewer does not have, as a rule, any connections to it other than the smaller tributary arrangement. These sewers vary in size from 2 to 10 feet in diameter and serve as a direct run to the disposal plant. The more recent installations of intercepting sewers are constructed in the shape of an inverted egg to assure a deeper and naturally a more self-cleansing flow. There are also sewers of oval and round design in common use. Fig. 1 illustrates a cross section of the three types of intercepting sewers.

The older type of intercepting sewers was constructed of brick and the interior was plastered with a rich cement mortar to make it water-tight. Some of the first public sewers were constructed of wood. Both of these early types of sewer installations are now obsolete.

Construction of an Intercepting Sewer. Before the contractor is invited to submit a figure for the construction of an intercepting sewer, he is given a very concise specification calling for certain kinds of materials, type of sewer, grades, data of soil test drillings and construction procedure, as well as many other important phases involved in the installation. Once the contract has been awarded, the procedure involved in construction is somewhat standard. Because the intercepting sewer usually is many feet below the surface of the

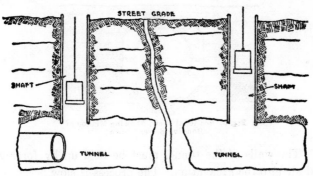

Fig. 2. Tunnel Construction of Intercepting Sewer

street, it would be impractical to attempt to lay it by the open trench method, hence it must be installed in a tunnel.

Shafts are dug at intervals of 750 to 1,500 feet or more to the depth specified by the engineering authority responsible for the work. Where the soil is of a substantial nature, lateral tunnels may be dug from the bottom of each shaft, Fig. 2, to meet at some point between them, or they may be dug from one shaft to the other. This is often referred to as the mining process. The tunnel is dug accurately, for its walls may serve as the outside surface of the finished concrete sewer. The tunnel may be made in the form of an octagon or, where unstable soil is encountered, it may be made rectangular in shape. Construction specifications are exact and require perfect grading and alignment.

The sewer proper usually is poured in sections. The length of the section which may be poured is determined by the stability of the soil. The bottom half of the sewer is framed first and concrete is then poured into it. A 4x4 timber is pressed into the soft concrete to form a depression, thus providing a more substantial bond for the top half of the sewer, which is poured later, Fig. 3.

After the lower half of the sewer sets, the top portion or crown of the sewer is framed. The concrete is forced or puddled into the space between the ceiling of the tunnel and the framing or rough shoring which forms the interior of the sewer. After the concrete sets, the interior walls of the sewer are waterproofed with cement mortar, brushed or troweled to a smooth finish.

When quicksand is encountered, a much more complex problem

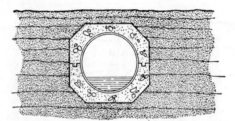

Fig. 3. Sectional View, Showing Method of Bonding Sewer Crown

arises. The walls of the tunnel must be held up by forcing compressed air into the excavation. The process is slow because of adverse conditions. Men must work under pressures which they ordinarily are unaccustomed to and they must be in splendid physical condition to endure the work for any length of time. Under these circumstances a series of locks or compartments is constructed at the bottom of the shaft to make entry and exit into the tunnels possible. The increased pressure must be taken on by the tunnel worker gradually and upon leaving the tunnel he must accustom himself to the normal atmospheric pressure in the same manner. In this way he averts sudden illness often referred to as the "bends" or "caisson" disease. Tunnel workers generally work in three shifts under continuous operation. The first shift tunnels, the second shift frames the sewer, and the third shift pours the concrete into the forms. These men are well paid because their work is unusually hazardous.

Intercepting sewers are thousands of feet in length. In larger cities the sewer may be required to extend a distance of many miles before it reaches its terminal. It would be unreasonable to suppose that a sewer could be placed an adequate distance from the surface of the street (10 feet) at its farthest point, be graded even slightly (one inch in 50 feet of run) and then be a reasonable depth below the ground at its terminal. A sewer 50,000 feet in length (approximately ten miles) under these conditions would be 1000 inches or 83 feet plus the depth at its farthest point under the level of the street. Sewers of greater length would be deeper in the ground, rendering them practically useless. To overcome this difficulty the lengths of the sewer are shortened, and the terminal of each division consists of a well, or lifting station, provided with a sewage ejector or pump to elevate the sewage to a higher level so it can flow by gravity to the

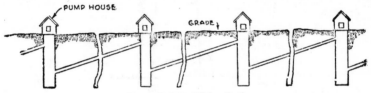

Fig. 4. Sewage Lifting Station

next lifting station and finally be discharged by the forces of gravity into the sewage plant, which is placed at some low point in the city. Fig. 4 illustrates this procedure.

Tributary Sewers. The intercepting sewer is provided with risers at various intervals at which points the smaller, tributary sewers terminate. The tributary sewers are of smaller diameter and are installed generally no more than 15 feet below street grade They are constructed of vitrified clay pipe cemented together and laid in an open trench.

Y-fittings are placed at such intervals that each piece of plotted property is provided with a house sewer terminal. Each connection is sealed and a record of its location is maintained by sewage authorities. Often the lateral house sewer is run to the curb line of the street to eliminate breaking up the finished street when it becomes necessary to make a connection for a new building. It is sealed at this point with a removable clay disc.

Manholes. Fig. 5 illustrates a manhole which may be constructed of brick or concrete and placed on the public sewer at intervals of about 250 to 500 feet. They serve as cleanouts and make the public sewer accessible for inspection and repair.

Manholes vary from 36 to 48 inches in diameter, and are equipped with iron rungs that form a ladder, enabling a man to reach the bottom or water table, as it is sometimes called. A well-fitting durable cover is then placed over the top, level with the road surface, Fig. 5. The depth and location of public sewers can be determined from the manholes.

Care must be exercised upon entering manholes for they may

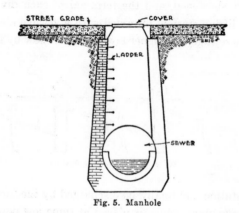

Fig. 5. Manhole

contain gases that are dangerous to the respiratory system. It is always good practice to aerate the sewer by removing two manhole covers, thus providing a circulation of pure air through the sewer.

MUNICIPAL SEWAGE TREATMENT

The treatment of municipal sewage is a complex problem and involves scientific aspects outside the sphere of plumbing. The design and construction of a modern sewage disposal plant requires engineering training in all the phases of natural science such as civil, for the design of the buildings, mechanical, for the construction of equipment, and an extended knowledge of chemistry, physics, and bacteriology.

A detailed scientific analysis of sewage treatment is not essential to this text, but it is important that the plumber be made familiar with the process involved in sewage treatment in a rather general manner.

There are a number of methods by which sewage may be treated, and those which are most commonly used are the activated sludge process and the trickling or sprinkling filter process. Fig. 6 gives an outline sketch of the activated sludge process used to correct the sewage wastes of the city of Milwaukee.

The bold lines with one arrow represent the route the sewage takes as it passes through the many stages of treatment. The lines with two arrows represent the heavy materials and their final disposition. The line with one arrow and one circle indicates duct through which air is administered to the sewage to effect its correction. The line with two arrows and one circle denotes vacuum applied for extracting the water from the organic material. The lines with two arrows and two circles show the course the organic material takes to the incinerator.

The Milwaukee sewage disposal plant serves approximately 650,000 people and is designed to correct 156,000,000 gallons of raw sewage every twenty-four hours. The sewage as it enters the plant contains much suspended material, such as paper, rags, leaves, garbage, ash, grit, grease and many other objectionable elements which must be removed before final disposition of the sewage can be made. It also contains much organic material from which a fertilizer high in nitrogen (N) content is manufactured.

The bacterial content in the raw sewage is approximately 2,500,-000 per cubic centimeter (c.c.) and it is the function of the plant to eliminate this serious condition and emit a stable effluent from its discharge terminal. The degree of purity attained is between 99 and 99.5 per cent, which represents a water practically fit for human consumption.

The motivating principle behind any activated sludge or trickling filter process is to oxidize the organic material with its bacterial content and reduce it to a nitrogen compound. The routine involved in this process is to aerate or administer oxygen either by mechanical or natural means, also to settle out the nonsoluble materials and remove them to some safe terminal. The entire process may be likened to a long river flowing over rocks and exposed to natural atmospheric conditions to produce oxidation of its bacterial content, its flow re-

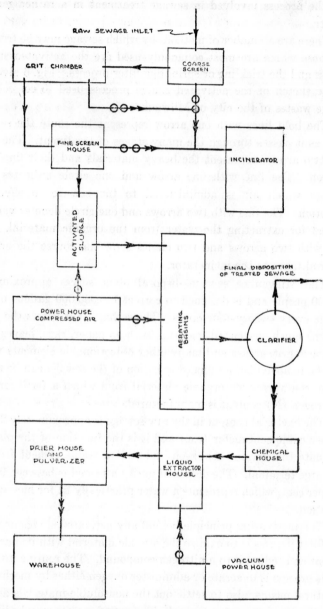

Fig. 6. Course of Travel of Sewage, Activated Sludge Process
(Large City Installation)

tarded to some extent at intervals to permit settling out of the suspended materials it would otherwise ultimately carry to its final destination.

Coarse Screen House. Referring to the sketch, Fig. 6, the reader will see that the first treatment the sewage undergoes is the separation of the larger particles of objectionable material. This is done by means of a large bar screen or grate through which the sewage must pass. The building is called the coarse screen house. The separated material has no value and is cut into small pieces and conveyed to the incinerator building where it is burned.

Grit Chamber. From the large screen house it enters the grit chamber where the finer particles—sand, ash, grit, and grease—are removed by a settling process. The velocity of the sewage is reduced tremendously at this stage to permit settling of these materials. The settled out material is removed to the incinerator.

Fine Screen House. The sewage now flows by gravity to the fine screen house, where the floating solids are removed by passing the sewage through mechanical equipment. This material is then conveyed by compressed air to the incinerator building, where it is burned.

Mixing Tank or Activated Sludge Tank. One of the essentials in the activated sludge process is to increase the bacterial activity to hasten decomposition and also produce a higher nitrogen content in the fertilizer. The sludge mixing tank, which is the next step in the treatment process, accomplishes this. A portion of the settled flocculent material from the clarifier unit, which is another part of the disposal system, is added to the raw sewage in this tank to increase the count of bacteria and change them from the anaerobic to the aerobic variety. Purified atmosphere taken from louvers located near the roof of the power house is passed into the sewage from the bottom of the tank through filter plates. The compressed air is under about 10 pounds of pressure. The oxygen content in the atmosphere, through some phenomenon of nature, oxidizes the anaerobic bacteria and forms compounds of nitrogen. These compounds are nitrites (NO_2) and nitrates (NO_3).

Aerating Tanks. The sewage next flows to the aerating tanks where it is retained for about a six-hour period. It is aerated continuously in the same manner as it was in the activated sludge tank.

The sewage has now changed into two distinct compounds, water and a flocculent material high in nitrogen content.

Clarifier Tanks. The clarification tanks are the next in the process of treatment. The function of these tanks is to settle the flocculent material and discharge the water, 99 to 99.5 per cent pure, back into its original source. The flocculent material is carried into a chamber where a portion of it is returned to the activated sludge tank and the remainder passes on to the chemical house.

Chemical House. This step in the process of sewage treatment is necessary to coagulate the flocculent material, which still carries with it a large volume of water. Ferric chloride ($FeCl_3$) is the chemical compound used in this operation.

Filter House. After the effluent has been thoroughly mixed with the ferric chloride compound, it runs through mechanical filters and the water is drawn from the mixture by a vacuum which has its origin in the vacuum equipment located in another building. The flocculent material is rolled in large sheets which are broken in rather coarse pieces and conveyed to the drier house.

Drier House. In the drier house the flocculent mass is carried into large revolving driers where the water content is evaporated and the residue is pulverized and prepared for use as fertilizer.

Warehouse. In the warehouse the product is divided and stored according to its nitrogen content. It is distributed to all sections of the United States and is considered a very good fertilizer for soils low in nitrogen content.

OTHER SYSTEMS OF THE ACTIVATED SLUDGE PROCESS

There are other systems of sewage treatment using the activated sludge and aeration process which are capable of serving smaller municipalities, towns, or individual institutions. One of the most popular and efficient installations of this variety is the product of the Chicago Pump Company.

The principle involved in this type of activated sludge process is identical with that of the larger sewage disposal plant previously discussed. The incoming sewage with its sludge content is activated by injecting into it a portion of that material which has already undergone treatment. The sewage is aerated to produce rapid precipitation, as it is in the larger plant. The method used, however, varies to some extent. Fig. 7 is an outline sketch of the activated sludge treatment of sewage as it passes through the various stages of the Chicago Pump Co.'s process.

The bold lines with arrows represent the route the sewage takes as it passes through the various stages of treatment.

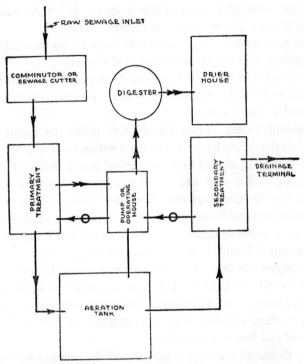

Fig. 7. Course of Travel of Sewage, Activated Sludge Process
(Small City Installation)

The lines with the two arrows indicate the route the sludge takes to the digester tank.

The lines with the arrow and circle indicate the route the activated sludge takes into the primary treatment tank.

Comminutor or Sewage Cutter. The comminutor, or sewage cutter, consists of an electrically driven device similar to a sewage ejector. The mechanism is made up of a stationary comb around which a drum with cutting edges revolves. The small pieces of sewage pass through the slotted drum into the primary settling tanks. The large particles are held against the drum by the flow of sewage while they are gradually reduced to smaller pieces, after which they are carried into the first stage of the treatment process.

Primary Clarifier. The primary clarifier of this type of plant is constructed of concrete and provided with agitators or sludge collectors of the drag variety. As the heavy materials precipitate, the sludge, or mass, is scraped into a sludge well from which it is pumped into the digester tank for further treatment.

The primary tank is provided with baffle plates which direct the liquid effluent into the aeration unit of the plant.

A quantity of sludge taken from the aeration unit and the sec ondary clarifier is injected into the primary clarifier to produce more rapid sludge precipitation.

Aeration Tanks. From the primary tanks the liquid effluent and a quantity of sludge pass into the aeration tank. In this opera- tion the sewage is aerated by a mechanical device which lifts it from the bottom of the tank and discharges it over the surface of the retained liquid. Constant circulation and subsequent exposure of the sewage to the atmosphere tend to reduce the once objectionable sewage to a stable effluent. A quantity of the settled sludge in the aerating unit is pumped into the primary unit.

Secondary Clarifier. From the aeration tanks the sewage passes into the secondary clarifier. This tank is a duplicate of the primary unit. The sewage is now a stable effluent or mixture which, when allowed to stand or when retarded sufficiently in flow, will separate into liquid and flocculent materials.

The flocculent material is passed into the primary treatment to activate the incoming sewage. The clear water is corrected to permit its discharge into a natural drainage terminal. Tests have proved it to be very close to 100 per cent pure.

Digester Tank. The heavy sludge is pumped from the sludge well, of the primary clarifier, into the digester, which is a large cylindrical tank 30 to 60 feet in diameter and 15 to 20 feet in depth. The size of the tank is determined by the volume of sewage to be treated. The sludge is retained in the digester for a given period to allow it to liquefy. The liquidation process is due to anaerobic bacteria which thrive in the tank content. To increase bacteria the tank is heated by means of a hot water coil to a temperature of about 70°F. The fuel which provides the heat to maintain this temperature is a gas given off by decomposing organic materials which is known scientifically as methane (CH_4). It is a compound of the elements carbon and hydrogen and is of a highly inflammable nature.

A liquid referred to as supernatant liquor must be drained periodically from the tank and is returned to the clear water well in the pump house. The heavy nonsoluble material is also drawn off and deposited on the drying filter bed.

Sludge Drying House and Bed. The drying filter bed consists of a well-ventilated glass-enclosed building, having a sand floor underlaid with drain tile, which discharges into the final disposal terminal. The nonsoluble material from the digester is spread over the sand bed and allowed to dry. When all the liquid content has been removed, the dried material may be used for fertilizer.

TRICKLING FILTER SEWAGE PLANTS

The trickling filter, sometimes called trickle, percolating, or sprinkling filter, is used extensively for secondary treatment of sewage. It is generally acknowledged to be the most efficient system in use today, attaining an 85 to 95 per cent range of correction. Its primary disadvantage is the large ground area required for its building and equipment.

Trickling filters have a relatively short history, the first installations being made in 1889 at the Lawrence experimental station in Massachusetts. The first municipal trickling filter was installed in Atlanta, Georgia in 1903. Other installations followed, although initially all were slow in disposing of sewage and were of a limited capacity. The development of faster methods of oxidation and the construction of more efficient filter beds have now made trickling filters a widely used method of secondary treatment.

Definition of Terms. In order to understand trickling filter operation it is necessary to know the meaning of the following terms. *B. O. D.* (Biochemical Oxygen Demand) is the amount of oxygen needed in sewage to maintain aerobic bacteria requirements during decomposition. It is expressed in pounds per acre foot (one acre of sewage one foot deep). *Media* is the material used in the filter, usually coarse crushed rock. *Volume load* is the amount of sewage applied to a filter on the basis of millions of gallons per acre of filter sewage surface per day. *Organic load* refers to the B. O. D. content of the sewage applied to the filter. *Primary treatment* consists of passing the sewage through settling tanks to permit sedimentation and clarification prior to its application to the trickling filter. *Filter bed blocks* are vitrified clay blocks laid on the floor of the filter to support the media, to provide drainage, and to supply ventilation. See Fig. 8A.

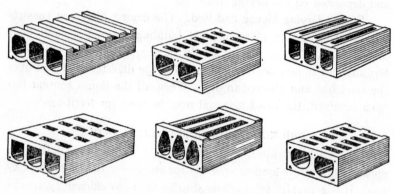

Fig. 8A. Types of Filter Bed Blocks

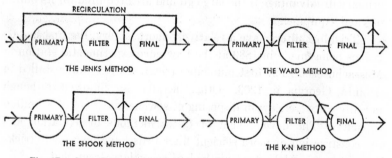

Fig. 8B. Some Recirculation Methods Commonly Used with Trickling Filters

Purpose and Operation. The main purpose of a trickling filter system is to provide the best possible environment for aerobic bacterial action, through an ample supply of oxygen. It differs from the activated sludge process, primarily, following that point at which the settled sludge has been separated from the liquid content of the raw sewage. The liquid content of the raw sewage is pumped from the clear-water well, under the pump-house floor, through rotating arms that sprinkle the sewage over the filter bed. The sewage filters through the media, the solids being removed and held by a fine film which forms on the surface of the media. Aerobic bacteria are present in the film and act on these solids to reduce B. O. D. and in other ways purify the effluent, which in turn filters through the media and drains into a large duct terminating in the secondary clarifier. (The processes preceding and following actual trickling filter application often vary in use and order of application.)

Trickling Filter Construction. Simply, a trickling filter consists of a circular cement tank, containing a bed of media resting on a floor of vitrified clay filter bed blocks constructed to include a drainage system. A *distributor,* two or more arms rotating around a center column, sprinkles the sewage evenly over the surface of the media. See Fig. 9. The sewage, sprayed into the atmosphere, oxidizes and is then subjected to the process described above.

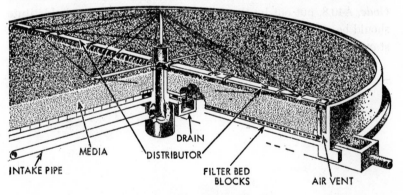

Fig. 9. Typical Sprinkling Type Filter

Types of Trickling Filters. In a *standard* or *low rate filter* the sewage goes through a sedimentation tank following the trickling filter treatment, for removal of those solids still present. It is rarely recirculated. This type of filter handles less sewage but produces a more stable effluent, low in B. O. D.

A *high rate filter* receives much more sewage, which is recirculated, usually continually, until the effluent is sufficiently purified for final disposal. See Fig. 8B.

The *roughing filter* is used for sewage having high B. O. D. concentrations such as industrial wastes. It is used as first stage treatment to reduce the load on existing or new sewage treatment plants.

There are other types of filters, such as the bio-filter, accelofilter, the aero-filter, using basically the same principles of trickling filter operation as described herein.

Chlorination. In most sprinkling and trickling filter sewage disposal plants, chlorine is injected to neutralize objectionable bacteria as an added precaution against water pollution.

MUNICIPAL AND STATE CODES AND REGULATIONS

Local codes and regulations should always be observed in the installation of all plumbing. All practices and suggested standards in this publication should be checked against local practices and standards. Any method for individual water supply or sewage disposal outlined in this text which is not permitted by a State or local health department should not be used. The *National Plumbing Code*, A40.8, put out by the American National Standards Institute should be consulted. Nevertheless, there is a wide variance in local standards and these also should be conformed to.

CHAPTER II

PRIVATE SEWAGE DISPOSAL

Modern transportation has made the country readily accessible to the urban resident, with a result that thousands of people spend week-ends, as well as extended vacations, in natural beauty spots. This migration presents a serious problem from the standpoint of sanitation. Facilities for the disposal of fecal and organic wastes left by the tourists have opened a new field for the plumber.

Thoughtless discharge of organic materials has in many instances polluted lakes, rivers, and streams to such an extent that aquatic life has suffered to the point of near extinction. As a protective measure many states now have conservation commissions cooperating with health boards in a regulatory capacity with respect to this problem. Rigid laws are being passed by state legislatures to give these boards power to specify and control sanitary requirements for individual homes, factory and business sites, tourist camps, and resorts.

Not alone does aquatic and animal life suffer from unregulated discharge of sewage wastes. Recent health statistics reveal the fact that most of the water-borne diseases, such as dysentery, typhoid, diarrhea and other intestinal disorders are more prevalent in rural communities than in the larger cities. In most instances these illnesses are caused by contaminated drinking water.

As a rule, water for domestic use in rural communities is *ground water,* usually from a *water table* (the upper limits of an underground portion of earth that is saturated with water), which can be contaminated with disease-spreading bacteria when close to the surface of the soil. The outside privy and cesspool are two formerly common methods of sewage disposal which are inadequate and even dangerous to health when located near the water supply. The organic material, with its extremely high bacterial content, is collected in the pit of the privy and carried into the drinking supply by surface water drainage through crevices in rock or loose formation of the soil. This is also true of the cesspool installation.

To avoid this dangerous condition, the organic materials must undergo treatment. They must be corrected or purified so that well or stream pollution will not result. The combination of septic tank and purification unit is the best means to obtain this end.

It is not uncommon for residents of summer homes closely associated with one another to build a community disposal system. These installations are generally supervised by state sanitary authorities and are a benefit because of their controlled terminals.

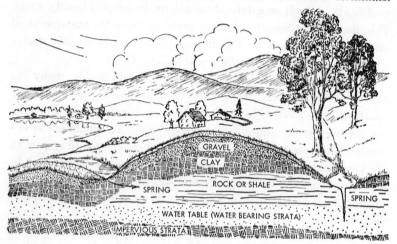

Fig. 10.   Source of Water Supply (Rural)

In dealing with the problems of private sewage disposal, as in every phase of plumbing, sewage, and water supply work, there are many general scientific principles, terms, and methods used daily. The actions of water and air, particularly as producing, and affected by, pressure, are the foundation of the sanitary sciences. In order to facilitate understanding, some of these are described where applicable throughout the book. Two such sections, describing phenomena relevant to most phases of plumbing, sewage, and water supply work are at the beginning of the VENTILATION chapter and the beginning of the WATER SUPPLY chapter. The first explains *atmosphere, atmospheric pressure, siphonage, vacuum, trap seal loss, back pressure,* and *capillary attraction.* The second defines terms commonly used in reference to heat and water, such as types of *pressure* (*static, critical,* etc.), types of *heat* (*latent, sensible,* etc.), and *temperatures.* Bearing the basic scientific prin-

ciples in mind greatly simplifies the job of understanding new material, which is usually a form of different application of one or more
of these basic principles.

Cesspool. So that the reader may be better informed on the
necessity of sewage treatment, the cesspool might be defined as a
hole in the ground curbed with stone, brick, or other material, laid in
such a manner as to allow raw contaminated sewage to leach into the
soil. (Some cesspools hold sewage until cleaned.) The wastes are
deposited in the cesspool by terminating the drains from the plumb-

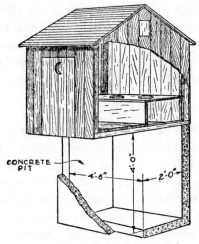

CONCRETE
PIT

4'-6"   2'-0"

Fig. 11. Improved Type of Outside
Privy

ing fixtures installed in the residence into it. The organic wastes
accumulate and are finally disposed of by a disintegration process.
Rain and surface water carry the decomposed organic materials with
their bacterial content into the soil. This is objectionable when
the water for domestic use is obtained from this source. The cesspool is also accessible to vermin and insects and therefore may be
an agency for the spread of disease.

Privy. The outside privy is the oldest form of disposal of
organic waste. It consists of a vault constructed of concrete for the
collection of raw sewage and a wooden shelter. The disintegration
of excrement is accomplished in the same manner as in a cesspool,
and is objectionable because of the danger of contaminating the

source of water supply. It should be used only where soil conditions do not permit a more sanitary method of disposal.

Because a modern septic system requires favorable soil conditions, not always present, the privy is still in common use. Much can be done to increase its sanitary quality by incorporating the following suggestions in its construction. The privy vault should be constructed of concrete and located at least 150 feet from the water supply. (If the vault is of water-tight design, 50 feet is considered a safe distance.) It is advisable that the privy vault be as dark as possible as well as protected against vermin. It should also be ventilated locally through the roof of the shelter house. The shelter house should be substantially built and kept scrupulously clean. It should be screened and fastened securely to the vault. Fig. 11 illustrates these essentials and gives approximate dimensions for a privy designed to serve a family of six members.

THE SEPTIC SYSTEM

The installation of a rural septic system requires careful and intelligent planning on the part of the plumber. The contour and character of the soil are the first important factors that must be considered. The soil most favorable for the disposal of effluent, which is the purified liquid waste of the septic system, is of a sandy nature. Sand or gravel is pervious to moisture and permits rapid leaching of the effluent into the subsoil. Another favorable condition is where the contour or slope of the soil is at such distance from the purification unit of the septic tank that a gravity flow in the discharge line is possible. In some instances, because the top soil is flat and moisture resistant, it becomes necessary to convey the waste to the undersoil which may be more favorable for leaching the effluent.

Design of the plumbing system is affected by a septic tank installation. It is important that the septic system be installed close to the surface of the soil, because correction of the effluent depends on oxidation, as well as aerobic bacteria. These bacteria are found only in the subsoil, not more than five feet below the surface. Oxidation of the effluent at greater depths becomes extremely difficult. In order to provide proper conditions, the house drain is generally suspended from the basement ceiling and discharged into the sewer, which is installed just below the surface of the soil.

Another important factor to consider is the drinking water supply. The purification unit, even though its function is to purify sewage, must be located a safe distance from the drinking water source. It is well, therefore, for the plumber to establish location of the well and septic system for the home owner or building designer, in order to secure this protection.

Septic Tank. The septic tank, Fig. 12, is a device used to expedite the decomposition of the elements contained in raw sewage

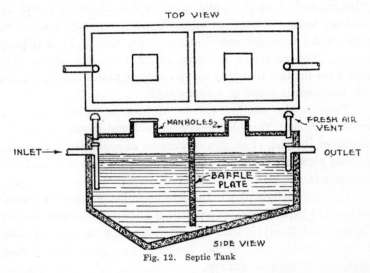

Fig. 12.   Septic Tank

wastes. Raw sewage consists of water, and settleable solid materials, such as grit, grease, fats, vegetable and animal materials in a suspended state, and non-settleable materials of a vegetable and mineral nature in solution. It has a milky color and may have an extremely offensive smell. The settleable solids, usually referred to as organic materials, can be precipitated in a septic tank in a very short time. About a 24-hour period of retention offers satisfactory results. The solid organic materials, being more dense than water, tend to settle to the bottom of the tank and are technically referred to as sludge. The lighter organic materials, being less dense than water, rise to the surface of the water and usually are referred to as scum. The suspended materials constitute only a small part of the raw sewage by actual weight. The organic materials in solution in the raw sewage offer a more difficult problem. They cannot be precipitated in the septic tank and must undergo further treatment by processes other than precipitation. The liquid content of raw sewage is referred to as effluent and does not present a problem other than that of its discharge into the subsoil or natural drainage terminal.

Decomposition of Settleable Solids. (Organic Materials.) The organic materials are composed largely of proteins, carbohydrates, ash, fats, and soap and are largely chemical compounds of nitrogen, hydrogen, carbon, sulphur, and phosphorous. Associated with the solids are countless numbers of bacteria of the anaerobic variety which, under favorable conditions, multiply rapidly by a splitting process. The anaerobic bacteria, called anaerobes, derive their name from the fact that they survive only in places lacking oxygen. This condition is found in a septic tank.

There is also another variety of bacteria called aerobes, or aerobic bacteria. These bacteria may be found in raw sewage, but because of the lack of oxygen or because of rapid association with chemical elements contained in the sewage, they do not survive in the septic tank.

The part bacteria play in the decomposition of organic materials is somewhat hypothetical. Very little is known of bacterial activity. The accepted theory is that the anaerobic bacteria consume much of the suspended material and change it into gases and chemical compounds of common variety.

The gases produced in the septic tank are known to chemical science. The most common gas formed in the decomposition of sewage is methane (CH_4)—a combination of hydrogen and carbon. It is odorless, highly inflammable, and is sometimes called swamp gas. Carbon dioxide (CO_2), a combination of carbon and oxygen, is odorless, tasteless, and colorless and it constitutes a part of the atmosphere as an impurity. Carbon monoxide (CO), hydrogen (H) hydrogen sulphide (H_2S), and sulphur dioxide (SO_2) are also formed in small quantities.

All of the gases formed in the septic tank are discharged into the atmosphere by means of septic tank ventilation. This may be accomplished in two ways. The septic tank may be equipped with a fresh air vent extended at least 24 inches above the surface of the soil, or it may be equipped with inverts, the tops of which are allowed to remain open. This practice is more favorable than the first, because the gases are carried through the plumbing system and expelled at the soil pipe roof terminal.

The heavy nonsoluble materials, called sludge, settle to the bottom of the tank. It is in the sludge formation that the anaerobic bacteria multiply. The fats and oils rise to the surface of the tank and serve as a source of bacterial growth. Both of these nonsoluble elements must be removed from the tank periodically. Frequency of removal depends on the thickness of the scum and the depth of the sludge. Vast quantities of this material tend to lower the retention period of the tank and reduce its capacity materially. Complete decomposition of the sewage cannot be accomplished under these circumstances and much soluble organic material is carried into the purification unit and tends to hinder its correction efficiency.

The nonsoluble sludge and scum must be removed manually. It is shoveled or dipped from the septic tank through the septic tank manhole. It is extremely offensive and must be disposed of by burning. Burying it in the soil is also a common practice. Care must be used in this case not to pollute the water supply.

The liquid content of the septic tank, which consists largely of water with or without chemical compounds in solution, is percolated into the soil and passes gradually into natural water tables under the surface. Purification of the discharged effluent will be discussed in later paragraphs of this chapter. Anaerobic bacteria are eliminated by the administration of oxygen which changes the bacteria into nitrogen compounds known as nitrites and nitrates.

To discharge large volumes of water into the septic tank, such as rainwater from roofs, surface water, or waste water from industial sources is objectionable. The function of the septic tank is to liquefy and precipitate solid materials. Water is already a liquid and therefore requires no further reduction. Large volumes of water tend to flood the tank and much solid material is forced into the purification unit. Bacterial activity is also affected by this process.

Wastes of an obnoxious nature, containing grease, oil, acid or gasoline in suspension, must also be excluded from the septic tank. Such wastes destroy useful bacteria and may be detrimental to the materials of which the septic tank is constructed

Construction of the Septic Tank. The septic tank usually is constructed of puddled concrete, 6 to 8 inches thick, well reinforced with steel rods to eliminate breakage caused by frost. It must be waterproofed by cement mortar applied to the inside walls with a plasterer's trowel.

The concrete septic tank usually is built longer than it is wide. This practice assures a retarded even flow, which is necessary to avoid disturbing the decomposition process within the tank. The minimum width of a septic tank should never be less than 2 feet 6 inches, and the minimum length at no time should be less than 5 feet.

The inlet and outlet inverts of the septic tank generally are constructed of a long turn sanitary tee and must be cast in the concrete structure of the tank. The invert should extend not more than 15 inches into the liquid content of the tank. This practice assures delivery of the incoming sewage, sometimes called affluent, into the liquid content of the tank below the scum line. The top opening of the inverts must be extended above the scum line of the tank. This is accomplished by calking a piece of cast-iron soil pipe 12 inches long into the tee. The extension must be left open to allow ventilation of the septic tank.

The inverts must be placed in the wall of the tank at least four feet from its bottom and it is advisable to space them equally from both sides. Four feet of liquid content in the tank usually is adequate for proper reduction of organic materials. It is not impractical, however, to build tanks of greater depth.

The bottom of the tank should slope to one low point. The purpose of this is to gather the settled organic materials into one mass and thus favor the propagation of anaerobic bacteria. Sludge removal is also facilitated by this practice.

The septic tank must be equipped with a manhole extended a few inches above the surface of the soil to overcome infiltration of surface water. The manhole is necessary to make the septic tank accessible for the purpose of cleaning, inspection, and repair.

Septic tanks for large plumbing installations are divided into compartments by means of baffle plates. Baffle plates are constructed of concrete and suspended into the tank from the ceiling slab of the septic tank.

The purpose of the baffle plate is to retard the large volume of flow from the plumbing installation. The plate must be slotted close to the top to assure a movement of air above the scum line. The baffle plate extends to within 18 inches of the bottom of the tank. When tanks are divided into separate compartments by means of baffle plates, each compartment must be equipped with a manhole, as shown in Fig. 12.

There are a number of commercially made septic tanks. Tanks of this variety usually are constructed of pre-cast concrete, vitrified clay tile, or sheet metal of no less than 12-gauge. These tanks should conform with the specifications of the tanks previously explained. Not always do commercially made tanks meet with the standards required by state or local governing agencies.

Size of the Septic Tank. Not much definite information is available for arriving at the proper size of a septic tank. It is agreed by sanitary authorities, however, that the minimum size of tank for a family of not more than six people should be no less than 2 feet 6 inches in width, 5 feet long, and not less than 4 feet deep. This affords a tank of approximately 50 cubic feet capacity, which should be adequate to serve the needs of a family of six. A septic tank of smaller capacity is impractical because some leeway must be allowed for storage of accumulated sludge. A tank of larger size is inadvisable because retarded bacterial activity is liable to result. For residence installations intended to serve a larger number of people it is accepted practice to allow 5 to 6 cubic feet of tank content per person. Should the installation be for service of twelve people, the tank should have a liquid capacity of not more than 72 cubic feet.

For school, commercial, or industrial purposes the cubical content of the tank per person, should be not less than 2, nor more than 3, cubic feet. Where large amounts of liquid waste from shower baths, industrial equipment, and laundries flow into the septic tank, the tank's capacity must be increased to accommodate them. It is not good practice to discharge these wastes into septic tanks. Some other means of disposal of these wastes should be provided.

Location of the Septic Tank. The septic tank must be a watertight as well as a gas-tight receptacle, hence not much difficulty is experienced in establishing its location. Good judgment should be used, however, in this phase of rural sewage disposal.

The septic tank may be located close to the building it serves, provided a minimum distance of five feet is allowed between septic tank and building. Care must also be exercised not to locate it close to a window or doorway for movements of air are likely to carry the obnoxious odors from the tank into the building.

It is also good practice to locate it a reasonable distance from lot lines and drinking water wells. Tanks should be at least 50 feet from any source of water supply, and further when possible.

Safety Precautions. An individual entering a septic tank for making repairs or cleaning purposes should be cautious. In most cases septic tanks are poorly aerated and there is a lack of free oxygen. This condition may cause almost instant death. Also the tank may contain very harmful and dangerous gases. When repair work or cleaning is to be done, the tank should be ventilated by removing the manhole cover several days in advance of operations, and, as an added precaution, some means of supplying fresh air while work is in progress should be provided. It should be remembered, too, that the tank may contain inflammable gases which, if ignited, may cause a terrific explosion. If light is needed to work in the tank, an electric light with properly insulated electric cord should be used.

Purification of Discharged Effluent. The second stage of rural sewage disposal, often referred to as secondary treatment, is much more complex than the first. This phase in the treatment process is designed to correct the effluent so it can be disposed of without danger of well pollution and contamination of lakes, rivers, and streams.

The effluent discharged from the septic tank is still in an objectionable condition. The settleable organic materials have been removed, and many of the objectionable gases have been eliminated. Nevertheless, it still contains countless numbers of harmful anaerobic bacteria and objectionable chemical compounds in solution which must be disposed of. Nature plays an important part in the removal of these objectionable elements.

The purification unit of a septic system accomplishes three objectives: (1) It serves as a fine screen, thus eliminating small particles of suspended materials contained in the discharged effluent; (2) the chemical compounds in solution and the anaerobic bacteria are reduced to nitrogen compounds which are beneficial to the soil and harmless to water supply; (3) the water which is the only remaining element is leached into the subsoil or discharged into a natural drainage terminal.

Types of Filtration Units. The dry well, filter trench, and distribution field are the three types of purification units usually used in connection with a septic tank. The use of any one of these units, or a combination of two units, is governed by the soil conditions and the available drainage terminal.

Dry Well. The dry well, sometimes called a seepage pit, is a hole curbed with stone, or other nonabsorbent material, that is so constructed as to allow liquid effluent to leach into the soil. See

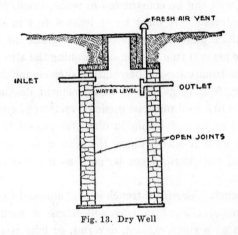

Fig. 13. Dry Well

Fig. 13. The dry well can be used as a discharge terminal for a septic tank installed in clay soil which has an under strata of gravel. It is not the most effective purification unit and should not be used unless the effluent it receives has undergone some preliminary treatment.

In the preceding paragraphs the objectives of the purification unit were discussed. The oxygen necessary to eliminate anaerobic bacteria and reduce the compounds in solution is found only within a few feet of the surface of the soil. Artificial means of aeration are provided by fresh air vents terminated above the surface of the soil. In the case of a dry well the gravel strata is found only at depths lacking oxygen, and to provide means of ventilation is practically impossible. The dry well serves in most instances only as a leaching or distributing unit with very low purification efficiency. It tends to deliver the objectionable sewage wastes into the soil in closer proximity to ground water tables. Although the gravel soil in which it may be installed serves as a fine screen eliminating suspended materials thoroughly, it becomes fouled rapidly and, when in this condition, loses much of its effectiveness.

The dry well must be located a safe distance from water supply. Fifty feet from a cased well 100 feet deep has been found to be adequate. Greater distances, however, are recommended. At no time should the dry well be located within 100 feet of a well 50 feet deep. Greater distances are also recommended in this instance.

The dry well can be constructed of stone, brick, cement blocks or similar materials. It must be at least 6 feet in diameter and equipped with a manhole extended above grade to make it accessible. The general rule used in determining the size of a dry well is to allow one square foot of percolating surface, including the bottom, for every 5 to 10 gallons of liquid effluent discharged into it The top of the dry well must be made of reinforced concrete with a fresh air vent at least 4 inches in diameter passed through it and terminated at least 20 inches above the ground, as shown in Fig. 13. An abandoned well must never be used as a dry well under any circumstances.

Filter Trench. The filter trench is best adapted for the disposal and treatment of effluent in clay soil where a natural drainage terminal such as a river, stream, dry run, or lake is available. It may be used as a preliminary treatment agency in connection with a dry well.

The filter trench is considered an efficient purification unit. It can be installed close to the surface of the soil containing free oxygen needed in the purification process. The filter trench can be artificially aerated with fresh air vents extended above the soil. The material between the tile lines used to remove suspended materials can be washed and replaced periodically, which tends to keep its purification efficiency at a very high level.

Good judgment must be used in locating the filter trench in relation to the drinking water supply. It is advisable to use the same minimum distances specified in the preceding paragraphs for dry wells

The filter trench consists of a ditch excavated in the soil to a depth of about 4 to 6 feet. The bottom of the trench must be graded slightly to the disposal terminal. Ordinary 4- to 6-inch drain tile is laid on the trench floor in such a manner as to serve as a collecting line for the purified sewage effluent. The end of the collecting line farthest from the disposal terminal must be extended vertically above the surface of the soil. This completes the circulation of atmosphere, so vitally important for the oxidation of the sewage effluent as it passes through the purification unit.

The collection line is then covered with from 30 to 36 inches of coarse gravel. The gravel fill forms a screen for the removal of sus-

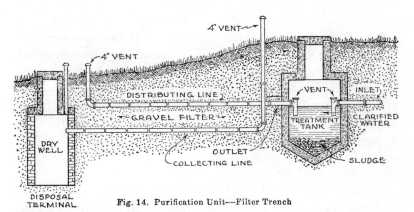

Fig. 14. Purification Unit—Filter Trench

pended materials as well as a source of aerobic bacteria. The aerobic bacteria consume much of the suspended materials and apparently are instrumental in changing them into nitrogen compounds.

A distributing tile line of the same diameter as the collecting line is laid on the gravel fill directly over the collecting line. The distributing tile line is also aerated by fresh air vents to help in the oxidation process. It must be covered with tarred paper or some other material to prevent earth or sand from entering it, thus causing stoppage of the line. The septic tank outlet invert is discharged into a cemented water-tight drain which connects with the distributing line, Fig. 14.

The number of feet of tile necessary to produce satisfactory results is determined by the number of people contributing to the septic tank. If a dosing device or siphon compartment is used ir connection with the septic tank the cubical content of the tile should be equal to that of the siphon unit. About 10 feet of tile per person contributing to the tank is adequate under normal conditions.

In extremely cold weather it becomes necessary to protect the filter trench against frost. This can be done by covering it with straw, tarred paper, or some material which will resist cold.

Distribution Field. In localities where the subsoil consists of sand, gravel, or sandy loam, the distribution field can be used as a means of disposing of the effluent of a septic tank. This type of purification unit is less costly to construct. It uses the earth as a means of purifying the discharged effluent.

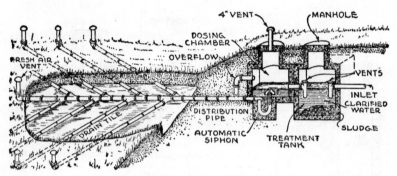

Fig. 15. Purification Unit—Distribution Field

The open jointed tile lines are constructed of field tile no less than 4 inches in diameter. Perforated, non-metallic pipe or 2—3 foot lengths of vitrified clay sewer pipe also may be used. The distribution field may be of any design the installer feels is best suited to the soil. The rows of tile may be laid parallel to one another with spaces of 10 feet between them, Fig. 15. They may be branched into a main run on a 45° angle to the main. The tile may be installed in the form of a large wagon wheel having a 36-inch tile set vertically at the distributing unit, Figure 16.

The drain tile must be placed as close to the surface of the soil as possible to take advantage of the oxygen and aerobic bacteria found there. The type of soil determines the amount of tile necessary to serve the installation. For sandy soil it is safe to allow 30 feet of tile per person contributing to the tank. For soil which does not leach water as readily as does sand, at least 50 feet of tile per person must be used.

The ends of the distribution field must be aerated by fresh air vents extended at least 20 inches above the ground level.

Siphon Compartment or Dosing Tank. Septic tanks which take care of a quantity of sewage exceeding the amount that can be disposed of in about 500 feet of tile should have a siphon

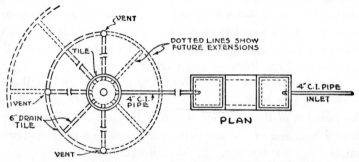

Fig. 16. Purification Unit—Distribution Field, Wheel Type

compartment, Fig. 17. The purpose of this added feature is to store a large volume of treated sewage and discharge it periodically into the purification unit. This practice assures the use of the entire tile

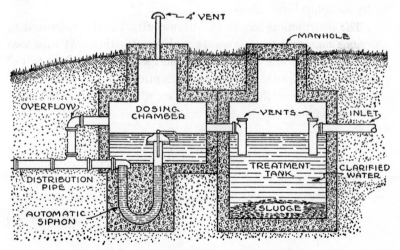

Fig. 17. Siphon Compartment or Dosing Device

line instead of only a few feet and permits the absorption field to dry out between dosings.

The siphon compartment usually is constructed of concrete and made a part of the septic tank. The siphon consists of a cast-iron trap of about 8-inch seal, the inlet leg of which is equipped with a bell forming another trap, Fig. 17. The double trap permits a compression of air between them, which can be overcome only when the water in the compartment rises above the top of the bell to a height equal to that of the lower trap seal. When this occurs, siphonage of the compartment's contents results. Siphonage is broken when the water recedes to a point below the air vent installed in the bell of the siphon. The siphon is often referred to as a dosing device.

Distribution Box. A distribution box is a device installed between the septic tank and the means used to dispose of its effluent. The function of the distribution box is to distribute the effluent so that all branches of the sub-surface disposal field receive approximately the same amount. It may also serve the same purpose in sewage disposal systems which use more than one seepage pit.

In many cases, particularly where the absorption field is in sloping ground, the distribution box prevents the flow of all of the septic-tank effluent into one branch of the field or into one seepage pit. It permits many variations in adapting the design of a system to topographical situations.

The distribution box should be watertight and constructed of the same materials used in septic tank construction. At least two outlets, placed at the same level, should lead from the box to the disposal field, with the inlet from the septic tank at a higher level. The line connecting the distribution box with the septic tank must have tight joints.

CHAPTER III

MATERIALS USED FOR SEWER PIPE AND FITTINGS

The plumber does not have a great variety of materials from which to choose in constructing a drainage system. The plumbing system consists of pipe and fittings assembled so as to waste and ventilate fixture traps. The system could not possibly function without the application of scientific principles relating to physical laws.

The kind of pipe generally used for sewers and drains, which is that part of the plumbing system installed in the ground, is limited to two materials. Vitrified clay pipe and cast-iron soil pipe. Either of these materials is permissible. Their installation, however, depends on local and state laws.

The materials most commonly used for waste, soil, and vent installations, which constitute the part of the plumbing system above the ground, are cast iron, galvanized steel and wrought iron, copper, brass, lead, or acid-resistant cast iron. Use of any one of these materials usually is determined by the architect's specifications. The factors that should be considered when choosing the materials for the drainage installation are the size and height of the building, the purpose for which the plumbing system is intended the materials in solution and suspension within the waste, etc.

The fittings used in connection with the pipe vary in design and construction. Fittings used in connection with waste lines must be of long radius and free from interior defect. Vent fittings do not require a long radius, although judgment should be used, because the air within the system is usually moving, and short radius fittings do offer resistance to free flow.

In the following paragraphs, kinds and variety of pipe and fittings are briefly discussed.

Vitrified Clay Pipe. Vitrified clay pipe and fittings generally are used for underground public sewers, house sewers, and drains. They are used commonly for storm as well as sanitary sewer installations.

The pipe is made of clay to which water has been added, and cast into lengths of 2 feet 6 inches. It is equipped with a bell and spigot end so it can be joined. After the pipe has been removed from the mold it is treated with glaze and fired in large kilns under temperatures of 2500°F. to make it impervious to moisture. All

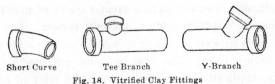

Short Curve   Tee Branch   Y-Branch

Fig. 18. Vitrified Clay Fittings

vitrified clay pipe used for plumbing installation must meet the requirements established by the American Society for Testing Materials (A.S.T.M.)

Vitrified clay pipe must be laid on substantial soil foundation.

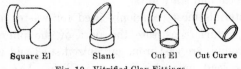

Square El   Slant   Cut El   Cut Curve

Fig. 19. Vitrified Clay Fittings

If the soil is of an unstable nature, clay pipe will sag and crack readily. This is partly due to the method used in joining the lengths. Should it become necessary to lay the pipe in swampy, unstable soil, an oak plank, or stone or concrete foundation must be provided.

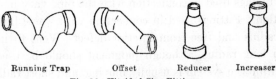

Running Trap   Offset   Reducer   Increaser

Fig. 20. Vitrified Clay Fittings

Vitrified clay pipe is resistant to most acids and is well fitted for underground work of this kind. The joints in this instance must also be made acid-resistant. The standard fittings used in connection with vitrified clay pipe are curves, **Y** branches, **T** branches, double **Y** branches, and short and long turn elbows.

The curve illustrated in Fig. 18 is used to complete changes of direction and offsets in the sewer. The **Y** and **T** branches shown in Fig. 18 are used for change of direction and connection of the branch drains to the main sewer.

The elbows and slant fittings in Fig. 19 are not used often. The traps, offsets, reducers, and increasers, Fig. 20, all have specific uses in sewer installations.

The following table gives the dimensions of clay sewer pipe as required by the A.S.T.M. It was taken from Recommended Minimum Requirements for Plumbing.

TABLE 1. Dimensions of Clay Sewer Pipe

Internal Diameter Inches	Laying Length Feet	Inside Diameter at Mouth of Socket Inches	Depth of Socket Inches	Minimum Taper of Socket	Thickness of Barrel Inches	Thickness of Socket
4	2	6	1½	1:20	9/16	The thickness
6	2	8¼	2	1:20	5/8	of the socket
8	2, 2½, 3	10¾	2¼	1:20	¾	¼ inch from
10	2, 2½, 3	13	2½	1:20	7/8	its outer end
12	2, 2½, 3	15¼	2½	1:20	1	shall not be less than
15	2, 2½, 3	18¾	2½	1:20	1¼	three-fourths
18	2, 2½, 3	22¼	3	1:20	1½	of the thick-
21	2, 2½, 3	26	3	1:20	1¾	ness of the
24	2, 2½, 3	29½	3	1:20	2	barrel of pipe.
27	2½, 3	33¼	3½	1:20	2¼	
30	2½, 3	37	3½	1:20	2½	
33	2½, 3	40¼	4	1:20	2 5/8	
36	2½, 3	44	4	1:20	2¾	
39	2½, 3	47¼	4	1:20	2 7/8	
42	2½, 3	51	4	1:20	3	

Cast=Iron Pipe. Cast-iron pipe has been used for drainage installations for many years. The material is well suited to the installation of house sewers, house drains, as well as soil, waste, and vent pipes in buildings of many types. The use of cast-iron pipe, however, should be limited to buildings less than twenty-five stories in height, and good judgment must be used as to its use in a building where constant vibration is present. Vibration affects the lead calk joint between pipe lengths, causing it to sag.

Cast-iron pipe is manufactured in lengths of 5 feet. See Fig. 21. The lengths are cast in two forms, single- and double-hub of standard and extra heavy specification.

Single-hub pipe is equipped with one hub and one spigot end. It is used, as a rule, in the installation of plumbing in its full length. Double-hub pipe is constructed with a hub on each end of it so it may be cut into two pieces when a short piece of pipe is needed.

Cast-iron pipe may be obtained either tarred or untarred. The tarred pipe is more acid-resistant, but the coating may cover cracks and defects in the pipe. Local regulations and practices vary as to type used.

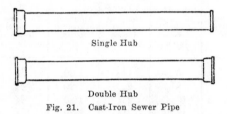

Single Hub

Double Hub

Fig. 21. Cast-Iron Sewer Pipe

Cast-iron pipe is affected to some extent by corrosion, which may be the result of chemical action in the system impossible to control. The air circulating in a plumbing system is always close

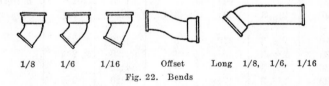

1/8　　1/6　　1/16　　Offset　　Long　1/8, 1/6, 1/16

Fig. 22. Bends

to the saturation point, with the result that carbon dioxide (CO_2), sulphur dioxide (SO_2), and methane (CH_4) gases, all acid in nature, are absorbed by it. These gases form solutions of carbonic (H_2CO_3) and sulphuric (H_2SO_4) acids, which attack cast iron and cause it to rust.

Cast-iron pipe is sometimes used to replace vitrified clay sewer pipe. It may be laid in unstable soil without danger of sagging. Common fittings used in connection with cast-iron pipe are **Y**'s, tees, sanitary tees, double **Y**'s and bends and elbows of 22½ to 90 degrees.

The bends shown in Fig. 22 are used to complete change of direction in soil, waste, and drain lines. The elbows in Fig. 23 are used at the base of soil and waste stacks and fixture openings, as

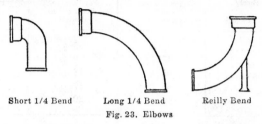

Short 1/4 Bend   Long 1/4 Bend   Reilly Bend
Fig. 23. Elbows

well as in change of direction on horizontal or vertical installations.

The **Y** and tee branches illustrated in Fig. 24 are used for change of direction and branch connections of soil, waste, and drain pipes. The fittings in Fig. 25 are used for special installations.

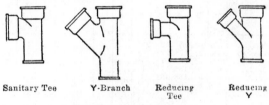

Sanitary Tee   Y-Branch   Reducing Tee   Reducing Y
Fig. 24 Y and Tee Branches

Tables **2** and **3**, taken from Recommended Minimum Requirements for Plumbing (A.S.T.M.) give the minimum weights of cast-iron pipe and fittings which may be used for plumbing installations. They are for the weight class known as extra heavy.

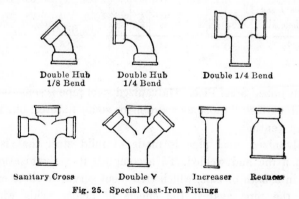

Double Hub 1/8 Bend   Double Hub 1/4 Bend   Double 1/4 Bend

Sanitary Cross   Double Y   Increaser   Reducer
Fig. 25. Special Cast-Iron Fittings

TABLE 2. Weights of Soil Pipe

Size (Inches)	Single Hub: Per 5-Foot Length Pounds	Single Hub: Per Foot Including Hub Pounds	Double Hub: Per 5-Foot Length Pounds
2	27½	5½	27½
3	47½	9½	47½
4	65	13	65
5	85	17	85
6	100	20	100

TABLE 3. Weights of Soil-Pipe Fittings

Only the staple fittings are shown. From the data herewith the weights of other fittings may be calculated. All values are in pounds

Fittings	Size of Fittings (Inches)														
	2	3	4	5	6	3 by 2	4 by 2	4 by 3	5 by 2	5 by 3	5 by 4	6 by 2	6 by 3	6 by 4	6 by 5
¼ bends, regular	6¾	10¼	15	19	23½	...	...	...	...	...	...	...	...	...	...
¼ bends, short sweep	8¼	12½	17¾	22½	27½	...	...	...	...	...	...	...	...	...	...
¼ bends, long sweep	10¼	15¾	22	27½	33½	...	...	...	...	...	...	...	...	...	...
⅕ bends	6¼	9¾	13¾	17½	21½	...	...	...	...	...	...	...	...	...	...
⅙ bends	6	9¼	13	16½	20	...	...	...	...	...	...	...	...	...	...
⅛ bends	5½	8½	12¼	15¼	18¼	...	...	...	...	...	...	...	...	...	...
1/16 bends	5	7¾	10¾	13¼	15¾	...	...	...	...	...	...	...	...	...	...
Return bends	8¾	14	20¼	26½	33½	...	...	...	...	...	...	...	...	...	...
Tees	10¼	15¼	21	26½	32½	13¼	16½	18¾	19¼	22	24½	22½	25	27½	30
Tapped tees (tapped up to 2 inches)	8¾	...	...	...	...	11¾	15	...	17¾	...	...	20½	...	...	...
Sanitary tees	11	16¼	22½	28	34½	14	17¼	19¾	20	23	25½	23	26	29	31½
Tapped sanitary tees (tapped up to 2 inches)	9	...	...	...	...	12	15¼	...	18	...	...	21¼	...	...	...
Y branch	11	17	24	31½	39½	14	17¼	20½	20	23½	27½	23	27	31	35
½ Y branch	10¼	15½	21½	27½	34	13½	16½	19	19½	22	25	22½	25½	28½	31
Tapped inverted Y branch (tapped up to 2 inches)	10¼	13¾	17½	21	24	...	...	...	...	...	...	...	...	...	...
Inverted Y branch	11½	18	25½	33	41½	15	18¾	22	22	25½	29½	25½	29½	33	37
Combination Y and ⅛ bend	12	18¾	27	35	44½	15	18¼	22½	21	25½	30½	24	29	34	39
Upright Y branch	12½	19½	28	36½	46	15¾	18¾	23	21½	26½	31½	24½	29½	35	40

Galvanized Steel Pipe. Galvanized steel pipe is sometimes used in plumbing installations as a drain for small fixtures, but its chief use is for venting.

Galvanized steel pipe is made of mild steel that is drawn through a die and welded. To accomplish its galvanization, it is dipped in a bath of zinc, which treatment serves to some extent to protect the pipe against the effects of acids. Acids which are harmful to cast iron will destroy galvanized steel. This is especially true if the steel pipe has not been thoroughly galvanized.

Galvanized Wrought=Iron Pipe. Galvanized wrought-iron pipe is better fitted for plumbing installations than steel pipe. It is constructed of wrought iron, dipped in molten zinc, and may be identified by its dull, greyish color. Tests have indicated that this type of pipe resists acid wastes more favorably than does steel, which is an important factor to consider when specifying plumbing materials.

The fittings used on steel and wrought-iron drainage pipe are the cast-iron recessed type, tapped with a standard pipe thread. Fittings for waste installations are of many varieties. Fig. 26 shows several standard patterns in use. Fittings of long radius or a

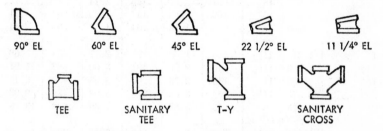

Fig. 26. Cast-Iron Recessed Drainage Fittings

combination of **Y**'s and elbows of 45 degrees should be used for making change of direction units. This practice overcomes stoppage, which usually is the result of short radius fittings or improperly planned work.

Table 4, on the page following, taken from Recommended Minimum Requirements for Plumbing, gives the specification of wrought-iron welded pipe.

Acid=Resistant Cast=Iron Pipe. Acid-resistant pipe usually is an alloy of cast iron and silicon. It should be used for installations which serve chemical laboratories, industries, or plumbing installations through which wastes of an acid nature are discharged.

Care must be exercised in handling acid-resistant pipe because it is brittle and cracks readily. Horizontal runs must be supported at 5-foot intervals to prevent breakage caused by sagging at the joint.

The pipe is cast in lengths of 5 feet and is of the single- and

TABLE 4.
Standard Weights and Dimensions of Welded Wrought-Iron Pipe

Size (Nominal Inside Diameter) Inches	"Standard Weight" Pipe				"Extra Strong" Pipe		"Double Extra Strong" Pipe	
	Outside Diameter Inches	Number of Threads Per Inch	Thickness, Inch	Weight of Pipe per Linear Foot, Threaded and with Couplings, Pounds	Thickness, Inch	Weight of Pipe per Linear Foot, Plain Ends, Pounds	Thickness, Inch	Weight of Pipe per Linear Foot, Plain Ends, Pounds
⅛	0.405	27	0.068	0.25	0.095	0.31	......	......
¼	.540	18	.038	.43	.119	.54	......	......
⅜	.675	18	.091	.57	.126	.74	......	......
½	.840	14	.109	.85	.147	1.09	0.294	1.71
¾	1.050	14	.113	1.13	.154	1.47	.308	2.44
1	1.315	11½	.133	1.68	.179	2.17	.358	3.66
1¼	1.660	11½	.140	2.28	.191	3.00	.382	5.21
1½	1.900	11½	.145	2.73	.200	3.63	.400	6.41
2	2.375	11½	.154	3.68	.218	5.02	.436	9.03
2½	2.875	8	.203	5.82	.276	7.66	.552	13.70
3	3.500	8	.216	7.62	.300	10.25	.600	18.583
3½	4.000	8	.226	9.20	.318	12.51	.636	22.85
4	4.500	8	.237	10.89	.337	14.98	.674	27.54
5	5.563	8	.258	14.81	.375	20.78	.750	38.55
6	6.625	8	.280	19.19	.432	28.57	.864	53.16
8	8.625	8	.277	25.00	......	......	.875	72.42
8	8.625	8	.322	28.81	.500	43.39	......	......
10	10.750	8	.279	32.00	......	......	......	......
10	10.750	8	.307	35.00	......	......	......	......
10	10.750	8	.365	41.13	.500	54.74	......	......
12	12.750	8	.330	45.00	......	......	......	......
12	12.750	8	.375	50.71	.500	65.42	......	......

double-hub variety. Acid-resistant pipe is difficult to cut because of its brittleness. Cutting can be accomplished with the use of a soil pipe cutter or by means of a small, cold chisel.

The fittings used in connection with acid-resistant pipe are of the same design as those used on cast-iron pipe. Y's, tees, and elbows of 1/16 to ¼ bend are those most commonly used.

Brass Pipe. Brass pipe is used extensively for waste installations in buildings constructed under federal specifications. It is a superior metal for this purpose, because of its smooth interior and its resistance against acids. Brass pipe is an alloy of zinc and copper, with the copper content as much as 85 per cent.

Brass pipe is drawn into lengths of about 20 feet and can be cut easily with a hack saw. It must be supported at intervals of from 8 to 10 feet.

The fittings used on brass pipe are of the cast brass, recessed drainage type, and are similar to those used on galvanized steel and wrought-iron pipe. The weight and quality and pipe specifications must be in accord with those established by the A.S.T.M.

Lead Pipe. Lead is probably the oldest material used for soil, waste, and vent pipe installations. Indications of lead wastes have been discovered by archeologists in their excavations of Roman, Egyptian, and Greek architecture of thousands of years ago. Do not use lead for water-supply pipes (particularly where soft water is used) as it can corrode and produce poisonous matter.

Lead pipe is very ductile, and, because of this feature, it can be placed readily. It must be well supported, however, because it deteriorates rapidly if permitted to sag. It is resistant to most acids and is commonly used on installations subjected to acid wastes. Its smoothness facilitates free passage of waste matter.

Lead pipe is drawn through a die and may be obtained in various lengths. Lead pipe of less weight than that indicated in the following table should not be used for wastes.

Inside Diameter, Inches	Weight per Foot
1	2 lb. 0 oz.
1¼	2 lb. 8 oz.
1½	3 lb. 8 oz.
2	4 lb. 0 oz.
3	6 lb. 0 oz.
4	8 lb. 0 oz.

From Wisconsin State Plumbing Code

Copper Pipe. Copper pipe and fittings may be used in the plumbing installation for waste, vent, and water pipes. Copper pipe is not commonly used for waste and vent pipe, because the cost is excessive. Ammonia contained in wastes from plumbing fixtures tends to dissolve the copper oxide, which is a protective coating occurring in the waste line, leaving the pipe exposed to the acids contained in the waste discharge.

Copper tube used for waste pipe installations should be rigid to

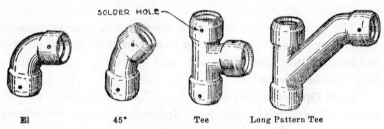

SOLDER HOLE

El   45°   Tee   Long Pattern Tee

Fig. 27. Drainage Fittings, Sweat Joint Type

overcome sagging. The pipe can be obtained in any desired length and is cut to size by means of a hack saw.

Fittings used on copper waste pipe are of the sweat joint variety. Standard fittings in use are **Y**'s, tees, and $11\frac{1}{4}$ to 90 degree short and long turn elbows, Fig. 27. The joint must be made of solder consisting of 50 per cent lead and 50 per cent tin. The process of making a sweat joint will be detailed in the following chapter.

CHAPTER IV

DRAINAGE SYSTEM PIPE JOINTS

The plumbing industry does not have as many skills associated with it as do other trades. It consists largely of joining pipe of various kinds and materials, incorporating in the completed system many scientific principles. A plumber must know what kind of joint to use for certain materials and be able to produce it and make it water- and air-tight.

The plumbing installation contains many noxious gases which would be detrimental to good health were they permitted to enter an occupied building through leaking joints. These gases affect the delicate breathing mechanism of the human body and therefore must be controlled.

Imperfect joints allow the liquid content of the waste pipe to escape and, in this event, property damage is likely to result. As a general rule soil and waste lines are concealed in partitions under the floor and, should a joint fail, painted walls and plastered ceiling will be damaged.

A leaking joint on a sewer or drain, installed below the surface of the soil, would allow sewage to enter the subsoil and contamination of drinking water might result. A joint of this kind also might allow soil to enter the drain and thereby cause stoppage of the line. One of the common causes of leakage of underground joints is the entrance of fine fibrous tree roots. Once the roots have gained entrance they spread quickly until the entire area of the sewer is filled. Correction of this difficulty can be accomplished only by digging up the drain and replacing it with new material.

All the joints which make up a plumbing system therefore must be made with care and when finished must represent careful workmanship. There are many kinds of joints that may be used for joining the pipe of a drainage system. The following are those which are used most commonly.

Cement Joint on Vitrified Clay Pipe. Fig. 28 illustrates a joint which is used to join vitrified clay pipe. The bell and spigot ends of the pipe end must be cleaned of foreign material and the spigot end of the pipe inserted into the bell of the pipe to be joined. It is necessary to align the pipe and space the joint evenly before the cement mortar is applied. The mortar consists of equal parts of

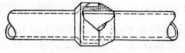

Fig. 28. Cement Joint

clean, sharp sand and cement, to which water is added until a sticky mass is formed. The mixture is squeezed into the bell until it is completely full, then it is banked on the pipe for about 2 inches and smoothed with a trowel.

Scraper   Swab   Trowel
Fig. 29. Tools Used in Making a Cement Joint

A swab or scraper should be used to remove the mortar which has squeezed into the pipe, in order to avoid obstruction of the sewer.

Mechanical Seal Joint. A new improved type of interlocking mechanical compression joint, made part of the vitrified pipe joint at the factory, is now replacing the cement joint wherever vitrified clay pipe is used. This "speed seal" is made of permanent polyvinyl chloride and called a plastisol joint connection.

Easily laid by one man, the joint is made by inserting the spigot end into the bell or hub, then giving the pipe a strong push to make the spigot lock into the hub seal. A solution of liquid soap, spread on the plastisol joint, will help the joint slip into place.

When connecting vitrified clay pipe to cast-iron soil pipe or soil pipe to vitrified pipe, special mechanical seal adapters are available.

The mechanical seal compression joint pipe ensures the home owner of tight joints that are root proof, infiltration proof, flexible, and collision resistant. They can be installed quickly and easily, and are manufactured in sufficient variety to meet any type of installation problem. There are other types of mechanical seal joints using practically the same method of installation.

Cast=Iron Pipe to Vitrified Clay Pipe. In localities where vitrified clay sewer is permitted, the base fittings for fixture connections and soil and waste pipes usually are of cast iron. It is rather difficult to scrape a joint of this kind and therefore an extra precaution is necessary.

The fitting must be inserted into the clay pipe and, when aligned properly, a ring of oakum is yarned into the bell of the clay pipe, Fig. 30. The cement mixture is then pressed into the bell. The ring

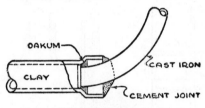

Fig. 30. Cement Joint, Showing Oakum
Packing

of oakum prevents the mortar from entering the sewer and thus eliminates the necessity of swabbing. The joint must be well banked and troweled to make it durable and water-tight.

In some instances, especially on municipal sewers, a wrapping of light canvas is applied to the soft cement joint. This practice assures additional strength and prevents the joint from cracking readily during the drying process.

Composition Joints on Vitrified Clay Pipe. There are a number of compounds manufactured which, when local or state authorities permit, may be used for joining vitrified clay pipe. The compounds generally are of a plastic nature and are applied to the joint in the same manner as the cement mixture.

Vertical Calk Joint. Cast-iron soil pipe is joined by means of a calk or percussion joint. This type of joint is made of molten lead and oakum and calked with irons to make it water-tight.

In making a vertical calk joint the first process involved is to wipe the hub and spigot ends of the pipe dry and also wipe them clean of foreign material. Moisture in the hub causes splitting of

the metal when the molten lead is poured, and because of the great
difference in temperature may result in a small explosion. The pipe
to be joined must then be aligned and the joint carefully spaced.
Oakum, which is hemp treated in pitch to make it moisture proof
and resistant to the elements contained in the waste, is yarned or
packed into the joint to a depth of about 1 inch from the top.

The lead is heated on a plumber's furnace and when it appears a
cherry red color it may be poured into the hub with the aid of a ladle.
Some mechanics use two pours of lead to fill the joint completely.

The calking irons, Fig. 31, are used to pack the oakum and lead,
making the joint watertight. These irons are used for both vertical

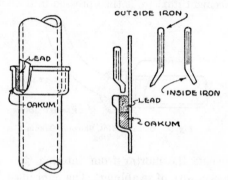

Fig. 31.   Vertical Calk Joint

and horizontal calk joints. As Fig. 31 illustrates, the angle of the
iron makes it possible to calk both inside and outside edges within
the hub. The iron is slowly moved around the joint and tapped
gently with a light ball peen hammer. This phase of the operation
requires great care. For if the calking iron is struck too hard, the
brittle cast-iron hub may crack, necessitating replacement of both
the pipe and hub.

When the calking is properly finished, the lead should be a dull
gray color. The next step is to smooth the lead, giving the pipe joint
a neat appearance, by using a wide iron.

In large plumbing shops, pipe joints are calked on a special
bench and pipe rack. The pipe is held rigid with a chain type clamp,
and a compressed air calking gun is used. The calking iron is
shaped like a broad screw driver bit, and the compressed air gun is
trigger operated. The primary advantage of this tool is increased

speed of operation. However, in many installations not suited for fabrication by a large plumbing shop, the plumber must calk joints by hand. There are fittings manufactured to fit practically every angle or bend.

Horizontal Calk Joint. The process involved in making a horizontal calk joint differs slightly from that used in making a vertical joint. After the oakum has been yarned into the hub, a pouring rope must be applied to the pipe, Fig. 32.

One type of pouring rope is made of asbestos, Fig. 32, and is held in place with a clamp. Another type, Fig. 33, is made of fiber and hard rubber, and fits snugly around the pipe and against the hub. It has a pouring groove on the upper side, making it easy to pour the molten lead.

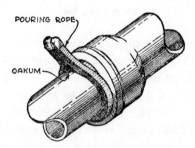

POURING ROPE

OAKUM

Fig. 32. Horizontal Calk Joint

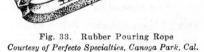

Fig. 33. Rubber Pouring Rope
Courtesy of Perfecto Specialties, Canoga Park, Cal.

Cutting Cast-Iron Soil Pipe. Cast-iron soil pipe frequently requires "cutting." To do this, all that is required is a hammer and a not too sharp cold chisel. Make a line around the pipe as a guide, to ensure making the piece the same length at all points. Then begin to hammer the chisel, pointing the chisel at a slight angle away from you, so that chips from the pipe will not fly into your face. Strike with quick moderate blows, moving the chisel a little forward on the line after each blow, to make a continuous dent all around the pipe. Continue working the dent until the pipe falls apart. The separating of soil pipe in this way is not a cutting process at all; it is simply packing the iron down in a line until the fiber of the iron is disturbed entirely through the wall, or at least sufficiently to wedge the pipe apart. Where the chisel strikes, the force tends to make the pipe longer, and the strain thus produced wedges it asunder. See page 52.

Acid=Resistant Calk Joints. Acid-resistant calk joints are made in the same manner as a calk joint on cast-iron pipe is made, except that in place of oakum, a material that is treated to make it acid-resistant is used. Care must be taken to calk the lead gently, because acid-resistant pipe is brittle and cracks readily.

Screw Thread Joint on Galvanized Steel and Wrought=Iron Pipe. Wrought-iron, steel, and brass pipe are joined by a screw thread joint. The screw-thread joint involves a number of definite

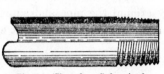

Fig. 34. Thread on Galvanized Iron Pipe

Fig. 35. Reaming Tool

processes which must be followed carefully to assure good results. After the pipe has been cut to length with the use of a pipe cutter, it must be wiped clean of oil and foreign materials.

The thread is cut with a stock and die designed to give the thread the necessary depth and taper, Fig. 34. The thread varies in length, depending on the diameter of the pipe. However, it is the

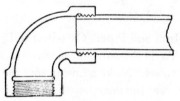

Fig. 36. Screw Thread Joint, Properly Shouldered

general rule to run the pipe completely through the die so that the thread will be long enough to butt against the recessed shoulder of the fitting. Before removing the stock and die from the pipe, the pipe must be reamed thoroughly with a reaming tool, Fig. 35. This is an important precaution and must be faithfully observed. An unreamed pipe produces an obstruction in the waste line. It actually reduces the diameter of the pipe about ¼ inch.

After the thread has been cut and the pipe well reamed, it must again be wiped free of oil, and a coating of white or red lead should be applied. The leaded thread can then be faced into the fitting and the pipe turned into it with a wrench until the leaded thread butts the shoulder, Fig. 36. Not more than three of the remaining threads must be exposed and these must be coated with iron paint of rust-proof quality.

Sweat or Solder Joint on Copper Pipe. Copper pipe with sweated joints frequently is used for waste pipe installation. The joint is made with string solder of about 50 per cent tin and 50 per cent lead. The pipe and fitting to be joined are first cleaned

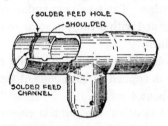

Fig. 37.   Solder or Sweat Joint on
Copper Pipe

thoroughly with sandpaper, and a non-acid flux applied to the ends to be soldered. Then both pipe and fitting are heated, with the solder being applied either directly to the surface or through the solder holes, as in Fig. 37. The solder flows quickly over the cleaned copper surface and will produce a joint capable of withstanding great pressure. Some localities have codes specifying the use of silver solder only, others make use of the 50–50 ratio of lead and tin. Advantages of the sweat joints are that they can be done faster than lead calked or cut and threaded pipe joints.

Wiped Joint on Lead Pipe. A wiped joint on lead pipe is the most difficult joint to accomplish of any used on the drainage system. Skill in this work requires many hours of practice and patience on the part of the apprentice.

There are many kinds of wiped joints used on lead waste pipe. The processes involved vary to a considerable extent. Every mechanic uses a method of his own to produce the desired result, though some elements of procedure are basic.

Because it would require many pages to illustrate the method involved in making each type of wiped joint, even though the text detailed clearly the procedure followed in the operation, the apprentice could not hope soon to become really efficient in this phase of plumbing practice. The only method of achieving this skill is actually to pour solder under the supervision of an experienced mechanic.

The most commonly used joints on a waste line are the horizontal round joint, the vertical round joint, and the 45° branch joint. In preparing a round joint of the horizontal type, the ends of the pipe to be joined must be squared with a rather coarse rasp. The spigot end must be tapered with the rasp for about ¼ of an inch, almost to a featheredge. The receiving end must be tapered in like manner and then opened with a turnpin so as to allow the spigot end to enter about ⅛ inch.

After the joint has been well fitted, a thin coat of soil or paste and ground pumice is applied to both ends of the pipe and extended on its surface for about 4 inches.

When the soil has dried, the joint is cleaned or shaved to the distance of 1 inch on each side of the union with the aid of a shave hook. It is not necessary to wipe a joint on any pipe, whether of small or large diameter, longer than 2 inches. The edges of the joint should be cut deeper than the tapered end by applying greater pressure to the tool. This practice allows for a smooth and substantial edge. The pipe must be supported substantially before starting the actual wiping process and be set sufficiently high to allow the mechanic to place his hand under the pipe to catch solder which may drop from the joint.

Wiping of the joint is accomplished with the aid of a cloth or pad made from a good quality of herringbone ticking folded into the desired thickness and size. A cloth 2¾ inches wide and 3 inches long, treated with wax or tallow to prevent the solder sticking to it, is adequate. The cloth is held in the left hand, with the stripes of the ticking at right angles to the length of the fingers. The solder consisting of about 37 per cent tin and 63 per cent lead is then heated on a plumber's furnace to a temperature that will just scorch a piece of paper.

The cloth is held under the cleaned surface and the solder is applied with a ladle. Drippings from the pipe are caught on the cloth and placed on the top of the cleaned surface to avoid burning of the pipe. The solder is poured and placed until a substantial amount of it has accumulated on the pipe. It can then be packed by cleaning the outside edges with the cloth, allowing it to remain high at the union of the pipe. By spreading the forefinger and the middle finger, after the edges have been cleaned, the solder can be shaped roughly into a convex form. Wiping of the joint is accomplished by putting pressure on the spread fingers to clean the edges, using the edge of the cloth to make the joint smooth. Some mechanics use two cloths to pack the joint and then use both hands in holding the wiping cloth to complete the joint.

Preparation of the vertical joint is identical with that of the horizontal joint. The wiping process varies to some extent however. The solder is splashed on the cleaned surface, the wiping cloth serving as a means of lifting accumulated solder to the union of the pipe. Once the heat required has been attained the joint is packed, using the cloth first in one hand and then in the other. The top edge must be wiped first, because the joint tends to cool at that point. After this has been done the forefinger and middle finger must be spread and the joint can be wiped in the same manner as is the horizontal one, except that both the hands must be used. A vertical joint requires less heat than does a horizontal one.

The branch joint on waste pipe is wiped in a vertical position. In some instances, however, this practice is impossible. The spigot end of a branch joint is prepared by rasping it diagonally (about ¼ of an inch should be sufficient) and then tapering the end in the same manner as for the spigot end of a horizontal joint.

A hole about the size of a dime must be made with a tap bore at the point of the branch. The bent end of a turning iron is inserted and the hole is increased in size by raising a shoulder of the lead to receive the spigot end of the pipe. The joint must be soiled in the same manner as is a horizontal joint. It can then be set temporarily to lay out the surface to be cleaned. Usually a 1-inch distance is laid out on the vertical spigot end and 1 inch on both sides of the horizontal piece.

A compass is used to scribe the horizontal portion of the joint and the two half circles are joined by a straight line on the side of the horizontal piece. The spigot end of the pipe and the horizontal piece are then shaved from the outside line to the union of the joint.

The joint must be well supported and the solder applied by the splashing method. After sufficient solder has accumulated the top edge of the spigot end can be wiped. Starting from the side farthest

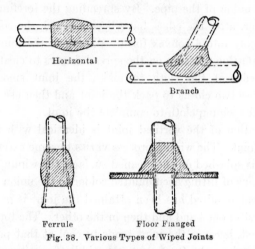

Fig. 38. Various Types of Wiped Joints

away from the mechanic, the left and then the right edge of the branch can be cleaned. Alternate hands must be used in this process. Fig. 38 shows the kinds of wiped joints most commonly used.

Chain-Type Cast-Iron Pipe Cutter. A special chain-type cast-iron pipe cutter is also available. It consists of an adjustable chain containing several cutting discs. One end of the chain is fastened to a handle, the other is looped around the pipe and tightened by means of a screw arrangement on the same handle. The handle is then moved backward and forward and the chain is gradually tightened, as the cutters cut into the pipe, until the pipe is cut completely through. This type of cutter is particularly useful in cutting cast-iron pipe which is already installed. It provides a smooth, even cut, with a minimum of disturbance to the installation.

CHAPTER V

THE HOUSE SEWER

The house sewer is that part of the drainage system beginning just outside the foundation wall and terminating at the main sewer in the street, Fig. 39. The terminal of a house sewer can also be a septic tank.

The house sewer is an important part of the plumbing system. Much of the efficiency of the drainage installation depends on this part of the plumber's work, and he must use care in producing it. Good judgment should be used in the selection of materials for house

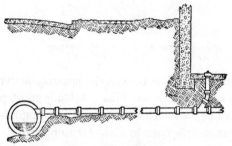

Fig. 39. House Sewer

sewer construction. When an installation is made in soil of an unstable nature, it is advisable to use cast-iron pipe. This is also true where the soil contains tree roots or is subjected to vibration of any kind. Cast-iron pipe is well suited to house sewer installations, and though the original cost is somewhat higher than for vitrified clay pipe its installation is recommended.

Vitrified clay pipe serves house sewer requirements adequately, but its use should be limited to substantial soil. It is unsatisfactory in swampy, loose, unstable ground.

One of the common difficulties which occurs in vitrified clay sewers, due largely to poor workmanship, is the entrance of fine tree roots which grow and eventually fill the entire inside area of the sewer. The mechanic must make the joints of the house sewer carefully to avoid this trouble. A well proportioned cement mortar and thorough troweling of the joint add to its sound construction.

Vitrified clay pipe should always be used in soil containing cinders or ashes. Water in contact with these elements produces an acid condition which is detrimental to cast-iron pipe. The joints in this case, however, must be of an acid-resistant nature.

Connection with the Main Sewer. In a preceding chapter, sewers constructed of concrete, brick, and vitrified clay pipe were discussed. Each of these types of sewers requires a special kind of house sewer connection. Efficiency of the house sewer can be increased by making a good connection at the main.

Fig. 40 illustrates the method of making a connection into a

Fig. 40. House Sewer Connection to Concrete Main

concrete main sewer. The hole in the concrete sewer is made with a well-sharpened point chisel. The mechanic must proceed slowly, making a small hole through the sewer wall, then gradually enlarging it until it is sufficiently large in diameter to receive the sleeve, which is cut from vitrified clay pipe. Precaution must be taken against breaking down the interior wall of the public sewer. This type of connection generally is supervised by local authority.

After the hole in the sewer has been completed, the wall thickness can be measured and a sleeve of vitrified clay pipe of this length may be cut. As the illustration indicates, the exterior surface of the bell rests on the outside wall of the sewer, and the sleeve does not extend into the sewer but is flush with the inside wall. This practice prevents the sleeve from falling away from the next pipe connection and also eliminates obstruction of the sewer.

As a rule the fitting used to start the house sewer into the building is a long sweep curve. The curve is inserted into the sleeve and a substantial cement joint is made to hold the sleeve in place and strengthen the damaged sewer wall.

The connection into a concrete sewer, as Fig. 40 illustrates, is made above the flow line of the sewer, entering it at a 45-degree angle or directly into the top.

Connections in Brick Sewers. Public sewers are no longer being constructed of brick. The connections into them, however, are identical with those of concrete sewers.

Connections in Vitrified Clay Sewers. Connections usually are provided in the vitrified clay type of public sewer for each piece of plotted property. They consist of **Y** fittings, usually of 6-inch

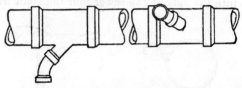

Fig. 41. Vitrified Clay Sewer Connection

diameter, sealed with a clay disc. Connection of the house sewer consists of the removal of the disc and the insertion of a clay curve of 45 degrees. A cement joint made of equal parts of sand and cement

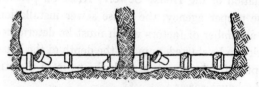

Fig. 42. Method of Inserting a New Y

mixed with water completes this type of sewer connection, Fig. 41.

Y connections in the public sewer generally are recorded by local administrative authority, and the plumber can obtain their location at the proper agency. Often, however, upon opening the street, the measurement indicated proves false and the **Y**'s branch cannot be located. In this event a new connection must be made.

The best procedure in this case is the insertion of a new **Y** in the public sewer. This can be accomplished in two ways. The most practical method of inserting a **Y** is to remove three lengths of pipe from the sewer and then fit in a new **Y**. Although this is difficult and involves expense it is the most advisable procedure.

The other method of inserting a new **Y** is accomplished by removing one length of pipe from the sewer. The top portion of the bell must be removed from the permanent installation to permit insertion of the **Y**. The **Y** fitting is dealt with similarly, using care to remove the top portion of the bell above the flow line of the sewer. The **Y** can then be dropped into place and turned in the direction of the house sewer. The removed parts of the bell are now cemented in place with a substantial application of cement mortar, Fig. 42.

In some localities, a slant connection can be used to make a new connection into a vitrified clay sewer, see Fig. 19. In this case a hole must be cut into the side of the sewer and the slant fitted carefully

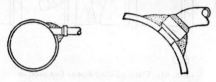

Fig. 43. Method of Banking Cement Mortar
at Public Sewer Terminal of House Drain

against it. The joint can be completed by banking a substantial amount of cement mortar around the connection, Fig. 43.

Installation of the House Sewer. After all permits have been paid to the proper agency, the house sewer installation can begin. There are a number of factors which must be determined before the actual work can be started, namely, the depth of the house drain outlet, the depth of the connection with the main in the street, and the grade of the house sewer.

The depth of the house drain outlet can be established by measuring the length of the longest branch of the house drain and multiplying it by the predetermined pitch per foot it is to be given. To the result must be added the required ground covering, which factor usually is established by local authority. Twelve inches of ground covering from the top of a concrete floor and 18 inches of ground covering without a concrete floor are considered adequate.

For example, from the architect's print, it is found that the length of a house drain is 240 feet and the predetermined pitch or grade is established at ¼ inch to each foot. Thus 240 feet x ¼ inch = 60 inches on this installation, and 60 inches plus 12 inches of required floor equals 72 inches, the depth of the house drain outlet below the finished basement floor. Fig. 44.

The depth of the connection with the main in the street must next be established. This can be done by removing the manhole covers on both sides of the proposed connection, and, with the use of a long stick, the actual depth of the sewer can be obtained.

Should the manholes be 200 feet apart and the tentative connection be midway between them, and, after obtaining the depth, it is

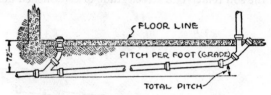

Fig. 44. Depth of the House Drain Outlet

found that the sewer is 10 feet 1 inch from the street grade in one manhole and 9 feet 11 inches in the other, the connection will then be 10 feet below the street grade.

The depth of the house drain outlet and the depth of the con-

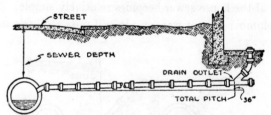

Fig. 45. Total Pitch of the House Sewer

nection in the street must be known to determine the grade of the house sewer. The total grade of the house sewer may be found in the difference between the depth of the house drain outlet and the depth of the connection in the street. For example, suppose the basement floor is 7 feet below the street grade and the house drain outlet is 6 feet below the basement floor, the house drain will then be 13 feet or 156 inches below the street grade. Should the public sewer be 16 feet below street grade, the total pitch of the house sewer would be 192 inches less 156 inches, or 36 inches of total fall, see Fig. 45.

In order to arrive at the pitch per foot the house sewer location must be given and its length determined. The total pitch is divided by this figure. For example, suppose the length of the house sewer is 144 feet, then 36 inches divided by 144 feet will equal ¼ of an inch grade to each foot.

A leveling instrument can be used to advantage. Very often the house sewer is installed before the basement floor grade has been laid out. Then it becomes necessary to make a general survey of the contour of the soil relative to street grade or established bench marks

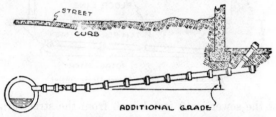

Fig. 46. House Sewer Graded More Than 1/4 Inch to the Foot

and in this way determine the grade of the house drain outlet. In most instances architects provide beginning marks. In this event, laying out of the house sewer becomes relatively simple.

The plumber cannot rely on hit-and-miss methods to establish

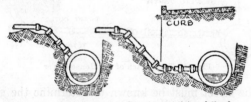

Fig. 47. Fittings Used to Provide Additional Grade

the pitch of the sewer. Uncertain procedure often results in house sewers being needlessly above basement floors. The usual pitch of a house sewer is not less than ¼ inch to each foot. However, should it be necessary because of building difficulties, less pitch per foot may be allowed.

Additional Grade. It is not unusual to pitch a house sewer more than ¼ inch per foot. This practice becomes necessary when the house drain outlet is sufficiently high. The general procedure under these circumstances is to grade the part of the house sewer between the main sewer and the curb not more than ¼ inch per foot and the remainder of the sewer, between the curb and the house drain outlet, not more than ¼ inch. This practice overcomes the necessity of deep, costly digging and adds to the installation materially. Fig. 46

Additional grade can also be made with the use of two curves or ⅛ bends, see Fig. 47. This is the most practical method of obtaining grade.

Size of the House Sewer. The size of the house sewer, for ordinary residence installation, has been established by sanitary authorities whose compilations are the result of installation tests and mathematical conclusions. Standard practice requires that the minimum size of a house sewer constructed of clay pipe be not less than 6 inches in diameter. It has been found that a sewer of this kind has ample discharge capacity and serves the installation adequately.

If cast-iron pipe is used, the size of the sewer can be reduced to a 4-inch diameter pipe, provided local authorities sanction this installation. In many instances a 6-inch diameter pipe is installed from the main in the street to the curb line and then may be reduced to a 4-inch pipe to its connection with the house drain. This practice averts the necessity of breaking up the street should additions be made to the original plumbing installation.

When sizing a house sewer for hotels, apartment houses, and industrial and commercial buildings, the accumulated fixture discharge must be taken into consideration. The problem becomes a complicated one involving a certain amount of conjecture. Overlapping of fixture discharge and simultaneous use of fixtures are also important. The task of ascertaining house sewer size, considering these factors, is beyond the capacity of the average installing mechanic. Fixture unit discharge values are usually found in local plumbing regulations, recommended codes, and similar publications. From this information it is possible for him to arrive at fairly accurate conclusions in determining house sewer size.

Trenches in Various Soils. *Clay Soil.* Very little difficulty is experienced in laying a house sewer in clay soil. The only precaution necessary is that the trench floor be graded properly and that the house sewer be laid in perfect alignment on the solid floor of the trench. See Fig. 48. Care should be exercised in backfilling the trench that no large pieces of ground be permitted to fall on the newly laid sewer.

Sandy or Unstable Soil. When sandy soil is encountered, the laying of the house sewer becomes more complex. The first difficulty the plumber may encounter is caving in of the trench walls. To overcome this, shoring is done. This is accomplished by driving in

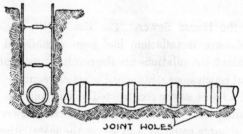

JOINT HOLES

Fig. 48. House Sewer Laid in Clay Soil

1x6 boards, bracing them at close intervals with horizontal **2x4's** **or** jacks of the wedge or screw variety, Fig. 49. If water is encountered, it becomes necessary to pack the shoring with hay or

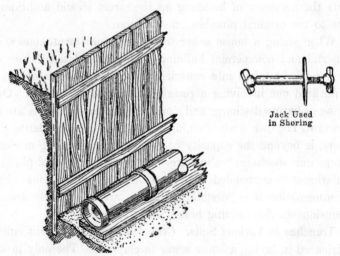

Jack Used in Shoring

Fig. 49. House Sewer Laid in Unstable Soil

straw, and possibly drive interlocking metal forms. In such case, **it** may become necessary to lay the house sewer on concrete or oak **plank** foundation, as illustrated in Fig. 49. The plumber should **work** **slowly** and carefully under these conditions.

House Sewer for Buildings on the Same Lot. Under certain special conditions, it may be necessary for one house sewer to serve two buildings located on the same lot. The buildings, however, should be one back of the other and constructed on a lot which cannot be divided by sale. See Fig. 50. It is inadvisable to permit one sewer to serve more than one building under other circumstances and, generally, plumbing codes prohibit this type of installation, except for unusual situations (for example, where a private sewer is not available and cannot be constructed for one of the buildings).

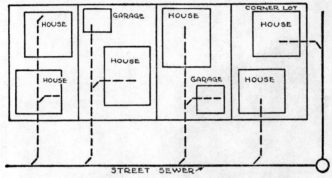

Fig. 50. Layout Showing How a Sewer May Serve Two Buildings

PROCESS OF APPLYING ENAMEL TO A PLUMBING FIXTURE

The iron casting is heated to a high temperature and the enamel, in powder form, is sifted on and fused by the heat into a hard, smooth, glossy covering in white or pastel tints.

Courtesy of the Kohler Co., Kohler. Wis.

CHAPTER VI

THE HOUSE DRAIN

The house drain is that part of the plumbing system which receives the discharge of all soil and waste stacks within the building and conveys it to the house sewer, Fig. 51. It may be installed under-

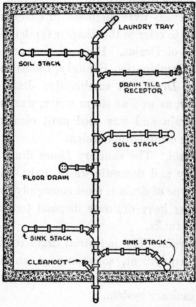

Fig. 51. House Drain Installation

ground, or it may be suspended from the basement ceiling. It is sometimes referred to as the collection line.

In most buildings, especially residences, the house drain can be installed under the basement floor. There are a number of advantages gained by this method of installation. It is more economical and does not reduce the headroom in the basement. Also, an unsightly network of piping is eliminated when the drain is placed below the floor.

In the larger types of buildings, because of their exceptionally deep basements, it may become necessary to suspend the house drair. from the basement ceiling. In doing this, the mechanic must use good judgment in its location, the manner in which it is suspended, grade, change of direction, and other important considerations.

Under all circumstances, however, the mechanic must use precision in the installation of the house drain, because the efficiency of the waste and soil system is dependent primarily on the accuracy of the house drain installation.

Materials Used. The materials which can be used in the construction of the house drain are vitrified clay pipe and cast-iron soil pipe. These materials were discussed in Chapter III. It may be well for the reader to refer to this chapter for detailed information.

Classification of Drains. House drains are classified in four ways as follows: Combination, Sanitary, Industrial, and Storm.

Combination Drain. A combination drain receives the discharge of sanitary, as well as storm water, wastes. It is the oldest form of house drain and was used most commonly when public sewers were of the combination design.

Sanitary Drain. The sanitary house drain receives the discharge of sanitary and domestic wastes only. All storm water is excluded. This type of drain is most commonly used today, because most modern cities have drainage disposal terminals consisting of sewage treatment tanks.

Industrial Drain. The industrial drain receives the discharge from industrial equipment, sometimes of an objectionable acid nature. It must be terminated into some drainage basin not associated with the sanitary system.

Storm Drain. The storm drain receives the discharge of all storm, clear water, or surface water wastes. All sanitary wastes must be excluded. Its terminal is usually a river, dry run, lake, or natural drainage basin.

Sizing the House Drain. To determine the proper size of a house drain, its classification must be known. If it is a drain of the combination type, still used, although not as widely, the roof area is the deciding factor in determining the proper size. Should the sewer be one of the sanitary variety, the fixture discharge is the governing factor. Because of the complexity of the problem, the responsibility of sizing the house drain is not generally assumed by the plumber. However, the sizing of the house drain is simplified by means of charts formulated for use in plumbing codes. Engineering data compiled over a long period, and the experience of installations giving satisfactory service, are the bases for these charts. State and local codes govern the house drain installation to a large extent.

It is impractical, even though the code and fixture discharge rating would permit, to install a vitrified pipe of 4-inch diameter. A 4-inch tile is impractical because it does not always run true to size, and it is difficult to lay without forming cocked and poor cement joints which may cause stoppage. It is also subject to breakage, due to settling and backfill of earth into the open trench. For this reason a clay drain should at no time be constructed of pipe less than 6 inches in diameter.

When iron pipe is used there is a somewhat better condition. The 4-inch pipe is generally full size in diameter and can be laid as satisfactorily as the 6-inch pipe. Codes and fixture load govern the installation of this size of pipe, and if these requirements come within the service possibilities of a 4-inch pipe, the installation is served efficiently.

Many mechanics believe that if they make the drain a little larger than is necessary they will increase its efficiency. However, this is not the case. Scouring action is not secured by increasing the size of the drain. The solids are carried along the bottom of the pipe and, because the water flow within the larger pipe is shallow and slow, they become separated from the water and remain in the drain. The result of this may be stoppage of the drain branch, and often the entire house drain is affected.

A drain, therefore, should be of proper size with a flow of about 50 per cent of the pipe diameter. This assures a scouring action and overcomes many house drain difficulties. On the other hand, a drain too small in size is overtaxed by flow and is apt to produce siphonage, back-pressure, and basement flooding.

The most practical method to use in determining the size of a house drain is the unit system. The unit system has been formulated from tests conducted by the Uniform Plumbing Code Committee, a body consisting of representatives of management, labor, and agencies of the Government. Standard plumbing fixtures were individually tested and the amount of liquid waste which could be discharged through their outlet orifices in a given interval was carefully measured. It was found that a wash basin, which is one of the smaller types of plumbing fixtures, would discharge through its waste approximately 7½ gallons of water in a 1-minute interval. This volume was so close to a cubic foot of water that the Committee decided to establish it as a basis of the unit system and called the discharge of the wash basin "one fixture unit." Hence, one fixture unit represents approximately 7½ gallons of water. Sinks, bathtubs and other fixtures that were tested revealed that their waste discharge was greater than that of a lavatory and, naturally, the unit values of these fixtures were larger. The following list presents the findings of the Uniform Plumbing Code Committee and may be used to determine the volume of waste, expressed in fixture units, that a soil pipe, waste, or drain may be subjected to.

Fixture Unit Values

	Units
Lavatory or wash basin	1
Kitchen sink	2
Bathtub	2
Laundry tub	2
Combination fixture	3
Urinal	5
Shower bath	2
Floor drain	1
Slop sink	3
Water closet	6
One bathroom group (consisting of water closet, lavatory, bathtub and overhead shower, or water closet, lavatory, and shower compartment)	8
180 square feet of roof drained	1

Further tests on horizontal drains indicated that given diameters of pipe would discharge up to a certain number of units without subjecting the system to minus or plus pressures. Tests were made of varied installations of standard design. Change of direction, materials, grade and many other factors were carefully considered, and the discharge capacities obtained may be regarded as authoritative. Tables 5 and 6, presented on the following page, were compiled as a result of experiments authorized by the committee which established Recommended Minimum Requirements for Plumbing.

TABLE 5. Sanitary Drain Sizes

Diameter of Pipe (Inches)	Maximum Number of Fixture Units			Diameter of Pipe (Inches)	Maximum Number of Fixture Units		
	Slope, 1/8 Inch Fall to 1 Foot	Slope, 1/4 Inch Fall to 1 Foot	Slope, 1/2 Inch Fall to 1 Foot		Slope, 1/8 Inch Fall to 1 Foot	Slope, 1/4 Inch Fall to 1 Foot	Slope, 1/2 Inch Fall to 1 Foot
1¼............	1	1	1	5............	162	216	264
1½............	2	2	3	6............	300	450	600
2.............	5	[1]6	[1]8	8............	990	1,392	2,220
3.............	[2]15	[2]18	[2]21	10............	1,800	2,520	3,900
4.............	84	96	114	12............	3,084	4,320	6,912

[1]No water-closet shall discharge into a drainpipe less than 3 inches in diameter.
[2]Not more than two water-closets shall discharge into any 3-inch horizontal branch, house drain, or house sewer.

Note: Table 6 deals only with storm drains. Methods in determining storm drain size will be included in Chapter VIII. However, it is essential to use Table 6 in estimating the size of the combination type of house sewer.

TABLE 6. Storm Drain Sizes

Diameter of Pipe (Inches)	Maximum Drained Roof Area (Square Feet)[1]			Diameter of Pipe (Inches)	Maximum Drained Roof Area (Square Feet)[1]		
	Slope, 1/8 Inch Fall to 1 Foot	Slope, 1/4 Inch Fall to 1 Foot	Slope, 1/2 Inch Fall to 1 Foot		Slope, 1/8 Inch Fall to 1 Foot	Slope, 1/4 Inch Fall to 1 Foot	Slope, 1/2 Inch Fall to 1 Foot
3..........	865	1,230	1,825	8..........	11,115	15,745	24,890
4..........	1,860	2,610	4,170	10.........	19,530	27,575	43,625
5..........	3,325	4,715	7,465	12.........	31,200	44,115	69,720
6..........	5,315	7,515	11,875	14.........	42,600	60,000	95,000

[1]The calculations in this table are based on a rate of rainfall of 4 inches per hour.

With the information contained in the preceding tables the plumber can establish the total discharge of all the fixtures in a building in units and can select a size of drain to serve the demand.

In order to familiarize the student with the unit system and its application to an installation of plumbing, the following example is offered.

Suppose a plumbing installation consisted of 30 water closets, 28 wash basins, 4 shower stalls, 3 urinals, 2 combination fixtures, 4 floor drains, and 2 slop sinks. The sum of the unit values of all the fixtures would be as given in the following summary:

Fixture Unit Summary

No.		Unit	Total
30	Water closets	× 6 =	180
28	Lavatories	× 1 =	28
4	Showers	× 2 =	8
3	Urinals	× 5 =	15
2	Combinations	× 3 =	6
4	Floor drains	× 1 =	4
2	Slop sinks	× 3 =	6
	Total fixture units		247

According to Table 5 (extending the diameter with a slope or pitch of $\frac{1}{4}$-inch to each foot) a 6-inch house drain would be required. This would apply only to a sanitary type of house drain.

The combination drain offers a more complex problem because of the additional roof area which must be served. A conversion table which simplifies the difficulty is here presented.

TABLE 7. Conversion Factors for Combined Storm and Sanitary System

Drained Roof Area, in Square Feet	Number of Fixture Units on Sanitary System														
	Up to 6	7 to 18	19 to 36	37 to 60	61 to 96	97 to 144	145 to 216	217 to 324	325 to 486	487 to 732	733 to 1,098	1,099 to 1,644	1,645 to 2,466	2,467 to 3,702	3,703 to 5,556	Over 5,556
Up to 120.......	180	105	60	45	30	22	18	15	12	10	9.2	8.4	8.2	8.0	7.9	7.8
121 to 240......	160	98	57	43	29	21	17.6	14.7	11.8	9.9	9.1	8.3	8.1	8.0	7.9	7.8
241 to 480......	120	75	50	39	27	20	16.9	14.3	11.5	9.7	8.8	8.2	8.0	7.9	7.8	7.7
481 to 720......	75	62	42	35	24	18	15.4	13.2	10.8	9.2	8.6	8.1	7.9	7.9	7.8	7.7
721 to 1,080.....	54	42	33	29	20	15	13.6	12.1	10.1	8.7	8.3	8.0	7.8	7.8	7.7	7.6
1,081 to 1,620...	30	18	16	15	12	11.5	11.1	10.4	9.8	8.4	8.1	7.9	7.7	7.7	7.6	7.5
1,621 to 2,430...	15	12	11	10.5	9.1	8.8	8.6	8.3	8.0	7.9	7.8	7.7	7.6	7.5	7.4	7.4
2,431 to 3,645....	7.5	7.2	7.0	6.9	6.6	6.5	6.4	6.3	6.2	6.3	6.4	6.4	6.8	7.0	7.1	7.2
3,646 to 5,460....	2.0	2.4	3	3.3	4.1	4.2	4.3	4.4	4.5	4.7	5.0	5.1	6.1	6.4	6.9	6.9
5,461 to 8,190....	0	2.0	2.1	2.2	2.3	2.4	2.5	2.6	2.8	3.2	3.7	4.6	5.0	5.6	6.2	6.4
8,191 to 12,285...	0	0	2.0	2.1	2.1	2.2	2.3	2.3	2.4	2.5	2.6	2.7	3.5	4.5	5.2	5.6
12,286 to 18,420..	0	0	0	2.0	2.1	2.1	2.2	2.2	2.3	2.3	2.4	2.4	2.6	3.2	4.2	4.7
18,421 to 27,630..	0	0	0	0	2.0	2.1	2.2	2.2	2.2	2.3	2.3	2.3	2.4	2.5	2.8	3.1
27,631 to 40,945..	0	0	0	0	0	2.0	2.1	2.2	2.2	2.2	2.2	2.2	2.2	2.2	2.3	2.4
40,946 to 61,520..	0	0	0	0	0	0	2.0	2.1	2.1	2.1	2.1	2.1	2.1	2.1	2.1	2.1
Over 61,520.....	0	0	0	0	0	0	0	2.0	2.0	2.0	2.0	2.0	2.0	2.0	2.0	2.0

The method of the table is simple. The total number of fixture units which the drain serves must first be determined. The area of the roof must also be calculated. After this information has been established, the conversion factor may be found readily by reference to the table.

Example: Suppose that 10,000 square feet of roof pitched at ¼-inch to the foot were to be drained (using 247 units of discharge). Then by referring to Table 7, it is found that the conversion factor would be 2.3.

The next step is to multiply the number of fixture units by the conversion factor. Hence, 247 units × 2.3 = 568.1 units.

The result of this multiplication is now added to the roof area as follows: 10,000 sq. ft. of roof + 568.1 = 10568.1 sq. ft.

With the use of Table 6 the size of the combination drain can now be ascertained. In this example an 8-inch drain would be required.

Grades. After the size of the house drain has been determined, other factors in laying a house drain must be considered. The grade

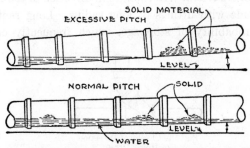

Above: Deposit of suspended bodies occurs when the drain is given excessive pitch, thus producing an unsanitary condition.

Below: Adequate depth of flow throughout the length of the drain assures thorough discharge of waste materials.

Fig. 52. Showing Proper and Improper Pitch of the House Drain

at which it is to be installed is probably of next importance. It is advisable under any circumstances to grade the house drain ¼-inch to each foot. Tests verify the soundness of the pitch. Horizontal pipes have been found to produce the necessary velocity and discharge capacity at this inclination to scour themselves properly and also to function without producing abnormal or subnormal pressures in the plumbing system. It may be necessary, however, because of the depth of the basement floor and inadequate depth of the sewer, to give the house sewer less than ¼-inch pitch per foot.

Also, an unusually long sewer would require less pitch per foot, because the accumulated or total pitch would result in an exceptionally deep house drain outlet. If the grade is slight, the plumber should use an instrument to level the sewer so that it is graded accurately and is free from sags or trapped piping.

A pitch of more than ¼-inch per foot increases the velocity and discharge capacity of the waste, but it might decrease the depth of waste necessary to provide a self-scouring condition. See Fig. 52. It might also account for a minus pressure if the drain were taxed to capacity flow.

The pitch per foot of a house drain can be calculated by dividing the total pitch in inches (which is the distance between the house sewer and the level of the basement floor) by the length in feet of the longest branch.

Example. Suppose the longest branch of a house drain were 48 feet and the total pitch were 12 inches, then $12 \div 48 = \frac{12}{48}$ or ¼ inch pitch per foot.

Change of Direction. All changes in direction of the house drain should be made with fittings of long radius. Long radius fittings lessen the probability of stoppage, which frequently occurs in a

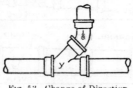

Fig. 53. Change of Direction
of House Drain

norizontal house drain. The branches should be run at right angles to the main, and fixture connections run at right angles to the branch. Fig. 53 shows the advisable method of making change of direction.

Cleanouts. The house drain must be equipped with an adequate number of cleanouts, so that the entire drain is accessible without breaking the basement floor should stoppage of the drain occur. The plumber must use good judgment in the location of cleanouts.

A cleanout should be installed on the house drain just inside the foundation wall of the building. A cleanout should be placed at every 75-foot interval as well as at the base of all soil and waste stacks. Any branch of the drain terminating at a floor drain or fixture also should be provided with cleanout facilities.

The cleanout should consist of a 4-inch cast-iron pipe extended at least 2 inches above the basement floor and inserted in a 45-degree Y branch in the direction of flow of the drain. It should be equipped with a standard brass screw thread cover that is provided with a raised head so it can be gripped with a wrench to facilitate removal.

A cleanout extended above the floor cannot be used as a floor drain. Fig. 54 illustrates a practical installation of a cleanout.

Installation of a Cast=Iron and Clay House Drain. After the plumber has completed the necessary details, such as size, grade, change of direction, and location of soil and waste terminals, he may begin to lay the pipe of the house drain. The pipe must be laid in trenches having solid foundation to prevent sagging of the completed drain. Digging of the trench usually is done by unskilled labor.

The general practice is to begin the house drain installation (from the connection with the house sewer outside of the foundation

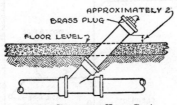

Fig. 54. Cleanout of House Drain

wall) by calking a **Y** fitting into a length of cast-iron pipe to be used as a cleanout for the finished house drain. It is necessary to test all house drain installations, hence a fitting of the tee pattern must be placed in the drain. This fitting is generally calked into the **Y** with its side opening facing upward. The **Y,** tee, and length of pipe may now be joined to the house sewer, using care not to exceed or diminish the grade, which was established previously. This much of every house drain installation is identical.

Fig. 55 gives a layout of a house drain which may be installed in the following manner. It is advisable to run the main line of the drain, from its beginning as shown at *A* to the waste pipe terminal *B*, leaving connections for the various branches at properly determined intervals. The pipe may be laid in the trench in separate lengths, or a number of lengths may be calked together on the surface and then put in place. Each branch connection must be planned beforehand, and may be made in many ways. When the main run of drain is moderately deep in the ground, connection may be made by placing a **Y** or sanitary tee on its back and using a ¼ bend as a riser to the level of the branch as shown in Figs. 56 and 57. For deeper drains a **Y,** or tee having pipe extension pieces between it and the fitting becomes necessary. This practice eliminates deep digging and therefore effects a saving. The branches of the house drain which

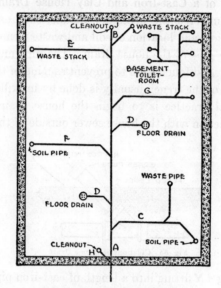

Fig. 55. House Drain Installation

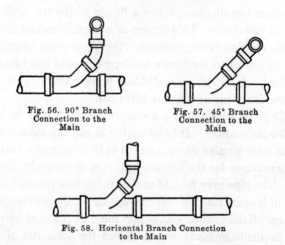

Fig. 56. 90° Branch Connection to the Main

Fig. 57. 45° Branch Connection to the Main

Fig. 58. Horizontal Branch Connection to the Main

are not deep in the ground may be connected to the main run by laying a **Y** fitting on its side, using a ⅛ bend to make the right angle, as shown in Fig. 58. After the main run has been completed the branch can be installed.

The branches should be run at right angles to the main and must be as direct and free from diagonal offsets as is practical. The changes of direction should consist of **Y** fittings and ⅛ bends laid on their sides to assure long radius turns. As illustrated in Fig. 55, branch *C* serves as a soil and waste pipe. The run has been centered between the two terminals to reduce long horizontal connections to it. This practice is considered efficient and signifies good workmanship. The terminals or base fittings of the soil and waste pipe which the

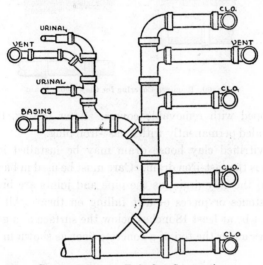

Fig. 59. Toiletroom Underfloor Construction

branch serves must be fittings of long radius, designed especially for this purpose.

The next connections to the main run serve the floor drains, *D*. These drains must be set at proper height to allow the water to drain into them. The trap of the floor drain should be placed not more than 2 feet under the finished floor so it may be cleaned readily in case of stoppage.

Branch *F* serves a soil pipe installation and is run in a direct line to the base of the soil pipe, as is branch *E*, which is a terminal for a waste.

Branch *G* is connected to the main run at right angles and accommodates a basement toiletroom consisting of four water closets, a battery of wash basins and two urinals. Fig. 59 illustrates how this underfloor work should be constructed. Care must be used in planning this toiletroom. The openings of each fixture must be measured accurately from tentative partitions or basement walls, for after the basement floor is completed not much can be done to change the underfloor work should it be incorrectly laid out.

The cleanouts *H* and *J*, Fig. 55, are now brought up to floor height

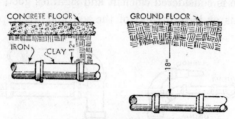

Fig. 60. Required Covering for Clay House Drain

and equipped with removable covers. After testing, the test tee must be sealed permanently with a cast-iron plug.

The vitrified clay house drain may be installed in the same manner as is the cast-iron drain. Care must be used in backfilling the trenches of the clay drain, as the pipe and joints are broken easily by large stones or pieces of dirt falling on them. All clay house drains must be at least 18 inches below the surface of a ground floor, or 12 inches under the top of a concrete floor as shown in Fig. 60.

CHAPTER VII

HOUSE DRAIN APPLIANCES

House Trap. The house trap is a device placed in the house drain immediately inside the foundation wall of the building. It serves as a barrier and prevents the gases which occur in public sewers from circulating through the plumbing system.

The advisability of the house trap has for many years been a problem which has divided sanitary authority. Some maintain that its value is negligible; others contend that its installation is a necessity for the protection of life. One point, however, that must be accepted as a certainty, and one which is a factor in favor of the house trap installation, is that public sewers are filled with various gases.

The gases which occur in public sewers are common to the science of chemistry. Oxygen (O), nitrogen (N), carbon dioxide (CO_2), hydrogen (H), hydrogen sulphide (H_2S), methane (CH_4), carbon monoxide (CO), and sulphur dioxide (SO_2), constitute the more common gases which sewers contain.

In making an analysis of the properties of each of these gases, one would find beneficial as well as detrimental characteristics so far as human welfare is concerned. It is the contention of those individuals who advocate the use of a trap, that whenever an element that is dangerous to life or health is present, even though in small volume, adequate protective measures must be taken. Because this condition prevails, where noxious gases are present, the house trap must be installed on the house drain. On the other hand, some sanitarians have established by actual test that sewer gases and the manner in which they occur are not detrimental to health provided the plumbing system is properly installed. They contend also that a sewer well aerated would not be a gas-producing agency.

A liberal attitude relative to the house trap and the advisability of its installation is recommended. In the older and more congested districts, where this device has been used for many years, causing an unaerated condition to prevail in the public sewer, its installation on a new building is made necessary because of the presence of gas.

This is also true in industrial areas, or where buildings are constructed in close proximity to hospitals; or for individual factories that may discharge noxious chemical compounds and mixtures which may form objectionable gases in the public sewers.

Public authorities, as a rule, advocate elimination of the house trap, for its use decidedly lessens the discharge capacity of the sewer. Because of the rapid growth of cities, many sewers are taxed to their limit of discharge. Elimination of the trap minimizes this problem materially.

House Trap Assembly. Fig. 61 illustrates the installation of a house trap and fresh air pipe. The trap usually is referred to as a

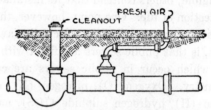

Fig. 61. House Trap and Fresh Air Pipe

running trap, having its inlet and outlet on the same level. The trap is equipped with a cleanout extended above the finished floor, as shown in Fig. 61.

The fresh air pipe is connected within 2 feet of the trap to assure completion of a movement of air in the system, which the installation of the house trap has impaired. The fresh air pipe must be extended to the outside of the building and be provided with a cap to prevent its obstruction. Two inches is the minimum diameter of the fresh air pipe. It is recommended, however, that at least a 4-inch pipe be used. The trap must be set level and installed in a workmanlike manner to assure efficiency.

Back=Flow Valves. Back-flow valves are devices used in a drainage system to prevent the reversal of flow. They are constructed in two patterns and are classified in the plumbing industry as either balanced or unbalanced.

Back-flow valves are used commonly on house drains, or branches of the house drain, which, when subject to reversal of flow, might flood and cause damage to the building and its contents.

Balanced Valve. The balanced valve is by far the most advisable installation. It is constructed in such a manner as not to interfere with the movement of air in the drainage system. Fig. 62 illustrates a valve of this type. The valve has a cast-iron body fitted with an air-tight removable cover. The body of the valve is equipped with a hub and spigot end so it can be calked into the drainage system. The interior mechanism consists of a brass seat into which is fitted a gate counterbalanced with an adjustable cast-iron weight.

Fig. 63 illustrates a balanced, rotating disc-type valve with standard iron pipe threaded connection. The valve has a cast-iron body and a removable top which sets flush with the floor. This valve incorporates a manually operated gate valve and an automatic swing check valve in one unit. Both valves are accessible from the floor line. The spade-shaped gate valve can cut through sewage easily. Its bronze metal non-rising stem is operated by means of a removable wheel handle.

No pit is required for this installation regardless of the depth of the sewer. The adjustable extension feature of the valve (*right*

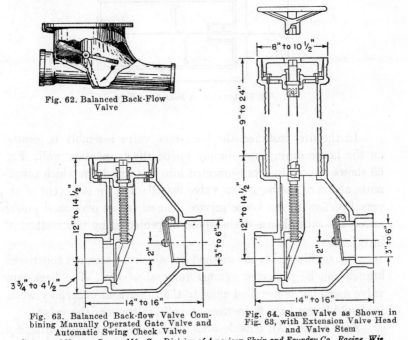

Fig. 62. Balanced Back-Flow Valve

Fig. 63. Balanced Back-flow Valve Combining Manually Operated Gate Valve and Automatic Swing Check Valve

Fig. 64. Same Valve as Shown in Fig. 63, with Extension Valve Head and Valve Stem

Courtesy of Norman Boosey Mfg. Co., Division of American Skein and Foundry Co., Racine, Wis.

view) enables the plumber to bring the floor plate to floor level for any floor line measurement between 18 and 36 inches.

Unbalanced Valve. The unbalanced back-flow valve (not illustrated) is, in outside appearance, similar to the balanced valve. Its body is constructed of cast iron and it is fitted with a removable air-tight cover. Its internal mechanism consists of a non-corrosive seat fitted with a swinging gate or flap. This type of valve remains closed until a flow of sewage strikes the gate. The velocity of flow is affected materially by this action as also is the movement of air through the drainage system. Because a circulation of air is essential in a drainage system, this type of valve should not be used unless a fresh air pipe is provided.

Location of Back-Flow Valve. A back-flow valve may be installed in such a manner as to protect an entire house drain. It also may be used to prevent reversal of flow from occurring in a branch of the house drain.

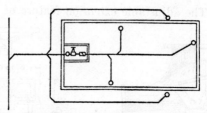

Fig. 65. Location of Back-Flow Valve

In the first instance, the back-flow valve assembly is located on the house drain, immediately inside the foundation wall. Fig. 65 shows the conductors connected into a storm drain which terminates ahead of the back-flow valve installation. In the event of reversal of flow in the house service, caused by an overtaxed public sewer, the entire house drain would be protected by this method of installation.

Fig. 66 illustrates the method by which a basement toiletroom branch can be protected against reversal of flow. The back-flow valve assembly is installed close to the toiletroom underfloor work. It should be set level to increase the effectiveness of the valve.

Area Drain. An outside area is a space constructed around a

basement window so that light may be admitted to the basement.
Because the area is below the surface grade, it may serve as a reservoir in case of a heavy rain. Should the accumulated water develop sufficient head to break the window, flooding of the basement
would result. This objectionable condition can be avoided by equipping the area with a drain.

Area Drain Assembly. Fig. 67 shows an installation of a drain
in an outside area. The drain assembly consists of a cast-iron
running trap installed under the basement floor to protect it from
freezing. The trap must be equipped with a cleanout to make it
accessible. The inlet side of the trap is extended with cast-iron pipe

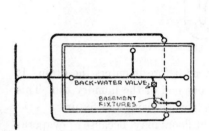

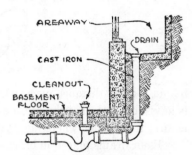

Fig. 66. Back-Flow Valve Installation on
Branch of House Drain

Fig. 67. Area Drain Installation

and fittings of long radius to the area floor, and is provided at this
point with a bar strainer. A bar strainer is preferable to a drilled
one, because it permits a greater quantity of water to pass through
it and thus eliminates the possibility of building up head in the area
proper. It does not clog as readily as a drilled strainer and therefore
is more effective.

The minimum diameter of pipe that should be used on an area
drain is a 4-inch cast-iron pipe. It may be constructed of larger
pipe should the area be of excessive size. It is inadvisable to construct it of smaller diameter, because the water must be drained as
rapidly as it enters the area. The area drain assembly can be used
for draining basement entry ways, loading platforms, or cemented
driveways.

Floor Drains. A floor drain is a receptacle used to receive water
to be drained from floors into the plumbing system. It is recog-

nized by sanitary authorities as a plumbing fixture and should be properly designed and located for the waste it will receive.

Building designers do not, as a rule, deem the location of a floor drain in a basement of much consequence. The average residence is provided with two floor drains, one located near the heating equipment and the other in the vicinity of the laundry. In some instances one floor drain is required to serve the entire basement. The usual result of this false economy is a wet floor.

Every room in which laundry equipment is used should be provided with adequate floor drainage. The drain proper must be located in such a manner that the overflowing water is not required

Fig. 68. Deep Seal Floor Drain

to run a great distance over the floor before it enters the drain. It is advisable to locate the drain at one end of the laundry tub. This assures a dry floor where one stands when using the fixture.

Floor drains placed in basement floors may, because of evaporation of their content, become a means of by-passing sewer air. This danger can be overcome by using the drain as a terminal for some clear water fixture, such as an ice box, soda fountain, or bubbler.

Every floor drain should be supplied with running water from a fixture located near by. If the fixture is less than 5 feet from the drain, it should be trapped but not necessarily vented. Fixture drains, which supply water to a floor drain, should be connected to the house side—never the sewer side—of the trap.

Construction of a Floor Drain. There are many types of manufactured floor drains. All are constructed on the same principle. A desirable type of drain consists of a 4-inch **P**-trap installed not more than 2 feet under the basement floor. The trap should be of the deep seal variety. To assure durability, a stub of 4-inch cast-iron pipe at least 2 feet 6 inches in length should be calked onto the outlet side of the trap. The trap also may be of the low inlet hub pattern. The inlet side of the trap should consist of a 4-inch pipe equipped with an 8-inch cast-iron or brass strainer set level and flush with the basement floor. The strainer must fit securely, as a precaution against accident. Fig. 68 illustrates a floor drain of approved design.

Yard Catch Basin. This receptacle is used to catch the surface water which drains from cemented courts, driveways, and yards. It also may be a terminal for drain tile installations used to drain water from athletic fields.

Yard catch basins are exposed to freezing temperatures, hence some precautions must be taken against their freezing. This is done by placing the invert, which serves to prevent sewer air from escap-

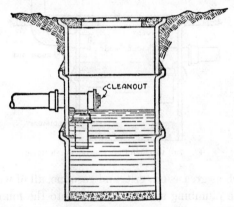

Fig. 69. Yard Catch Basin

ing, a safe distance below frost line of the earth. A distance of 4 feet 6 inches usually is considered adequate to guard against freezing.

Construction of a Yard Catch Basin. The yard catch basin, illustrated in Fig. 69, is constructed of three lengths of vitrified clay pipe, 36 inches in diameter, joined together with cement mortar. This diameter permits cleansing of the basin without difficulty. The bottom is of poured concrete, at least 6 inches in thickness, troweled to make it thoroughly water-tight. It is equipped with a substantial cast-iron bar strainer set flush with the grade.

The outlet invert consists of a 4-inch cast-iron sanitary tee, provided with a stub of pipe calked to its inlet side to provide at least a 10-inch water level. The outlet side of the tee extending through the clay tile must be equipped with a 2-foot 6-inch cast-iron stub to assure durability. The top opening of the invert should be pro-

vided with a standard brass screw thread cleanout. The invert of
the basin must be placed at least 4 feet 6 inches below the strainer.
The damaged tile around the cast-iron invert is replaced with a
substantial cement joint. At least 1 foot 6 inches of space between
the inlet of the invert and the bottom of the basin must be allowed
to provide for accumulation of sediment.

Garage Catch Basin. A garage catch basin is a device used to
convey wastes from garage wash racks, grease pits, and repair floors
into the house drain. These wastes contain many objectionable

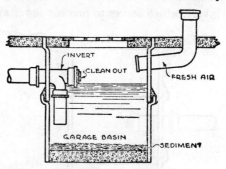

Fig. 70. Vitrified Clay Garage Catch Basin

elements, such as grease, oil, grit, and gasoline, all of which are detrimental to the plumbing system as well as to the municipal sewage
disposal system.

Oil, grease, and sediment tend to produce stoppage and affect
the operation of the sewage disposal plant. Large quantities of oil
and grease adhere to the mechanical devices used in sewage treatment and may reduce the bacterial activity necessary to the process.
Gasoline discharged into the public sewer vaporizes and is often
responsible for serious explosions if it is ignited. The function of a
garage basin is to retain these noxious materials and discharge only
the associated water into the house drain.

The efficiency of a garage catch basin depends on how often it is
cleaned. This usually is the responsibility of the garage operator
and often is either entirely neglected or done in such a manner that
much waste material is forced into the plumbing system. Therefore
it is essential that garage workers be instructed in the proper method
of cleaning the basin and also informed of the dangers attending
neglect.

Types of Garage Catch Basins. There are two types of garage catch basins. The concrete or vitrified clay basin and the manufactured cast-iron catch basin. The vitrified clay or concrete basin is constructed on the job by the plumber. The cast-iron receptacle is only placed, and is then connected to the house drain.

Vitrified Clay Basin. The vitrified clay garage basin, Fig. 70, should be constructed of two lengths of pipe at least 30 inches in diameter. The lengths must be joined together with cement mortar

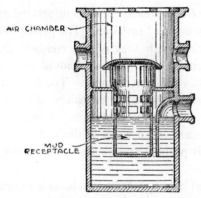

Fig. 71. Cast-Iron Garage Catch
Basin

made of equal parts of cement and clean, dry sand. They must be provided with a water-tight concrete bottom at least 6 inches thick.

The invert may be constructed of a 4- or 6-inch cast-iron sanitary tee, to which a stub of cast-iron pipe has been calked to provide a substantial connection to the branch of the house drain. The invert should be extended downward in the catch basin at least 10 inches to provide an adequate seal. As shown in Fig. 70, it must possess a cleanout to make the drain branch accessible. The invert should be placed close to the top of the garage basin to permit evaporation of gasoline through the open strainer. Ten inches from the top of the strainer is considered practical. If it is found impossible to keep the water line of the basin close to the floor level, it becomes necessary to ventilate the basin locally. This can be done by using a fresh air pipe, extended to the outside of the building, or by connecting a rain water leader to the basin. A rain water leader can be used only when the basin is connected to a combination sewer.

The space between the bottom of the invert and the floor of the garage catch basin must be at least 2 feet 6 inches, to provide for accumulated sediment. It should be equipped with a durable cast-iron bar strainer of the same diameter as the body of the catch basin.

Concrete Garage Catch Basin. The concrete garage basin is constructed of 4-inch concrete walls and bottom and is essentially of the same dimensions as the vitrified clay receptacle shown in Fig. 70.

Cast-Iron Garage Catch Basin. The manufactured garage catch basin, Fig. 71, is constructed of cast iron. The principle involved in its operation is identical with that of the concrete and vitrified clay types. The sediment, however, is removed from the water-carried waste in a little different manner. The basin is provided with a cast-iron removable bucket or container, through which the water-carried waste must pass. Sediment is permitted to accumulate in the container. The water is drained from it, through holes provided for this purpose and located close to the top of the unit.

The basin is sealed by a trap built into it, as is indicated in Fig. 71. Cleaning this type of catch basin is a simple matter and may be accomplished by removing the sediment container.

Steam Boiler Blow=off Basin. Water used to produce steam has a varied mineral and chemical content which, when heated, may be precipitated and prove detrimental to the heating boiler and its equipment. Constant heating of water also affects the steaming qualities of the boiler. Scale and disintegrated water must be ex- changed periodically for new water. The process of removal is re- ferred to as "blowing off" the boiler.

When a boiler is blown off, approximately one gauge of its entire content is removed. One gauge of the boiler is the volume of water contained between the two highest drain cocks located on its water column.

Boilers operating under high pressures contain water heated above the boiling temperature of water subjected to heat under atmospheric pressure. For each pound of pressure there is a corresponding temperature. For example, heat applied to water exposed to atmospheric pressure (14.72 lb.) attains a temperature of

212°F. before it disintegrates into vapor. Water heated under 105 pounds of pressure, as in power plants operating under high pressures, vaporizes at a temperature of 331.4°F. before disintegration occurs. Expansion of the molecules of water under these conditions is tremendous. Scientific authority has proved that from 1 cubic inch of water a maximum of 1,700 cubic inches of steam or vapor can be produced.

To the layman a steam heating or power unit contains only vapor. This is partly true, but it also contains a boiler partly filled with water that is heated to excessive temperature. The only reason it remains a liquid is that the tensile strength of the material which comprises the heating unit is sufficient to withstand the pressure

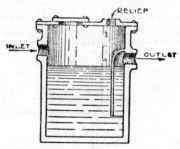

Fig. 72. Steam Boiler Blow-off Basin

under which it is operated. If the operating pressure of the system is reduced, the water changes from a liquid to a vapor and, with this change, excessive expansion occurs. This condition prevails when a steam boiler is blown off, and, unless it is controlled in a practical manner, may result in injury to any person in close proximity to the plant.

The most logical place in which a steam boiler can be blown off, is the house drain. The operation can be accomplished safely with the use of a steam boiler blow-off basin, the function of which is to eliminate the expanded vapor and permit the condensate to drain by gravity into the house drain.

Installation of the Steam Boiler Blow-off Basin. The steam boiler blow-off basin illustrated in Fig. 72 is manufactured of cast-iron. It is equipped with an invert, which is cast into the basin to seal the basin against the entrance of sewer air and also to maintain a water content within the device to cool the steam wastes below a temperature of 212°F. The discharge of the steam boiler is connected to the basin above its water line. This phase of the installation generally is a steamfitter's job. A ventilation pipe of adequate size must be extended through the roof with the least possible number of fittings to prevent the condensate from slugging as the boiler is being blown off. The ventilation pipe is that part of the installation which makes the connection of a blow-off basin to the house drain permissible. The purpose of this installation is to reduce the high pressure to one of atmospheric condition, so that the seal content of traps installed on the fixtures will not be affected. The branch of the house drain between the blow-off basin and the drain should

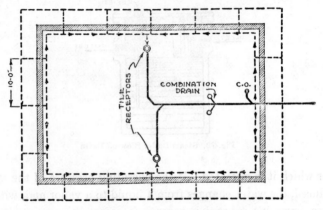

Fig. 73. Drain Tile Installation

be built of cast-iron pipe, as the high temperature of the steam wastes deteriorates joints made of other materials.

Steam boiler blow-off basins vary in size from 18 to 60 inches in diameter and may be from 24 to 72 inches in length. The size of basin required is determined by heating authorities by the size of the boiler and the pressure under which the system is to operate.

The size of the ventilation pipe is fixed by the manufacturer and should never be less than the diameter of the tapping provided for this purpose in the cover of the blow-off basin. The house drain branch usually is of 4-inch diameter.

Drain=Tile Receptor. One of the factors contributing to comfort in a home or apartment building is a dry basement. To effect this condition the surface or ground water must be prevented from seeping through foundation walls or basement floors. To this end, drain tile, which is properly connected with the house drain, must be placed around the outside and inside of the building footings. The device used to connect the tile to the house drain is called a drain-tile receptor.

Drain Tile Installation. Fig. 73 illustrates the manner in which the tile may be placed around the foundation walls of a small residence. Tile pipe of 3-inch diameter is used customarily. The outside drain may be laid level or graded to the bleeders or connectors

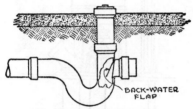

Fig. 74. Drain-Tile Receptor

placed at 10-foot intervals through the foundation proper. The inside or collecting tile line may be laid slightly lower than the footing and pitched to one or two established terminals where the drain-tile receptor is installed.

The drain-tile receptor, Fig. 74, consists of a 4-inch **P**-trap of deep seal, into which is calked an ordinary cast-iron tee. The top opening of the tee may be provided with either a cleanout plug or floor-drain strainer. The side opening of the tee must be provided with a flap trap or back-water valve to exclude the entry of sanitary waste from the tile installation should reversal of flow occur in the house drain. The materials used to construct the drain-tile receptor usually are of 4-inch diameter.

Sewage Ejectors. Public sewers usually are placed at a depth of from 8 to 15 feet under the street grade. This is done for practical reasons and also to avoid deep excavation when making connection to the sewer. The depth at which the public sewer is installed usually is inadequate to provide drainage by gravity for large buildings located in business or downtown areas. Because of the high rental value of property in these areas, all available space is utilized and multistory buildings having sub-basements that extend many feet below street grade are erected in the interests of investment.

The basements are used for building equipment and storage and must be provided with sanitary plumbing facilities. Because these fixtures are below the level of the public sewers their wastes must

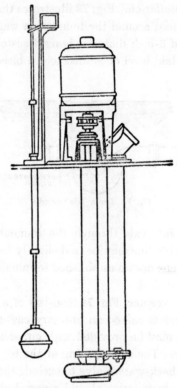

Fig. 75. Sewage Ejector

be discharged into a sump and pumped to the house drain installed overhead. The pump used is termed a sewage ejector.

Types of Sewage Ejectors. The most commonly used sewage ejector consists of a horizontal centrifugal pump that is driven electrically, Fig. 75. Ejectors operated by compressed air are now obsolete. The impeller of the modern sewage ejector is made of noncorrosive metal such as bronze or brass. It is encased in a cast-iron housing and driven with a steel shaft, also encased, and is connected to an electric motor mounted vertically on a steel plate, which serves as a cover for the sump pit. The motor is started and stopped automatically by an electric switch, mounted on the sump cover, and is connected to a hollow copper ball which floats on the surface of the liquid content of the sump.

The outlet side of the centrifugal pump is connected directly with the house drain and must be equipped with a back-flow valve to avoid filling the basement with water should reversal of flow occur in the house drain.

The suspended materials which raw sewage contains are detri-

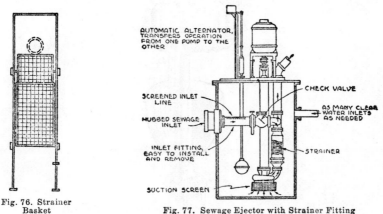

Fig. 76. Strainer Basket

Fig. 77. Sewage Ejector with Strainer Fitting

mental to the operation of the sewage ejector and should be removed before they come in contact with the pump impeller. In the older types of sewage ejector installations, a wire mesh basket was placed over the house drain inlet, as shown in Fig. 76, for the purpose of arresting all the nonsoluble materials. When the basket became full it had to be removed from the sump and its content burned or disposed of in some other manner. This practice is unsanitary and is therefore objectionable.

Recent types of sewage ejectors are equipped with a device installed on the discharge outlet which consists of a strainer and a single check valve, as may be seen in Fig. 77. The house drain outlet is connected to the strainer fitting in such a manner that its content must pass through it, the nonsoluble materials being retained. When the sewage level in the sump pit attains its peak, the pump begins to function and ejects the content of the strainer fitting into the house drain. The check valves prevent the sewage discharge from backing up into the house drain.

Sump Pit. The sump pit usually is constructed of concrete, cast iron, or vitrified clay tile, the material depending upon the size of the installation. The pit must be water-tight and of a size sufficient to accommodate the actual volume of fixture discharge, which can be determined by making a count of the total number of fixture units. It is well to allow about 50 per cent of the actual discharge volume

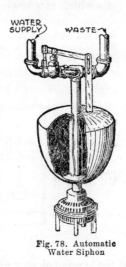

Fig. 78. Automatic
Water Siphon

for reserve storage in case the pump should be taxed to its discharge efficiency.

The sump pit should be provided with a manhole, so it can be cleaned frequently and also to make the submerged portion of the ejector accessible. The pit may be provided with a ladder cast into its concrete walls.

Ventilation Requirements of the Sump Pit. The soil pipe connected to the house drain which enters the sump pit should be extended through the roof of the building and be no less than 4 inches in diameter. The pit must be provided with a fresh air pipe connected to its cover and extended to the vent pipe system or to the outside of the building. Its terminal should be a reasonable distance (not less than 4 feet) from any door, window, or fresh air intake

Ventilation of the sump pit, in the manner described, insures that the entire drainage installation will have an adequate movement of fresh air.

Automatic Water Siphon. The automatic water siphon is a type of ejector once commonly used on small clear-water installations, Fig. 78. The use of this type of ejector unit is not recommended because it causes a cross-connection between contaminated wastes and domestic water supply.

Grease Basins. A large percentage of stoppage in a plumbing system is caused by grease contained in the waste discharge. To overcome this difficulty, a device known as a grease trap must be installed on the wastes from large kitchens serving hotels, dining rooms, clubhouses, and restaurants. Special forms of grease receptors are used in connection with slaughter houses, packing plants, and sausage factories. Grease traps can be used in connection with a residence kitchen, but their installation is not recommended under normal conditions.

The efficiency of a grease trap is dependent on the attention it receives. Almost daily removal of the grease is necessary to obtain the full benefit of the trap. This is true primarily of the smaller variety of manufactured grease traps which do not, as a rule, have an adequate amount of space for grease storage. Removal of the grease is a disagreeable task, and, in most instances, is done only when the trap ceases to function. The percentage of grease separated is probably very low, because the trap slowly stops up, allowing more grease to pass through it into the waste pipe with each fixture discharge.

The grease trap should be installed as close to the fixture as possible. More than one fixture can terminate into the same trap, provided that the trap is of sufficient size and that the waste pipe is not too long. Cleanouts at frequent intervals are necessary to prevent stoppage of the waste line.

The same principles apply in accomplishing separation of grease in all types of grease traps. Grease suspended in a waste discharge is in tiny globule form. It is less dense than water and floats on the water's surface. The greasy wastes may be 100° to 200°F., and require cooling to a temperature of 95 degrees to solidify the grease particles; or the trap must be provided with baffle plates and compartments which retain the grease globules.

Types of Grease Traps. Grease interceptors of approved design are of two types, each designated according to the method used to congeal the warm grease. Earth cooling of the waste is accomplished by placing the grease trap in the ground. This type of trap is used on larger installations and is the most desirable form of grease separation. The earth cooling method is not always feasible, however, because of a lack of ground space. In other cases, the type of trap used on small single fixture installations may be of the air-cooled variety. Water-cooled grease traps are no longer recommended because of the danger of contamination of the water supply.

Earth-cooled Grease Basin. The earth-cooled type of grease trap usually is constructed of poured concrete and is of rectangular design. It is inadvisable to make the basin less than 2 feet in width. Its length should be from three to four times its width to provide smooth and unagitated flow. Minimum depth of a concrete grease trap should be no less than 4 feet below the outlet invert.

The waste branch may be connected to the concrete receptor in two ways. It may consist of an inverted tee extended below the water line not more than 12 inches, or it can be a straight piece of pipe, cast in the wall at least 12 inches above the water level of the basin. The first method offers advantage in that the velocity of flow into the basin through the submerged inlet is reduced materially. Whenever possible, this form of connection should be made.

The outlet invert also should consist of an invert submerged at least 30 inches into the trap's liquid content. When inverts are used on the inlet and outlet of the trap, they may be set at the same level. Both inverts must be left open at the top and extended at least 12 inches above the water level to allow for accumulation of the separated grease. The larger variety of earth-cooled grease receptors may be equipped with baffle plates, which provide smaller, individual compartments in the tank. Each of the baffle plates must be slotted near the top to permit movement of air. Each grease trap, or the compartment of each trap, must be provided with a manhole extended above grade to make it accessible.

The size of a grease trap of this variety usually is determined by the amount of fixture discharge. It may be sized according to the number of meals served, allowing from 4 to 5 gallons of liquid capacity for each meal. It is good practice to double the actual volume of waste to which the trap will be subjected. This practice assures sufficient time to separate the grease from the liquid wastes.

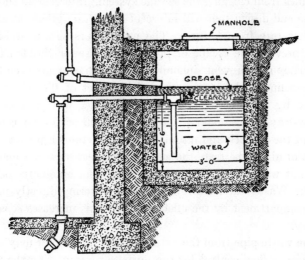

Fig. 79. Earth-cooled Grease Basin

Fig. 79 illustrates an earth-cooled basin and gives minimum dimensions.

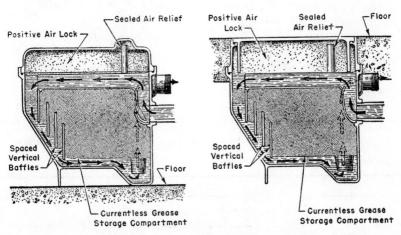

Fig. 80. Grease Interceptor
Resting on Floor

Fig. 81. Grease Interceptor
Installed Flush with Floor

Courtesy of Norman Boosey Manufacturing Co., Division,
American Skein and Foundry Co., Racine, Wis.

Trap with Water-Sealed Inlet. The main reason for using grease traps or interceptors is to prevent grease discharged into waste lines from clogging the sewage system. Interceptors should be of a size and design that will intercept and retain grease at a maximum flow rate from fixtures. The interceptors shown in Figs. 80 and 81 prevent the escape of grease to the sewer through the air relief. They also prevent siphoning of the trap contents even though the waste line vent becomes clogged. Neglect of the trap is compensated by the fact that accumulated grease will gradually clog the flow-ways, thus making cleaning compulsory. The combined action of the air lock and internal air relief retains a predetermined amount of air in the trap, the cushioning effect of which maintains a constant water level. The water-sealed inlet eliminates odors at the sink. Waste water is prevented from entering directly into the grease compartment by the inlet scupper. This increases separating efficiency.

The waste pipe from the fixtures to the interceptor may be provided with a flow control for the purpose of regulating the rate of discharge of the fixtures into the interceptor so that the established flow rate of the interceptor cannot be exceeded. If the orifice type of control is used, it should be vented locally to the outer air.

CHAPTER VIII

STORM DRAINAGE

The storm drain is that unit of the plumbing system which conveys storm water to a satisfactory terminal. The disposal of surface and storm water, up to a few years ago, was accomplished by discharging it into the drain which served the plumbing fixtures of the installation. This practice was considered satisfactory, because the public sewer was drained into a river, lake, or natural drainage basin, and the discharge of a large volume of clear water waste did not create a serious problem. As cities grew in population, the need for sewage treatment became more apparent. At present many large municipalities maintain sewage disposal plants. One of the processes of sewage treatment is the liquefying of suspended organic materials. Rain water passing through a disposal plant affects this process, hence it is advisable to separate the clear water from the water-carried organic waste before it enters the plant for treatment. Storm water is relatively pure and may be discharged into some natural drainage terminal without material effect. Most municipalities have constructed storm sewers which serve privately owned buildings, and the connection of storm drains to them is compulsory. This practice is a good one, for it eliminates the discharge of rain water leaders into gutters and over sidewalks where it may become a nuisance to pedestrians.

Materials Used. The materials used for storm drain construction are identical with those of house sewers and house drains. Vitrified clay pipe and fittings with cemented joints are used for this purpose. It may be well for the reader to refer to Chapters III and IV, which describe materials used in drainage systems.

INSTALLATION OF THE STORM DRAIN

Inside Storm Drain. Storm drains usually are classified as inside, outside, and overhead drains. The inside storm drain is located under the basement floor within the walls of the building, see

Fig. 82. This type of drain is used in buildings located in congested business areas.

The plumber must take care to plan the installation of the inside storm drain so it will not interfere with the sanitary drain of the system in service. The greatest difficulty encountered is in crossing over the main runs of the sanitary drain with branches of the storm

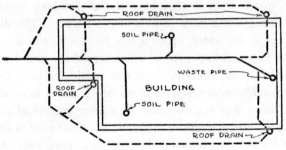

Fig. 82. Combination House Drain

drain. Fig. 83 illustrates the proper procedure where this problem arises.

Outside Storm Drain. The outside storm drain is located around the outer foundation wall of the building and is placed below the frost line of the ground, as shown in Fig. 84, a depth of 4 feet being considered adequate.

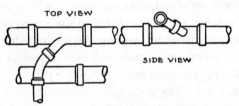

Fig. 83. Method of Crossing the Sanitary Drain with
Conductor Branches

This type of drain is advisable on buildings where an ample amount of ground space between buildings is available. It should be laid on a substantial bed of earth and may be constructed of vitrified clay or cast-iron pipe.

Overhead Storm Drain. The overhead storm drain is suspended from the basement ceiling by substantial hangers, placed at close intervals, and generally is adapted to buildings where the public storm sewer is not sufficiently deep in the street to permit gravity drainage of the underfloor inside drain.

Careful planning of the overhead storm drain is necessary. The drain should not conflict with heating mains, ventilation ducts, windows, or beams. It must be installed in such a manner as to allow sufficient headroom and so as not to mar the appearance of basement areas.

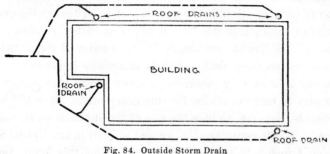

Fig. 84. Outside Storm Drain

It may be constructed of cast-iron pipe of the calk-joint type, or of galvanized steel with screw thread fittings. The overhead storm drain should be labeled to identify it from the sanitary installation.

Grade. All storm drains should be graded at least ¼ inch per foot. It has been found that this amount of pitch provides an unobstructed and self-cleansing flow. It is not considered impracti-

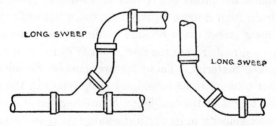

Fig. 85. Change of Direction of Storm Drain

cable, however, to give it more grade than ¼ of an inch per foot, since fixture traps which might lose their seal and allow sewer air to enter the building are not associated with it.

Change of Direction. Change of direction of the storm drain must be made with fittings of long radius as these diminish the possibility of stoppage. A **Y** fitting and ⅛ bend, Fig. 85, is advisable

SIZING THE STORM DRAIN

In determining the size of a storm drain, a number of factors must be taken into consideration. It is difficult to arrive at definite conclusions because of the nature of some of these elements, hence the process of determining storm drain size involves a certain amount of conjecture. Practical experience is of great value.

One factor which makes sizing the storm drain difficult is the matter of gauging rainfall over a given period. Although the average of rainfall throughout the United States is rather constant, some regions are subject to exceedingly heavy showers of short duration. In these regions storm drains must be materially larger in size. Because any locality may experience a severe storm occasionally, it is customary to make provision for emergency needs. It would be impracticable, however, to base the size of the storm drain on extreme demand. A safe estimate of maximum rainfall in the United States is about 1 inch in a 10-minute interval. Using this figure, the approximate volume of water that will fall on a roof in one minute's time can be determined readily.

Roofs vary in size, pitch, and the distance water must travel over them to reach the conductors or roof leaders. Even though the amount of water in gallons or cubic feet can be computed accurately, these elements produce conditions which affect the drainage and therefore determine the problem. A roof of 45-degree pitch allows the water to drain from it more rapidly than one which is flat and therefore requires a larger drain because of the greater velocity of flow. Gravel-covered roofs retard the flow of water owing to the roughness of the surface material. These factors must be considered.

Another element to be considered in determining the size of the drain is the height of the building. A building ten stories high develops greater velocity in its vertical conductors than does a building only two stories high. This accelerated flow taxes the storm drain to capacity and requires a larger installation. Entrance loss caused by water spilling over the conductor terminal varies the discharge capacity of the pipe to some extent. This occurs as a result of air accumulating in a pocket under the fall of the water—a circumstance which tends theoretically to reduce the area of the drain. Short offsets and indiscriminate use of fittings also affect the installation.

There are other factors to be considered. The discharge capacity of the pipe varies according to its length and grade per foot. It would be impossible for the mechanic to develop this information on the job because much of it is established by actual test.

The mechanic can do a great deal to increase the efficiency of the drain by devising proper means of support, and by constructing the branches with fittings of long radius.

Sanitary authorities realize that determination of the foregoing factors is difficult and, in view of this, have compiled data to assist the plumber in making his calculations. All of the elements discussed have been taken into consideration in establishing these guides.

Table 6, presented in Chapter VI, page 67, is based on the assumption that a roof of certain area and slope can be drained by a pipe of given size.

To familiarize the reader with the method of Table 6, the following problem is offered.

Example: How large a drain would be necessary to serve a roof graded at ¼ inch per foot whose length is 150 feet and whose width is 100 feet?

Solution: The area of the roof is 150 x 100 or 15,000 square feet.

By referring to Table 6, the reader will find that 15,000 square feet of roof graded or sloped at ¼ inch per foot require an 8-inch storm drain.

Formulas which do not require the use of a table and are more practical because they offer a method by which the installing mechanic may graduate the size of the storm drain are as follows:

$$S = \sqrt{\frac{A}{B}} \times 1.128 \text{ in. or } \quad S = \sqrt{\frac{A}{B} \div .7854} \text{ sq. in.}$$

Either formula may be used and it is up to the mechanic to select the one which suits him best. The results are the same in both instances.

Formulas of this kind may be confusing unless one has some knowledge of the use of constants and the values of the letters which appear in the formulas.

The letter S represents the unknown size of the sewer. The letter A is the area of the roof. The letter B is the area which may be drained with 1 square inch of pipe and its value usually is placed at from 200 to 300 square feet, depending on the pitch of the roof

The given 1.128 inches represents the diameter of a circle the area of which is 1 square inch, and the given .7854 square inches, given in the second formula, represents the area of a circle whose diameter is 1 inch. These figures (constants) are used to change a square pipe to a round one. The radical sign $\sqrt{\ }$ which embraces $\frac{A}{B}$ is the symbol indicating square root.

Example: How large a drain would be necessary to drain a roof whose area is 15,000 square feet?

Solution: Substituting the values for the letters of the formula we have

$$S = \sqrt{\frac{15000 \text{ sq. ft. x } 1.128 \text{ in.}}{300 \text{ sq. ft.}}}$$

Solving the equation

$$\frac{15000 \text{ sq. ft.}}{300 \text{ sq. ft.}} = 50$$

This means that 50 square inches of pipe are required to drain the roof. The equation now is

$$S = \sqrt{50} \text{ sq. in. x } 1.128 \text{ in.}$$

In order to find the length of the sides of the square pipe we extract the square root of 50.

```
          _____
         |50.0000  |7.07 inches length of sides
         |49
     1407| 1 0000
         |   9849
```

To change a square pipe to a round one the length of one side of the square pipe may be multiplied by 1.128 inches. Then

7.07 in. x 1.128 in. = 7.97 inches

Therefore a 7.97-inch pipe would be necessary to drain a roof of this size. Because a pipe of this diameter is not manufactured, an 8-inch pipe would be substituted.

STORM DRAIN TERMINAL

A storm drain may be terminated at any of the locations listed below:

1. The combination house sewer
2. The storm sewer in the street
3. The curb
4. A natural drainage basin

Combination House Sewer. Where a storm sewer is not available, the drain may be terminated into the combination house sewer, as illustrated in Fig. 82. This type of terminal should be used only where the public sewer discharges its content into a terminal other than a disposal plant. As explained in the preceding paragraphs, rain water does not require correction and is detrimental to the function of a septic system.

Storm Sewer in the Street. Wherever local sanitary engineers

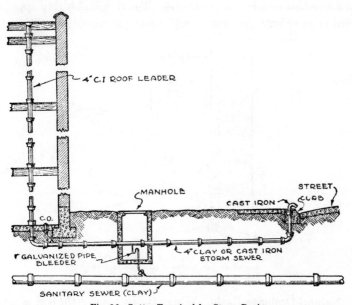

Fig. 86. Gutter Terminal for Storm Drain

construct a storm sewer in the street, they are required by ordinance to connect the storm drain to the sewer. The storm drain is connected to the sewer at right angles by means of a 45-degree **Y** and curve, assembled in exactly the same manner as a sanitary house sewer connection.

Curb Terminal. Not all municipalities have storm sewer installations, yet nowhere are storm drains allowed to discharge into the sanitary sewer. Therefore the plumber is obliged to terminate the drain at the curb. A storm drain thus terminated is exposed to freezing because it does not have sufficient ground covering and, therefore it must be installed below frost line (approximately at a depth of 4 feet). Fig. 86 illustrates an installation of this kind. This procedure traps the drain, permitting it to fill with water as the terminal is situated at a higher level.

To avoid freezing of terminal, the trapped line must be equipped with a bleeder or drain constructed of galvanized pipe connected to the sanitary sewer. The drain should be equipped with a trap to prevent the passage of sewer gas into the storm drain installation. It is advisable to make the trapped bleeder accessible by building it into a manhole of adequate diameter. The terminal fitting may be a return bend or any device that will direct the water into the gutter

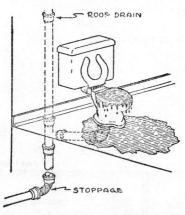

Fig. 87. When the Roof Leader Is Connected to the Waste Pipe, Flooding Is Likely to Result Should Stoppage Occur

Natural Drainage Basin. Where the contour of the soil is favorable, the storm drain may be discharged into a natural drainage basin This practice is one that is adapted to factory buildings located in outlying, sparsely populated areas.

ROOF LEADERS

A roof leader is often referred to as a conductor or downspout. It is that portion of the storm drainage system which extends between the storm drain and the roof terminal. The roof leader is as important, and requires as much care in its construction, as any other unit of the plumbing system. A defective roof leader may be instrumental in causing property damage or may permit objectionable odors to enter the building.

The roof leader must never be used as a soil or waste pipe for a plumbing fixture. It is subject to stoppage and, in this case, overflow of the fixture is likely to occur. Fig. 87 illustrates a probable result. Fixture trap seal loss also may occur as a result of this practice.

Location of Roof Leaders. There are two varieties of roof leaders, each designated according to its location in the building. The outside roof leader is located on the outside wall of the building and the inside roof leader is installed within the building walls.

Outside Roof Leader. The outside roof leader generally is constructed of corrugated sheet metal or copper and is placed by the tinner or sheet metal man. The plumber is responsible for the base fitting installation and its connection to the storm drain.

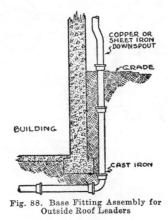

Fig. 88. Base Fitting Assembly for
Outside Roof Leaders

The base fitting, Fig. 88, consists of an extra heavy cast-iron ¼ bend which is calked into a half length of extra heavy cast-iron pipe of 4-inch diameter. The hub end of the ¼ bend must be extended to a point at least 12 inches above the finished grade to serve as a connection for the outside roof leader.

The base fitting is made up away from the job. When construction has progressed sufficiently, its exact location must be determined from the building plan. Care should be used to plumb the vertical portion of the fitting so its appearance will be in keeping with the remainder of the building.

Inside Roof Leader. The inside roof leader may be constructed of galvanized steel or wrought-iron pipe, cast-iron soil pipe, brass, or copper. Fittings of the common variety or recessed drainage type may be used.

The leader generally is started vertically from its connection with the house drain, which is a fitting of long radius. Some means for testing the conductor must be provided at its base. A test tee is the best device for this purpose. When the final test has been completed, the test tee serves the storm drain as a cleanout.

The roof leader is extended vertically through the floors of the building to a point just below the roof, and is then extended horizontally to reduce the danger of breakage which may result because

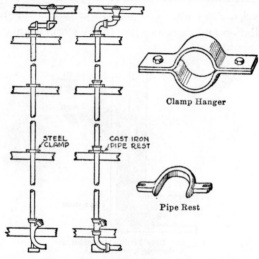

STEEL CLAMP

CAST IRON PIPE REST

Clamp Hanger

Pipe Rest

Fig. 89. Construction of Inside Roof Leader

of expansion and contraction of the roof. The change of direction at the highest point of the conductor should be made by means of an elbow and 45-degree fitting, as shown in Fig. 89. Fittings installed in this manner form a swing joint and take up any movement of the roof.

Changes in direction of the roof leader must always be of long radius so the flow of water in the conductor will not be retarded, except where the change is from horizontal to vertical direction. Branches from the installation may be made with 45-degree **Y**'s and ⅛ bends or 90-degree short pattern tees. The former are recommended.

If the roof leader is constructed of galvanized steel it must be supported at every floor with a pipe rest, Fig. 89. Horizontal runs may be suspended from band iron hangers or steel ring hangers anchored in the structure. The base of the roof leader must be provided with a concrete or stone foundation.

Roof Terminals. There are many types of manufactured roof terminals for rain water leaders. All are constructed on the same principle, the only difference, as a rule, being the size and design of the castings.

Fig. 90 illustrates a roof terminal commonly used today. It consists of a cast-iron body into which a screw thread galvanized pipe can be inserted as a means of connecting it to a cast-iron or gal-

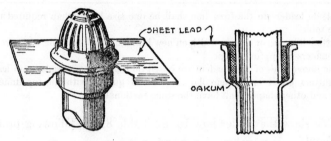

Fig. 90. Roof Terminals of Conductors or Roof Leaders

vanized steel roof leader. Roof drains are provided with a copper flashing or, in some instances, with a cast-iron clamp ring so the joint between the roof and the conductor may be water-tight. A cast-iron strainer basket is attached to the drain to prevent stones, leaves, and other materials from entering the conductor.

A roof leader may be terminated by placing the hub of a cast-iron soil pipe flush with the roof, as shown in Fig. 90. A piece of sheet lead 2 feet long by 2 feet wide may be laid on the roof and dressed into the soil pipe hub. A small sleeve about 2 inches long, cut from a cast-iron pipe, must be calked into the hub to make the connection watertight. It is inadvisable to nail the sheet lead flashing to the roof. A wire mesh strainer may be inserted in the hub to prevent the entrance of foreign materials.

Sizing the Roof Leader. The same problems which make sizing the storm drain a difficult task are involved in determining roof leader size. Therefore the mechanic must use available data.

The following table and specifications are from Recommended Minimum Requirements and may be used to determine roof leader size.

TABLE 8. Roof Leader Specifications

Area of Roof (in Square Feet)	Gutter Inches	Leader Inches
Up to 90	3	1½
91 to 270	4	2
271 to 810	4	3
811 to 1,800	5	3
1,801 to 3,600	6	4
3,601 to 5,500	8	5
5,501 to 9,600	10	6

Outside leaders to the frost line shall be one size larger than required in the above table.

Gutters 8 inches or over in width on new buildings shall be hung with wrought iron hangers of approved type.

The above sizes of rain leaders are based on diameter of circular rain leaders, and gutters based on semicircular sheet-metal gutters with the top dimension given and other shapes shall have the same sectional area.

To show the reader how to use Table 8, the following problem is offered.

Example: How large a roof leader would be needed to drain a roof whose length is 55 feet and whose width is 55 feet?

Solution: The area of the roof is length times width. Then

$$55 \times 55 = 3025$$

By referring to Table 8, the reader will find that 3600 square feet of roof require a 4-inch roof leader.

Should the mechanic wish to use the formulas which were presented earlier in the chapter under the heading "Sizing the Storm Drain" he may do so. The final conclusions are practically identical with the table. The number of square feet of roof which can be drained by 1 square inch of pipe can be increased to 350 square feet, because of the vertical position of the roof leader.

CHAPTER IX

SOIL PIPE

The soil pipe is that portion of the plumbing system which receives the discharge of water closets, with or without additional

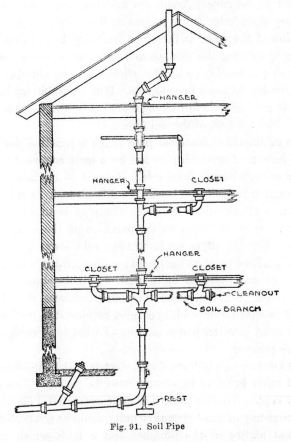

Fig. 91. Soil Pipe

fixtures, and conveys the wastes to the house drain, Fig. 91. It is used where stack, wet, and flat venting are permitted. Concealed in partitions, careful workmanship is required to install it, as otherwise the efficiency of the entire plumbing system could be impaired.

Before the plumber is able to begin the roughing in of the soil pipe, he must lay out the entire building installation. This includes the location of fixtures, the size of partitions, and the relation of these to each other. Location of windows and doors must be considered, as well as lowered ceilings which often serve to conceal soil and waste pipe branches.

The soil pipe generally is placed in such a manner that the branches which serve the water closets of a toiletroom are as direct and short as possible. Architects and building designers do not always use good judgment in indicating the location of the soil pipe. Relocation of the stack is therefore necessary to accommodate the plumber in making the runs of associated vent, waste, and water pipes more practical. To do this effectively, the plumber's knowledge of the layout must be complete. It is advisable for him to see the building designer before changes are made, and consult with him as to the best location of the soil pipe.

The minimum diameter of pipe which is practical for soil pipe use is 3 inches. Four-inch pipe has been most commonly used and is still considered most favorable. Because of its large diameter, concealment of the soil pipe becomes a difficult task. For this reason the location of partitions in which the pipe is to be installed is important. In some instances the partition on the second or third floor is not directly above the partition on the lower floors. A condition of this kind requires the use of an offset, which is often difficult to make, or too short to build with available standard fittings. Furring of the partition, under these circumstances, may be necessary. The carpenter or bricklayer must be consulted, and each trade involved must have an understanding of what is to be done before work may proceed.

The location of windows, doors, electric outlets and boxes, cabinets and other building necessities must be set up. The soil pipe generally is placed before partitions are constructed. Passing the soil pipe through any of these necessary units would be a reflection on the mechanical ability of the plumber and a fault which would be exceedingly hard to correct.

Materials Used. There are two materials most commonly used in the construction of a soil stack. The first is cast-iron soil pipe; the second, galvanized steel or wrought-iron screw thread pipe. The building design, in many cases, governs the material used for soil pipe purposes.

Cast-iron pipe is better adapted for soil and waste pipe in buildings of wood construction or in buildings of concrete and steel not subjected to vibration. Flats, apartment houses, hotels, residences, can be served efficiently with cast-iron soil pipe. Buildings used for ir.dustrial plants, such as machine shops, stamping factories, or where heavy machine work is done; or packing houses, subjected to varying temperatures; or buildings of great height, are best served with steel or wrought-iron pipe installation. It is reasonable to assume that steel or wrought-iron pipe will stand up under these conditions more satisfactorily than will cast-iron pipe.

For detailed information as to quality, weight of pipe, and fittings which may be used, it is well for the reader to review Chapter III.

Joints. Joints on the soil pipe must be made in a substantial manner. Leaking joints permit sewer air to escape into the building and are responsible for much property damage. The process of making a joint on soil, waste, and vent pipe installations was discussed in Chapter IV.

Supports and Hangers. The soil pipe must be well hung and properly supported, for much of the efficiency and durability of the installation depends on this part of the work. Various types of supports, such as clamps, brackets, or rests, have been designed to hold either cast-iron or galvanized steel soil pipe. In supporting a soil stack, the mechanic should use good judgment in placing the rests properly, allowing for a certain amount of expansion and vibration.

The pipe rests, designed to support vertical cast-iron soil pipe, are usually made of cast iron. They are round in form and cast in diameters to fit the various sizes of pipe. They are made with 4-inch projections on opposite sides. These rests can be placed around the pipe and located under the hub with the projections or ears fastened to concrete floors, wooden joists, or whatever solid footing can be found, as shown in Fig. 92.

Pipe clamps are much the same in design and purpose as pipe rests. The clamp generally is made of iron, heated and bent to fit the outside diameter of the pipe. It is made in two parts, which are bolted tightly together. The clamps have ears or projections which can be fastened to concrete floors or wooden joists. Clamps support steel pipe exceptionally well when they are installed properly. Soil pipe stacks can be supported at their bases by means of concrete piers, rest bends, or similar contrivances. Horizontal runs of soil pipe generally are hung or suspended by means of a band or cast hanger and band iron or steel rods.

Change in Direction. Change in direction is a term applied to the various turns which may be required in a soil pipe stack between its base and its terminal at the roof. One can understand readily

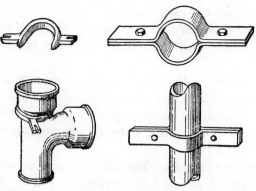

Fig. 92. Method of Supporting Cast-Iron and Galvanized
Iron Pipe

why these changes of direction are important factors to consider and must be installed correctly. The stack may rise vertically for one or two floors, then change to a horizontal position, and again rise vertically. It may also run north to south horizontally and then change to an east to west direction.

Care must be taken in selecting change of direction units. Short turn ¼ bends may be used for change of direction from the horizontal to the vertical. Long sweep ¼ bends, two ⅛ bends, or a combination **Y** and ⅛ bend may be used where the change of direction is from the vertical to the horizontal.

Accessible cleanouts, whether located in the basement or under the first floor, should be installed for every change of direction and for all horizontal lines two or more feet long. It is not recommended that cleanouts be installed between floor and ceiling unless they can be placed on the outside, and usually this arrangement detracts from the appearance of the building. Fig. 93 illustrates changes of direction of soil pipe.

Sizing of the Soil Pipe. The same problems occur in sizing the soil pipe stack as are encountered in determining the size of the house drain, vent system, and water supply. No definite mathematical procedure is possible, because of variable conditions.

It would be impossible for anyone to foretell how often he is going to use a plumbing fixture in a given interval. It would be even more difficult for a plumber to ascertain how often and at what times a plumbing fixture might be used.

To base the size of the soil pipe on the total maximum discharge of the fixtures connected to it during a one minute or one hour

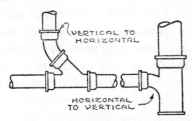

Fig 93. Change of Direction of Soil Pipe

interval, would be unreasonable and impractical. It is certain that all the fixtures could never be used at one time, and at no time would the soil pipe be required to handle a maximum load of this kind. If this reasoning is correct, it then becomes the task of someone to determine how many fixtures might be used simultaneously, and what size the soil pipe would have to be to accommodate the volume of discharge.

The plumbing industry has changed from one of a traditional aspect to one involving scientific principles which the mechanic must be able to apply in his work. At one time (not more than twenty years ago) six water closets, six wash basins, and six bathtubs were considered the maximum load a 4-inch soil pipe could possibly accommodate. In case one more closet was added to the installation, a 5-inch soil pipe was required. The regulation was changed, in later years, and ten closets were allowed on a 4-inch pipe. The older mechanics, being traditionally minded, were not ready to accept this change and many adverse opinions were expressed. After much practical testing, and with the aid of mathematics, it was found that a soil pipe of 4-inch diameter would accommodate 144 water closets without being overtaxed inasmuch as simultaneous use of this number of fixtures would be highly improbable. It was found, moreover, that proper ventilation of the fixture traps, and careful installation, did more to promote the efficiency of the system than installation of pipe of increased diameter. Under certain conditions more fixtures may be added to the installation without harm.

Of immense value to the plumbing profession is the *Report of the Uniform Plumbing Code Committee,* a publication issued jointly by the Department of Commerce and the Housing and Home Finance Agency, Washington, D.C. The *Report,* issued in 1949, is a reflection of the opinions of the Committee members based on practical experience and on research conducted at the National Bureau of Standards under sponsorship of the Housing and Home Finance Agency. Thousands of tests were made on a three-story installation having the usual number of fixtures on each of its floors. Back to back, the fixtures were discharged singly and simultaneously and the results recorded. It was upon the outcome of these tests that the writers of the Uniform Code formulated their rulings respecting disposal systems and the fixtures and fittings appurtenant thereto.

The conditions governing each of the tests conducted varied. Discharge was simultaneous into different types of fittings, and the fall or pitch of the drain not always the same. Trap siphonage was shown for the different installations tested. It was proved that wet venting is safe under certain conditions. Stack venting on a one-story installation, or on the top floor of a building more than one story tall, was shown to be satisfactory. Long drain wet vented bathtub drains were proved to be safe provided that a fall no greater than ¼ inch per foot was used.

The tests also proved that the arm on a basin drain, if more than 12 inches long, is unsafe when the fitting in the stack consists of a drainage **T-Y** and the fall from trap to stack in the arm is ½ inch per foot. However, by using a basin tee in the stack, and not over ¼ inch fall, an arm as long as 30 inches can be used safely.

Information such as this is available to any person interested in plumbing and desirous of obtaining sound knowledge of all phases of plumbing work. The regulations of the National Uniform Plumbing Code are based on findings established by scientific test. In the main, these regulations are as proper to one region as another.

The several factors involved in sizing soil pipe have been considered by sanitary authorities in a general way, and with the benefit of experience, and by making practical tests, a logical method of determining size has been worked out.

Some pertinent information is presented in Table 9, which shows the maximum number of fixture units that may be connected to a given size of building sewer, building drain, horizontal branch or vertical soil or waste stack.

TABLE 9. Pipe Sizes

	Maximum number of fixture units that may be connected to—							
Diameter of pipe (inches)	Building drain or sewer—Fall per foot				1 horizontal branch	Not over 3 branch intervals	Stacks with 3 or more branch intervals*	
	$\frac{1}{16}''$	$\frac{1}{8}''$	$\frac{1}{4}''$	$\frac{1}{2}''$			In 1 branch interval	Total in stack
1	2	3	4	5	6	7	8	9
1¼......	......	......	......	......	1	2	1	2
1½......	......	......	......	......	3	4	2	8
2	......	......	21	26	6	10	6	24
2½......	......	......	24	31	12	20	9	42
3	......	......	†27	†36	†20	30	11	60
4	......	180	216	250	160	240	90	500
5	......	390	480	575	360	540	200	1,100
6	......	700	840	1,000	620	960	350	1,900
8	1,400	1,600	1,920	2,300	1,400	2,200	660	3,600
10	2,500	2,900	3,500	4,200	2,500	3,800	1,000	5,600
12......	3,900	4,600	5,600	6,700	3,900	6,000	1,500	8,400
15	7,000	8,300	10,000	12,000	7,000	......	......	......

* Column 7 and columns 8 and 9 will, ordinarily, be found most economical for stack heights listed; that is, 3-branch intervals or less for column 7, or 3-branch intervals or more for columns 8 and 9. However, in certain circumstances the use of column 7 will give more economical sizing of the taller stacks than will the use of columns 8 and 9. Where such is the case, the use of column 7 on a stack of any height is entirely permissible.
† Not over 2 water closets.

To familiarize the reader with the method of Table 9, the following example is offered.

Example. How large a soil pipe would be required to serve the following installation?

14 Water closets	3 Urinals
22 Wash basins	2 Slop sinks
4 Shower baths	8 Bathtubs

By referring to the unit value of each fixture as set up in Chapter VI, we find the following:

Fixtures	Units
14 Water closets	84
22 Lavatories	22
4 Shower baths	8
3 Urinals	15
2 Slop sinks	6
8 Bathtubs	16
Total	151

Consulting Table 9, we find that 500 fixture units are permitted on a 4-inch vertical pipe. Therefore 151 fixture units would require a 4-inch soil pipe.

Installation of a 4=Inch Cast=Iron Soil Pipe. Fig. 91 illustrates a cast-iron soil pipe extended vertically through three floors and the roof of a building. The soil pipe may be started vertically from the base fitting, with the aid of a test tee, which can be sealed and used as a cleanout plug after the soil pipe has been put in service. This practice assures a foundation and saves considerable time and expense. If this method cannot be applied, the soil stack may be started by extending a half length of pipe through the first floor of the building. Connection of a stub to the house drain is made after the soil pipe has been tested. The mechanic may now add piping to the stack, leaving connections at their proper location to take care of the horizontal soil and waste branches. The stack generally is run vertically through the roof before the lateral branches are installed. It is advisable to support the soil pipe at every floor by placing a pipe rest, Fig. 92, on the rough floor joist or concrete construction of the building. A tee for ventilation purposes must be left 3 feet above the last floor or in the attic space of the building. The vent pipe system terminates at this connection. As a rule the soil pipe is terminated through the highest peak of the roof next to the ridge board. When the vertical run of soil pipe is completed, the plumber may proceed to run the horizontal branches.

Galvanized Soil Pipe Installation. Galvanized soil pipe stacks are installed in the same manner as are cast iron pipe stacks except that the joints are made with screw thread fittings. They must be supported in the same way as are cast-iron stacks. The branches require a little more judgment, in that room must be allowed to swing or screw them into the vertical pipe. This generally is accomplished by making the connection with a swing joint, consisting of a sanitary or 90-degree drainage tee and a 45-degree drainage ell.

Roof Terminals. A soil pipe extending through the roof must be flashed with sheet lead or some other durable material to make the roof water-tight. The sheet lead is of 4 pound weight. There are many roof flashings of satisfactory design manufactured and

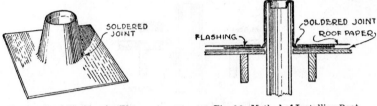

Fig. 94. Roof Flashing for Flat
Roof

Fig. 95. Method of Installing Roof
Flashing on Flat Roof

available to the plumber at a lower cost than any he could make himself.

The most common form of roof flashing in use at the present time is shown in Fig. 94. It consists of a piece of sheet lead about 18 inches long and 18 inches wide. Mounted on the sheet lead is a

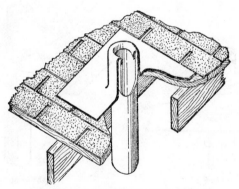

Fig. 96. Roof Flashing for Pitched Roof

cylindrical sleeve of sufficient diameter to fit around the outside of the pipe extended through the roof. The joint between the two pieces of lead may be soldered or of the burned variety.

On a flat roof this type of flashing is placed between the layers of roofing paper. The sleeve is dressed over the top of the soil pipe stack, Fig. 95. Flashings are also constructed and designed for pitched roofs, as illustrated in Fig. 96. This type of flashing generally is placed over the soil pipe terminal before the roof has been shingled. The roofer, as a rule, fits it into place as indicated in Fig. 96. It becomes the duty of the plumber, however, to dress the flashing over the top of the soil pipe. When this type of flashing is used, the soil pipe terminal should be increased in diameter because

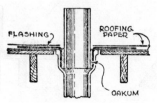

Fig. 97. Calk-Joint Method of
Flashing a Flat Roof

the diameter of the soil stack has been decreased slightly, and on large installations the atmospheric condition within the soil pipe might be affected.

A flat roof may be flashed as shown in Fig. 97. The illustration also shows the hub of a cast-iron pipe placed flush with the roof. A

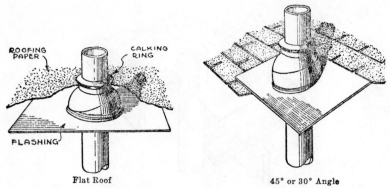

Flat Roof　　　　　　　　　　　45° or 30° Angle
Fig. 98. Adjustable Types of Roof Flashings

piece of sheet lead at least 2 feet square has been dressed into the hub. A cast-iron sleeve at least 18 inches long is calked into the hub to complete the terminal. This method of flashing the soil pipe may be used on pitched roofs, but it is not recommended.

There are other types of roof flashings constructed of copper and sheet metal. Some are of the adjustable design which permits them to be installed on any type of roof. See Fig. 98.

These flashings are provided with a lead ring which may be calked against the pipe to make it water-tight. Flashings of sheet metal are subject to corrosion and unless they are painted regularly they deteriorate rapidly.

Protection of Roof Terminal Against Freezing. In the colder

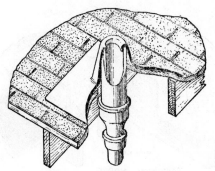

Fig. 99. Increased Roof Terminal

climates frost closure of the soil pipe terminal is a common difficulty and may be responsible for trap seal loss, which permits sewer air to enter the building. The air within the plumbing system usually is close to saturation point, and when this humid atmosphere is emitted

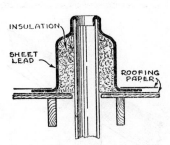

Fig. 100. Insulated Roof Terminal

through the soil pipe terminal condensation occurs. The condensate freezes rapidly.

Precautionary measures may be taken to eliminate this difficulty. One of the most advisable methods is to increase the size of the terminal, as shown in Fig. 99. This practice provides area too large to freeze solid, and because temperature varies, frost closure is averted. Another method used to overcome this defect is to cover the terminal with frost proof covering, tarring it well to make it waterproof. Fig. 100 illustrates this procedure. The roof terminal, especially on flat roofs, may be enclosed in a box packed with some kind of insulating material, as shown in Fig. 101. This practice is quite effective.

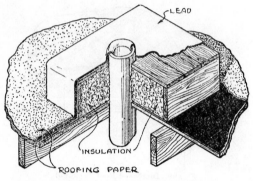

Fig. 101. Boxed Roof Terminal

A roof flashing may be used which allows an air space between the pipe and the sleeve of the flashing, Fig. 102. Air, being a poor conductor of heat or cold, serves as the protective element to overcome freezing.

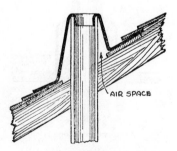

Fig. 102. Roof Terminal with Air Space

Connection of Soil Pipe to Sewer. After the soil pipe has been tested it must be connected to the house drain. This can be done in two ways.

Fig. 103 illustrates how three pieces of pipe tilted at proper angles may be used to complete the connection. The pieces of pipe should be of equal length and when fitted together take up the entire distance between the soil pipe stub and the house drain connection. Once the pipe is in place, the calk joints must be substantially made.

Fig. 104 shows the connection of the soil pipe to the house drain, using an insertable fitting to complete the work. The fitting is constructed of cast iron with a spigot and elongated hub end. After the pipe has been cut to proper length, the hub end of the fitting is passed over the spigot end of the pipe and pushed upward so the pieces can be set in place. The joints are made in the same manner.

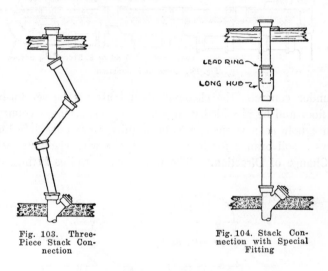

Fig. 103. Three-Piece Stack Connection

Fig. 104. Stack Connection with Special Fitting

SOIL BRANCH

The soil branch, installed horizontally with lateral or vertical connections, is that portion of the pipe which receives the discharge of water closets, with or without additional plumbing fixtures. See Fig. 91. It is usually concealed in floors, partitions, or lowered ceilings and unless it is carefully installed and thoroughly tested may be a source of trouble.

Cleanouts. The soil branch should be made accessible by equipping it with an adequate number of cleanouts. It frequently is subject to stoppage, and unless it is provided with a sufficient number of cleanouts, placed at such locations that the entire branch may be reached with a cleaning wire, the removal of a fixture or breaking into the branch may be necessary. This practice incurs expense, inasmuch as painted walls or ceilings must be cut into to get at the concealed branch.

Cleanouts should be installed wherever change of direction of the soil branch is made, as well as on the end of the branch farthest away from the vertical soil pipe installation.

Make the cleanout the same diameter as the branch, for a heavy wire must often be used to remove the stoppage, and a small cleanout may obstruct free movement of the wire. When inaccessible

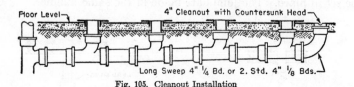

Fig. 105. Cleanout Installation

or under concrete, the cleanout should always be set flush with the floor and be provided with a brass cover having a countersunk square hole so a wrench may be applied to remove it. Fig. 105 shows a soil branch for four water closets with cleanout installed.

Change of Direction.   The use of short radius fittings on soil

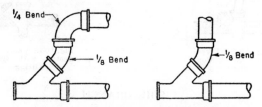

Fig. 106. Change of Direction of Soil Branch, *Left*,
Horizontal to Horizontal; *Right*, Vertical to
Horizontal

branches for making change of direction is impractical. Short sanitary tees, ¼ bends, and short turn ells are all objectionable. With their use the soil branch is subject to stoppage (which must be removed at the owner's expense) and to low rate of discharge as well.

Change of direction from horizontal to horizontal or vertical to horizontal should be made with long radius fittings.

It is permissible to use short radius fittings on changes of direction from horizontal to vertical, because this installation is somewhat different. However it is not advisable to use them. Fig. 106 illustrates the method of making these connections.

Supports and Hangers. The soil branch must be supported at close intervals to prevent sagging, which usually results in breakage.

Horizontal cast-iron runs of piping must be suspended by means of substantial iron ring hangers, placed approximately 5 feet apart, Fig. 107. It is advisable to locate the support as close to the calk joint as possible.

Galvanized steel or wrought-iron pipe branches can be suspended from eight- to ten-foot intervals. It is more durable than cast iron and therefore requires less support.

Runs of branches installed between wooden joists or steel joists must be supported on substantial wood or pipe headers spiked or otherwise fastened solidly to the building framework.

Care always must be exercised when installing soil branches in

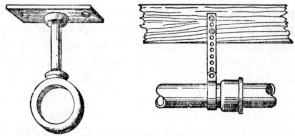

Fig. 107. Soil Branch Hangers

a frame building. Because of their large diameter, joists must be cut to provide for the installation of piping, and unless due care is exercised, the building may be weakened.

Grade and Alignment. The soil branch must be extended at a grade of ¼ inch per foot. Under this condition the waste content of the branch is given a flow of proper velocity to assure a self-scouring condition and the possibility of trap seal loss, stoppage, or retarded flow is materially lessened.

When a soil branch is given more than ¼ inch per foot grade, separation of waste generally results. The heavier suspended material becomes deposited on the bottom of the waste line because the flow does not have sufficient depth and velocity to scour the pipe.

A waste with grade less than ¼ inch pitch per foot also is likely to give trouble. Stoppage, retarded flow, crooked threads, and low efficiency are the results of this practice.

The branch must be aligned carefully. Cocked fittings and joints must be overcome, as they tend to create stoppage and affect branch efficiency generally.

Sizing the Soil Branch. The flow of waste in a soil branch differs from that in a vertical soil pipe. The movement of the waste is the result of only slight grade, hence it does not attain the velocity developed in vertical pipes. There is likely to be a decided separation of the water from the suspended organic materials, consequently, its interior may become fouled rapidly. This is especially true of long soil branches.

The efficiency desired in a horizontal waste installation is a scouring or self-cleansing action with each discharge of waste. If this action could be developed in every installation, the problem of stoppage would be eliminated. A favorable situation of this kind is difficult of achievement. Fixture groups vary in number and design, and to construct a branch installation of a size to serve each combination of fixtures is entirely impractical. Tests have also proved that a self-scouring flow is out of the question in many plumbing installations, and the only means by which satisfactory results may be obtained is to use a pipe of ample diameter and provide it with a minimum and maximum fixture limit.

To familiarize the reader with the use of Table 9,[1] the following example is offered.

Example. How large a soil branch would be required to serve an installation of six water closets?

By referring to the unit value of each fixture as set up in Chapter VI, page 66, we find that 6 water closets total 36 fixture units.

Consulting Table 9, we find that 160 fixture units are permitted on a 4 inch horizontal pipe in one branch interval.

Therefore, 36 fixture units would require a horizontal pipe of 4-inch diameter.

Installation. Fig. 108 gives a top view of a cast-iron soil branch which serves an installation of six water closets and several other fixture connections. As was previously stated, the connection for a soil branch to the soil pipe consists of a sanitary tee placed at the

---

[1] Taken from the *Report of the Uniform Plumbing Code Committee*, issued jointly by the Department of Commerce and the Housing and Home Finance Agency, Washington, D.C.

correct height and installed at the time of construction of the vertical
stack.

There are several ways in which a waste branch may be
installed. It may be necessary to calk the entire assembly of fittings
together on the floor and then hoist or lift the assembly into place.
This is a common procedure and saves much time and expense. The
awkwardness of handling the assembly because of its weight may be
eliminated by casting a hook into the ceiling of the floor above and
extending the sling of a chain block through a hole (located so the
entire pipe unit will balance) in the floor of the toiletroom which
the branch is to serve.

When there is a sufficient amount of space between the ceiling
and the branch to pour the joints, the underfloor work may be in-

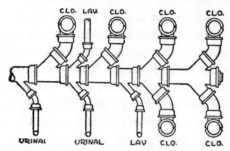

Fig. 108. Change in Direction of Waste Line

stalled in pieces. Several fittings can be calked together, or the
work may be installed in two parts; or the entire branch can be suspended, except the stubs which serve the closet installations. Procedure depends entirely on building conditions and the mechanical
skill of the plumber.

It is essential when installing heavy pipe overhead, to build
substantial scaffolds. Make sure, however, that ropes and equipment are in the best of condition. Serious accidents may result from
carelessness in this respect.

A galvanized underfloor work of the same design usually is
partially assembled on the floor. This especially is true of the main
run of pipe. Galvanized screw thread pipe can be provided with
more swing joints, or a thread may be backed up to permit a fitting
to clear a building obstruction during the assembly procedure which

allows the underfloor branch to be installed in smaller pieces than a cast-iron installation. This type of branch and the method of placing it also depends on the design or layout of the toiletroom and the judgment of the plumber. No concise rule can be applied.

It often becomes necessary for a mechanic to run the soil pipe and its branches between joists or in partitions subjected to cold drafts. The architect's building plan sometimes requires the plumber to place a soil pipe in a cold outside wall. In this case the plumber must be sure to insulate the pipe thoroughly against frost. Even if the wastes do not freeze, the greases and fats which they carry become chilled, thus solidifying and causing stoppage.

Any insulating material of good quality can be used as a protection against cold. Hair felt, mineral wool, or frost proof covering are some which serve this purpose satisfactorily. A circulation of warm air may be provided in some cases by allowing air from the basement to circulate through a partition by means of a register or opening concealed behind a fixture. This has proved to be a successful method to overcome freezing. When installing soil pipes the plumber should put headers between joists and in walls to close off drafts as an additional safeguard against freezing.

The noise caused by water rushing through the soil lines within the walls is still another objectionable feature common to carelessly installed pipe. Soil lines must not be allowed to touch plastered walls or ceilings, for such contact will cause the sound to be magnified. Hair felt or mineral wool packed around waste lines assures a quiet installation. Insulation of soil lines to avoid these difficulties is important and should be considered carefully when the soil pipe is being installed.

Horizontal runs of soil pipe may be subject to condensation in event a fixture flushing device leaks or passes water into the fixture continuously. In a short time the condensation may drip on to a ceiling over which the soil pipe branch has been installed and cause damage. This condition always must be guarded against and may be overcome by applying a good quality of anti-sweat covering to the soil-pipe installation.

CHAPTER X

WASTE PIPE

The waste pipe is that part of the drainage system which conveys the discharge of fixtures other than water closets, such as sinks, lavatories, urinals, bathtubs and similar fixtures to the soil-pipe soil branch or house drain. The waste pipe usually is smaller in diameter than the soil-pipe installation, and the materials suspended in the waste also differ, hence it must be separately classified.

The materials the waste pipe must carry are varied. The suspended materials in the water waste are grease, lint, matches, hair, garbage, and many other substances. Then there are the materials in solution with the acids, salts, and other elements which are soluble in water. Plumbing fixtures are too often misused. Household refuse of all kinds is carelessly disposed of by flushing it through the plumbing system. Improper use of plumbing fixtures can only result in waste line stoppage and deterioration of the piping, unless the mechanic uses exceptional judgment and skill in its installation. Some of the important factors are the *selection of proper material, conservative use of fittings, manner in which the waste line is supported, location of cleanouts, pitch, size, properly made joints,* and many other factors of lesser importance.

Materials Used for Waste Pipe Installation. The materials of which waste pipes may be constructed are cast-iron pipe of the calk joint and screw thread type, galvanized steel and wrought-iron pipe, brass, copper and lead.

The character of the matter to be drained should govern the selection of material for the waste pipe. In a preceding chapter the nature of the material as well as the effect various elements have on it, were briefly presented. Any waste line which conveys large quantities of acids should be constructed of acid-resistant material. All fixtures serving chemical laboratories, plating, engraving and photographic establishments, or departments in industry which use acids for various purposes, should be provided in this manner. It is permissible to use lead pipe with burned joints, and even cast iron under some circumstances, but it is inadvisable to use these materials whenever pipe that has been designed for this purpose can be obtained.

It is true that refuse from domestic kitchens contains acids of various kinds. However, the acids are in such small quantities that they are negligible and do not require acid-resistant pipe to withstand them. For domestic or commercial kitchens, cast-iron, galvanized, steel, wrought-iron, lead, brass, or copper pipe can be used with safety. Some of these materials offer advantages over others, but generally speaking, any one of them is suitable if correctly installed.

For ordinary domestic, commercial, and industrial toiletroom installations, lead, brass, copper, cast-iron, steel, or wrought-iron pipe may be used. Fixtures serving toiletrooms ordinarily are not subjected to matter which is detrimental to waste pipe of common variety. The kind of material which may be used, however, is often specified by local or state ordinance, and where such codes are in effect the plumber must use the materials specified.

Use of Fittings. A large amount of failure in the waste pipe system can be attributed to unwarranted use of fittings. There are many mechanics who build an installation piecemeal, and do not hesitate to use an extra fitting to help them out of difficulty. Stoppage of the waste line under these circumstances is common and it is always a reflection on the mechanic's work, as well as a source of annoyance and expense. Fittings are intended to make change of direction, turns, and offsets, but their use must be planned carefully so the waste within the line will have free passage. Injudicious use of fittings should not be tolerated on plumbing installations.

There has been perpetual argument between mechanics of the old and new schools as to whether or not lead pipe is superior to iron, brass, or copper, because the former does not require fittings. There is no reason to believe that a waste line of the latter type is inferior owing to the use of fittings. Those which fail, break down because of using too many fittings, or because of using the wrong type of fittings in a given location.

Change of direction of the waste line should be made in the same manner as a change of direction in the soil pipe. It is unwise to use fittings of short radius on horizontal or vertical-to-horizontal

changes. These must be made with long sweep fittings, as illustrated
in Fig. 109. Experience favors the use of **Y** and 45-degree fittings
to accomplish a turn of this kind. Some difficulty is experienced on
change of direction when fittings of the tee pattern are used, because
the side or branch opening is tapped in such a manner as to give

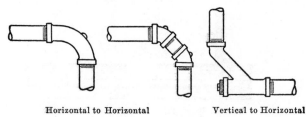

Horizontal to Horizontal   Vertical to Horizontal
Fig. 109. Long Sweep Fittings

the horizontal run a slight grade. Tee fittings are intended for use
in vertical runs having lateral branches only, Fig. 110. Their use
on horizontal installations produces a situation which requires a
crooked thread to correct, Fig. 111, and it is certain that careless
practice of this kind results in premature waste line deficiencies.

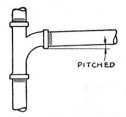

Fig. 110. Change of Di-
rection, Horizontal to
Vertical

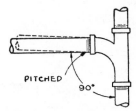

Fig. 111. Impractical Change
of Direction, Showing
Crooked Threads

Cleanouts. The waste line must be provided with an ample
number of cleanouts, so placed that the entire waste installation can
be made accessible. It often is difficult so to equip an installation
because usually it is concealed in partitions, ceilings, or floors.
Under these circumstances it is necessary to extend the cleanout to
the surface of the floor or ceiling and provide it with a plug or flush
plate so as not to impair the appearance of the room. Fig. 112 illustrates installations of this kind.

On exposed runs of pipe no difficulties are encountered, and cleanout openings can be placed at random. The location and number of cleanouts needed are factors the installing mechanic must decide. The installation should be provided with as many cleanouts as may be needed to simplify repair should obstruction occur.

Cleanouts should always be a size equal to the diameter of the

Fig. 112. Waste Pipe Cleanouts. *Left*, Vertical to Horizontal; *Right*, Horizontal

waste line so the rodding or cleaning out of the waste pipe is not interfered with.

Supports and Hangers. Vertical runs of galvanized pipe must be supported on each floor, using a pipe clamp for this purpose. A coupling of the drainage variety or a fitting should be used at about

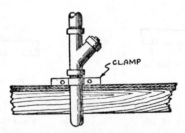

Fig. 113. Riser Support

every fifth floor interval so the weight of the line is not supported entirely by the split clamp, Fig. 113. It is not permissible to use the branch connection as a means of supporting a vertical line of pipe. A severe strain occurs at the threaded connection, and because the pipe has been materially weakened by cutting away stock to form the thread, breakage at the junction point is unavoidable.

Horizontal runs of galvanized pipe must be suspended from ceilings by ring hangers of various design. They must be placed no more than 10 feet apart to overcome all possibility of the waste sagging. Fig. 114 illustrates hangers in common use as waste pipe supporters.

Brass and Copper Pipe. Brass and copper pipe may be suspended in the same manner as galvanized steel pipe. Vertical runs are supported with pipe clamps. Because the clamp is usually of a different metal than the waste pipe, it is good practice to coat the pipe with acid-resistant paint. There is the possibility of an acid condition at the point of contact sufficient in energy to cause corrosion.

Horizontal runs of brass and copper waste pipe should be suspended at intervals closer than 10 feet. Pipe made of this material is less rigid than iron and sags more readily.

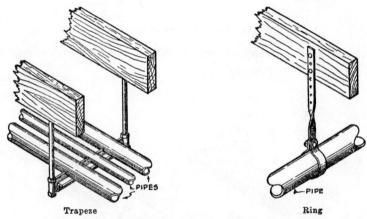

Trapeze   Ring

Fig. 114. Methods of Suspending Waste Line

Cast-Iron Pipe. Cast-iron waste pipe is supported in the same manner as cast-iron soil pipe. Regardless of diameter, it must be suspended at 5-foot intervals. Vertical runs must be supported on every floor.

Lead Pipe. Waste lines constructed of lead pipe are the most difficult to support. Vertical runs generally are provided with lead tags which may be soldered to pipe. These tags are provided with holes so they can be screwed to structural parts. In concrete buildings it may become necessary to use a lead pipe clamp or wipe a flange joint on the floor to give support to the line. See Fig. 115.

Horizontal runs must be thoroughly suspended. It may become necessary to hang a board or an angle iron from the ceiling and lay the horizontal run of lead pipe on it. Lead is very ductile and sags readily unless adequate precaution is taken. Fig. 116 illustrates this procedure.

Grades and Sizes of Waste Pipe. Waste lines should be extended at a grade of ¼ inch per foot and must be of sufficient

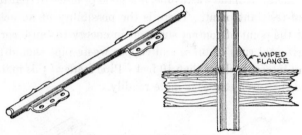

WIPED FLANGE

Fig. 115. Lead Pipe Supports

diameter to afford adequate velocity of flow to make them as nearly self-scouring as is practical.

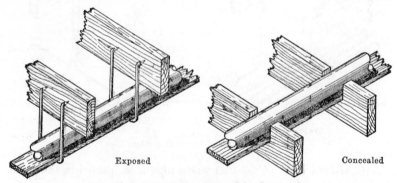

Exposed   Concealed

Fig. 116. Lead Pipe Supports on Joist Construction

Until recently there was much division of opinion among sanitary authorities as to what constituted satisfactory grade and size. This is no longer the case. The elements of doubt and uncertainty have been removed from many practices that formerly relied upon hit-and-miss methods for success. Matters such as the number of fixtures involved, simultaneous discharge of fixtures, design, ventilation methods, and other, less tangible, considerations no longer present problems that are difficult to solve. Scientific methods applied to the solution of these matters materially reduce the probability of error. Plumbers who desire to keep abreast of developments in the trade investigate and study these methods.

The preceding chapter told of the thousands of tests conducted by the National Bureau of Standards, under sponsorship of the Housing and Home Finance Agency, to determine conditions affecting fixture discharge under different conditions of use and in different types of fixtures. These tests were the means of supplying the information needed to determine the sizes and grades of waste pipe to be used in specific applications.

Inasmuch as transparent plastic pipe was used in the tests conducted, the action taking place could be plainly observed. Results of the tests were tabulated and the data made available in pamphlets and reports. It was found, for example, that wet venting and stack venting are safe in certain types of installations. An unexpected disclosure was that revealing the trap seal loss that occurs when the grade is increased from $\frac{1}{4}$ to $\frac{1}{2}$ inch per foot of fall.

Plumbing systems frequently suffer abuse of function. The human element is the factor of account here. It often occurs that waste lines identically graded and of the same size and design installed in different residences and used by different individuals function oppositely. The methods and materials used in preparing food and the habits of the housewife account for this. Some people use plumbing fixtures as a means of getting rid of almost any kind of unwanted waste, such as garbage, grease, hair, lint, matches, paper, etc. These materials can be found in most clogged waste lines. Plumbing installations are not intended to convey materials of this kind. People who are careful in the disposition of refuse seldom experience waste pipe difficulties.

METHOD OF SIZING THE WASTE PIPE

By analyzing the factors which affect the size of a waste pipe one can readily conclude that the waste pipe diameter must be adequate to serve the installation of fixtures in a general way. It would be impossible to build a specific diameter of pipe for each and every fixture arrangement. The best that can be done is to fit the diameters of pipe now manufactured into the fixture pattern in the most efficient manner.

As in the case of soil pipe and drains, the Federal Subcommittee on Plumbing has compiled a unit system for waste pipe size founded on laboratory and installation tests. Table 9, in Chapter IX, offers assistance to the plumber when he is confronted with a difficult sizing problem. The method of arriving at the size of a waste pipe is identical with that of sizing soil pipes and house drains.

To show how Table 9 may be used to determine waste pipe size, the following problem and its solution is offered.

Example. What size of horizontal waste pipe is necessary for three urinals, two lavatories, six shower baths and one slop sink?

By referring to the unit value data presented in Chapter VI on "The House Drain," we summarize as follows:

	Units
Three urinals	15
Two lavatories	2
Six showers	12
One slop sink	3
Total	32

Referring to Table 9, we find that 160 fixture units are permitted on a 4-inch horizontal waste pipe.

Then 32 fixture units would require a 4-inch horizontal pipe. Comparing the required 32 fixture units with the 160 units allowed in a 4-inch horizontal pipe, it would seem that a 4-inch pipe would be too large. However, this is within the limits of Table 9, and as nearly accurate as may be determined. The waste pipe may be graduated in size up to 4-inch diameter, depending on how the plumbing fixtures are placed.

WASTE INSTALLATIONS FOR VARIOUS FIXTURES

Waste pipe is classified in two ways, according to the type of fixtures it serves.

Direct Waste

A direct waste is one which has a terminal solidly joined to the plumbing system. An example of this is shown in Fig. 125, where all pipes are rigidly connected to the soil stack.

Indirect Waste

An indirect waste is one which has a terminal joined to the plumbing system locally, as in Fig. 121, where the drain pipe simply empties into the plumbing system.

Direct wastes usually serve the following fixtures:

1. Sinks
- Kitchen
- Pantry
- Scullery
- Slop—Shop

2. Bathtub
- Sitz
- Foot
- Bidet

3. Lavatories
- Wall hung
- Pedestal
- Two-piece

4. Showers
- Single stall
- Gang

5. Urinals
- Pedestal
- Stall
- Trough

6. Laundry tubs
7. Drinking fountains
8. Laboratory equipment
9. Hospital fixtures

Indirect wastes are used on the following fixtures:

1. Soda fountains
2. Bar wastes

3. Refrigerator wastes
4. Drinking fountains

Sink Waste. Sinks are constructed and designed to serve **many purposes.** The ordinary kitchen sink is the one which a plumber comes in contact with most frequently. This fixture requires a waste pipe of 1½-inch diameter as a minimum. The general trend is to increase the size to 2 inches because of the bulk and nature of the materials suspended in it. The waste must be as short and free from offsets as possible and provided with an adequate number of cleanouts to make it accessible.

Slop sinks are patterned in two styles, trap standard to the floor, and trap to the wall. This sink is used as a rule for janitorial service in a building. It consists of a deep basin permitting a pail to be placed under the water spigots. The trap standard to the floor fixture requires a 3-inch waste pipe. The trap to the wall fixture requires a pipe of 2-inch diameter.

Scullery sinks are constructed of galvanized sheet metal and are used in restaurant kitchens. They are provided with plugged strainers, so water may be retained in them and are used for washing dishes and vegetables, and for general kitchen use. The scullery sink requires a 2-inch waste pipe and must have cleanouts.

Pantry sinks are used in connection with butlers' pantries and are designed to retain water for washing dishes, glassware, and cutlery. A 1½-inch waste pipe is adequate for this fixture.

Factory wash-up sinks are installed in shop toiletrooms for use in washing the hands and face. Usually the water is discharged by an inverted shower spray into a circular basin. Because of the size of the fixture, a waste pipe of 2-inch diameter is recommended as a minimum.

Bathtubs. Bathtubs are constructed in many designs and their wastes do not offer trouble. The materials suspended in the wastes are few and, as a rule, any materials in solution are so thoroughly diluted that they are carried into the waste pipe system without difficulty. The waste pipe of a bathtub is generally scoured thoroughly because of the head of water in the tub. Its efficiency, however, can be increased by eliminating long runs of horizontal pipe and offsets. The minimum diameter of waste for a bathtub is a 1½-inch or 2-inch pipe. There are other types of bathtubs used to wash various parts of the body, such as sitz baths, foot tubs, and bidets. The size of the waste for these fixtures must also be 1½ or 2 inches in diameter.

Lavatories. Lavatories, like kitchen sinks, are used as depositories for much objectionable waste matter, and the waste pipe from these fixtures becomes stopped up frequently. The only factor which makes it not quite so objectionable as a sink waste is that the run of waste is much shorter. The plumber can add much to the installation by the proper use of fittings and by eliminating long runs of horizontal pipe. The minimum diameter of pipe which may be used as a waste for this fixture is 1¼ inches; but it has been found that 1½ or 2 inches is more satisfactory.

Shower Baths. Shower waste lines do not give trouble often, largely because they are a clear water waste and, as a rule, no objectionable materials are conveyed by them. The important factor in this installation is to provide the fixture with a waste pipe of adequate size. Experience has proved that a 2-inch diameter pipe is adequate for this purpose. As in every phase of plumbing, careful planning is essential to ensure correct operation.

Urinals. The waste pipe for a urinal should be at least 2 inches in diameter and must be increased in size with the addition of other fixtures. The waste pipe serving this fixture becomes stopped easily because the urinal is used in most instances as a drain for toiletroom floors. Much foreign material, such as floor sweepings, gum, cigar and cigarette stubs, and matches are passed into it and become lodged in the short turns of the waste pipe. The action of the acid content of urine on these materials seems to produce a soft gelatinous substance that soon becomes foul and is extremely difficult to remove.

Laundry Tubs. A laundry tub is used in the average residence for the washing of clothes and its waste pipe does not give a great deal of trouble as it is relatively short and usually of vertical design. It is discharged directly into a connection of the house drain. Pipe 1½ to 2 inches in diameter is proper for a laundry tub installation.

Drinking Fountains. Some drinking fountains are clear water fixtures requiring a trap no greater in diameter than 1¼ inches. The fixture may be connected directly to the plumbing system or, should fixtures of this kind be conveniently grouped, they may be discharged into an indirect terminal. For screw pipe drains, a riser of 1½ inch diameter is used; for soil pipe, the riser to the fixture and drain should be 2 inches in diameter.

Laboratory Waste. The waste pipe from chemical laboratory equipment need not be more than 2 inches in diameter unless manufacturers' specifications call for a larger pipe. This type of waste is subjected to objectionable acid compounds, and the self-scouring flow obtained in a pipe of small diameter adds to the installation materially. Acid-resistant pipe should be used.

Hospital Fixtures. Hospital and special type fixtures do not have a fixed size of waste. The size of the unit, as well as manufacturers' specifications, generally determine the diameter of the waste line. As a rule no less than a 2-inch pipe should be used for this purpose because generally rapid draining of the hospital fixture is necessary. It is also well to provide the waste pipe with an adequate number of cleanouts because of the nature of the materials discharged into it. The action of the atmosphere on the waste material tends to coagulate it and cause stoppage which is difficult to remove.

Indirect Wastes. Indirect waste pipe installations differ from those designed for direct waste, because they are clear water fixtures that generally are made up of many small units or compartments, making direct waste installations impractical. The terminal of an indirect waste may be a funneled drum trap, properly ventilated, and connected directly with the plumbing system, Fig. 117. The terminal may be a floor drain or any other plumbing fixture, provided the waste materials discharged into it do not become a nuisance. See Figs. 118 and 119.

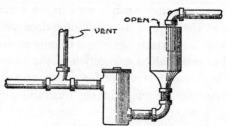

Fig. 117. Funneled Drum Trap

Discharging wastes from fixtures of this design into basement floor drains is considered good practice, because the discharged water tends to maintain a constant seal in the floor drain and prevents sewer air from entering the building.

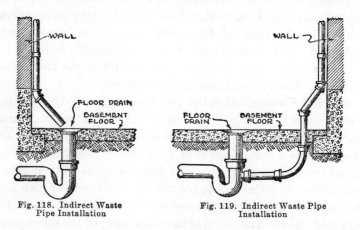

Fig. 118. Indirect Waste   Fig. 119. Indirect Waste Pipe
      Pipe Installation   Installation

Waste pipes indirectly connected to the drainage system often are referred to as a local waste and their installation does not require the rigid detail that direct wastes do. Indirect wastes are constructed of the same material and fittings as are direct wastes. Cleanouts, grade, and suspension also enter into their design, but because the waste usually is exposed not much difficulty is encountered in removing stoppage.

Soda Fountain and Bar Waste. Soda fountains and bars are connected to the plumbing system in the same manner. Both fixtures consist of many small compartments, such as rinsing sinks, steam tables, beverage units, drain boards, ice chambers and other devices. These fixtures may be connected to a common waste pipe run in such a manner that all service units may be accommodated.

It is an added responsibility to trap and ventilate each compartment, because the design of the fixture does not offer space to conceal the pipe. As a rule the local waste pipe is suspended from

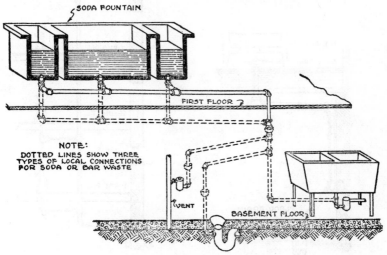

Fig. 120. Indirect Bar Waste

the basement ceiling or laid on the floor directly underneath the fixture.

The location of the waste usually is determined by the design of the fixture and becomes the responsibility of the mechanic. It is advisable to use a pipe of at least 1½-inch diameter for a soda fountain or bar waste and provide it with cleanouts spaced in such a manner as to make it accessible. Fig. 120 illustrates the installation of an indirect waste for a soda fountain or bar with possible indirect terminals.

Icebox Waste. Icebox wastes also are connected indirectly to the plumbing system. It would be objectionable to make a rigid connection, because foods stored in an icebox may be affected by odors emanating from the drainage system should the protecting trap's seal be destroyed.

The icebox must be protected from basement odors as well as from those of the plumbing system. This is accomplished by means of a bell trap or a trap composed of fittings installed at the base of the waste line.

The terminal for an icebox waste may be a basement floor

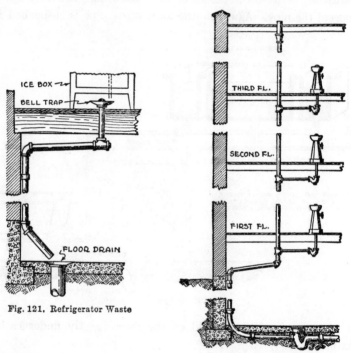

Fig. 121. Refrigerator Waste

Fig. 122. Bubbler Waste
Installation

drain, the local side of a fixture trap, a basement fixture, or any other convenient terminal. Fig. 121 illustrates the installation of an icebox waste properly dripped into a floor drain terminal. It is not good practice to drip the waste pipe onto the basement floor because it wastes continuously and the wet floor becomes a nuisance.

Drinking Fountain Waste. Fountains or bubblers, when they are conveniently located in the building, may be connected to a common waste that terminates with an indirect connection. Fig. 122

shows the installation of a drinking fountain on three floors of a building, the waste terminating into a floor drain.

Each fixture must be trapped individually so the waste pipe does not serve as a local vent for the basement. Individuals using the fixture are protected by this means. It is advisable to continue the waste pipe through the roof as a means of local ventilation and in this way avoid trap seal loss. The extended vent pipe, however, must never be connected to any part of the drainage system because of possible danger to health.

Industrial Equipment. Creameries, laundries, canning factories, breweries, slaughterhouses and other industries use large volumes of water in the manufacture of their products. These wastes must be disposed of rapidly, because industrial equipment must function

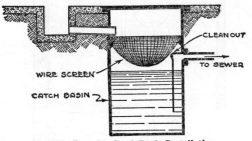

Fig. 123. Laundry Catch Basin Installation

almost continuously to maintain low operating costs. The time element involved in draining this equipment through small waste orifices, would raise the cost of producing the product tremendously. To avoid this difficulty industrial equipment is provided with large waste openings, operated mechanically, which allow the used water to escape in a very short interval. To connect these wastes directly to the plumbing system might overbalance it to the point where sanitary fixtures would not function properly, and minus or plus atmospheric conditions within the system would unquestionably result. Industrial wastes, therefore, must be connected indirectly to the plumbing installation.

Industrial devices usually are discharged into waterproof concrete gutters or troughs which terminate in a cast-iron receptor of the catch basin type. The troughs are designed to accommodate the volume of water discharged from the equipment and permit it to drain slowly into the receptacle connected with the drainage system.

The catch basin is usually provided with a strainer basket to separate from the water-carried waste suspended materials which may interfere with the working of the plumbing installation. Fig. 123 illustrates laundry washing equipment which is connected by means of troughs to the plumbing system. This installation is typical of that used in other industries.

UNDERFLOOR WASTE PIPE FOR DOMESTIC BATHROOM INSTALLATIONS

In the preceding paragraphs waste pipe installations for in-

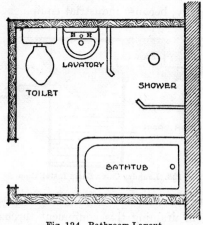

Fig. 124. Bathroom Layout

dividual fixtures were discussed. These installations are common in practice, but group installations constitute the greater part of all fixture work. Presentation of a large number of such installations would require a great amount of space and still might fail of the purpose intended owing to variability of existing conditions. The competent plumber has knowledge of what constitutes good waste pipe practice, and he has the ability to apply that knowledge.

Some details of the work are the same in all waste pipe installations. It is necessary, nevertheless, that the student have specific knowledge of underfloor construction. To provide this knowledge, the layout of a small bathroom consisting of water closet, lavatory, bathtub, and shower is shown in Fig. 124.

Galvanized Iron Underfloor Work (Soil Pipe Terminal).[1] The underfloor work, Fig. 125, terminates directly into a 4x2 **Y**, placed in the soil pipe underneath the sanitary tee which serves the closet bend. A calk joint **Y** is the best fitting for this purpose, although a tapped **Y** can be used. The first connection into the **Y** consists of a spigot into which is screwed a short nipple and a 45-degree drainage ell and long pattern 90-degree elbow to provide an opening for

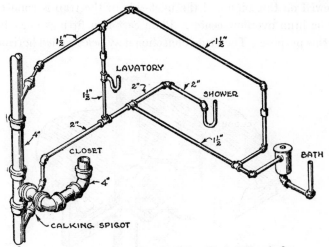

Fig. 125. Galvanized Iron Underfloor Work, **Y**-Terminal

the horizontal waste pipe. From the side opening of the elbow a piece of pipe is cut to proper length and screwed into a long turn 90-degree tee to serve as the terminal for the vertical basin waste. The vertical basin waste is cut to proper length and screwed into the long turn 90-degree tee. The top opening of the basin tee may be joined to the vent pipe installation by means of a long screw or calk joint fitting.

The shower bath is the next fixture connection to be made. A 45-degree ell is screwed onto a piece of pipe which is cut to a length that will locate the shower strainer in the center of the stall. This connection is then screwed into the fitting. The length of the pipe between the strainer and 45-degree ell must be measured accurately

[1] The underfloor work shown in Fig. 125 represents minimum requirements for the installation described.

and be provided with a **P**-trap for the shower bath. A 45-degree
Y-connection for the bathtub wastes must be included in this run.

The next step is to complete the bathtub branch. The branch
consists of a piece of pipe and a 1½-inch short pattern 90-degree tee
for the bath and shower vent pipe terminal. A nipple and a
45-degree ell must be screwed into the vent tee for the purpose of
leveling the bathtub drum trap. The bath waste is continued by
screwing a piece of pipe into the 45-degree ell. A 4×5 drum trap is
screwed on this pipe and the inlet side of the trap is connected to
the bathtub overflow center. Plain cast-iron fittings must be used
for this purpose. The vent connection may be extended horizontally,

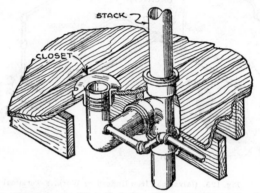

Fig. 126. Galvanized Iron Underfloor Work, Closet Bend Terminal

underneath the bathtub, to the partition, and then may be joined to
the vertical vent by a long screw or calk joint fitting.

The underfloor work must be checked for grade, support, and
fixture openings and all exposed threads should be coated with rust-
proof paint before the rough floor is laid.

Galvanized Underfloor Work (Closet Bend Terminal). The
underfloor work, Fig. 126, terminates into a closet bend provided
with tapped side openings for that purpose. This type of underfloor
terminal is most commonly used because it requires less cutting
away of the wooden framework of the building. After the closet
bend has been calked into place, two 60-degree drainage ells with
nipples are screwed into the side tapping. This practice allows the
runs of waste pipe to lie close to the top of the joist, and by turning

the 60-degree ells off center, the waste line is provided with pitch. Some mechanics use 45-degree fittings at this point, but this method requires the cutting of crooked threads to grade the underfloor work properly. Crooked threads represent inferior workmanship and often produce faulty installations.

The piece of pipe between the basin connection and the closet bend may now be cut to length and provided with a 45-degree **Y** and elbow, or long turn 90-degree tee, for the branch connection into the partition. The base of the vertical basin waste consists of a 90-degree long pattern ell which is screwed into the **Y** connection and extended to the basin trap center where a short pattern 90-degree tee may be used. The vertical basin waste can be connected to the vent pipe either by means of a long screw or a calk joint fitting.

From the basin connection the horizontal run is extended to the bathtub branch and then to the shower bath center, where it is provided with a **P**-trap. The **P**-trap can be leveled with a 45-degree drainage elbow.

The bathtub branch connection consists of a 45-degree **Y** and a 45-degree elbow installed in the shower run and extended to the drum trap location in the tub panel. The drum trap must also be leveled with a 45-degree fitting and its inlet side can be run to the tub overflow center by using plain cast-iron elbows of a type suited to the purpose.

Within a few inches of the drum trap a short pattern 90-degree tee must be installed and extended to the partition to act as a vent for the shower and bathtub trap. The vertical vent pipe may be connected to the vent pipe system in the usual manner.

Methods of waste installation often vary, being dependent upon building conditions. The choice of method is the responsibility of the mechanic, who can best weigh all relevant factors.

Brass and Copper Underfloor. The underfloor works of brass and copper pipe may be made in the same manner as that of galvanized pipe, except that on copper pipe, fittings designed for this material must be used. In some instances, especially when copper pipe is used, the underfloor waste pipe may be assembled above the floor and then set in place. This practice necessitates the making of only a few joints in awkward locations and does eliminate the danger of fire. When copper or brass underfloor work is installed, care must be exercised not to kink it or drive nails through it when laying back the rough floor.

Lead Underfloor Work. Lead pipe may be used to construct underfloor work, and is considered an outstanding material for this service. In many parts of the United States, the working of lead has become a lost art, due to the general practice of installing iron material. To build a lead underfloor work requires mechanical ability which can be attained only through experience and diligent

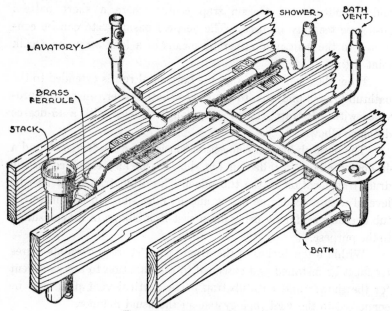

Fig. 127. Lead Underfloor Layout

practice. The young mechanic does not have the opportunity to perfect himself in this skill.

It would be impossible to formulate a set of rules as to how a lead underfloor work should be installed because of the high degree of skill associated with it and also because of differences of opinion held by mechanics. What might be an awkward position for one operator might be comparatively easy for another. There are, however, a few fundamental processes which may be presented as a beginning for the installation of a lead underfloor. Fig. 127 shows lead underfloor work for practically the same layout for which a galvanized waste (Fig. 125) was installed.

The first step in the process of building a lead installation is to select a room of sufficient area and then clean the floor thoroughly. The older mechanics always carried a piece of carpet with them on which to work the lead. After the room has been thoroughly swept, the bathroom partitions may be laid out on the rough floor, indicating in the layout the centers of all the fixtures. Lead underfloors generally are wiped up out of place (or at least the major portions) and then carried into the bathroom and placed in their permanent location.

The plumber now drifts all the lead waste and dresses it until it is free of any kinks. He also tins the brass fittings, and, after determining just where he is going to wipe the joints, he starts to make the necessary bends, disregarding the lengths of the runs.

Making a Lead Bend. The practice of making a lead bend is a problem to mechanics who have not had the opportunity to become efficient in it. Making a bend in lead pipe is a simple and easy

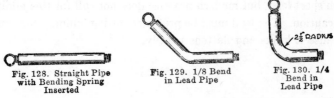

Fig. 128. Straight Pipe with Bending Spring Inserted

Fig. 129. 1/8 Bend in Lead Pipe

Fig. 130. 1/4 Bend in Lead Pipe

operation. It is made by means of a bending spring inserted into the lead pipe. It can be made with clean, dry sand, but this method is obsolete.

After the lead pipe has been cut to proper length and all kinks have been drifted from it, the bending spring is screwed into it. This can be done by inserting the pointed end and turning the opposite end of the spring to the right. After the spring has been inserted into the pipe its full length, it is turned to the left in order to expand it and take up the entire diameter of the pipe. See Fig. 128. The inside radius of the lead pipe to be bent must be heated slightly and the pipe placed over the knee and given approximately a 1/8 bend. See Fig. 129. The inside of the bend is again heated and the bend pulled around quickly to the proper radius. See Fig. 130. Care should be taken not to allow the pipe to become too hot, for this causes buckling of the inside radius, which is likely to split the outside of the bend.

Once the lead bends have been completed and the location of the joints determined, the pipe is laid out on the marked floor and cut to length. When this has been completed, the joints are prepared for wiping and the underfloor work is propped on blocks to permit this operation. When it is impossible to wipe the entire underfloor out of place, the sections can be laid out, so the joint that is to be wiped in between the floors will be in a convenient and accessible location.

After the work has been completed, the waste installation may be placed in the rough floor. Any slight variation in measurement can be corrected readily because of the ductility of the lead.

It is advisable to terminate the lead underfloor directly into the soil pipe stack. Past experience has proved that terminating a lead waste into a lead closet bend is impractical.

The lead underfloor work must be well supported on 7/8 x 4-inch lumber nailed to 2x4 headers located between the building joists, as the illustration indicates. The entire underfloor work may be safed with sheet lead, but modern practice does not call for this additional precaution. The lead must be protected during building construction because it kinks and flattens readily.

CHAPTER XI

TRAPS USED ON PLUMBING SYSTEMS

A trap used on the plumbing system is a device so constructed as to prevent the passage of sewer air through it and yet not affect the fixture discharge to an appreciable extent. The study of traps is an interesting one and has caused sanitary authorities interested in investigation of plumbing problems no end of difficulty. Since the innovation of the first patented trap used on plumbing fixtures in the United States (dating back to the year 1856), manufacturers have designed and offered to the plumbing industry hundreds of these devices, each one varying in construction, some cumbersome and bulky, others plain and simple, but each said to be the most efficient trap made. Testing of these traps demonstrated that some offered advantages over others, under certain conditions, but every one failed to come up to the standard expected of it when subjected to actual installation conditions.

The gases which occur in public sewage systems caused by the decomposition of organic materials within the sewers have been discussed briefly in previous chapters. It is improbable that water-borne diseases, such as dysentery, typhoid, cholera, etc., can be transmitted through the gases of the public sewers. Recent tests and experiments have indicated that the bacteria responsible for these diseases may be carried into the body through a faulty cross-connected drinking water supply.

This fact, however, does not lessen the importance of trap installations on a plumbing system. The properties, both physical and chemical, of the many gases found in sewage systems are known, and their effect on the human body is often serious. No individual could maintain health if he were required to breathe large quantities of hydrogen, hydrogen sulphide, methane, or carbon dioxide; and even a small amount of carbon monoxide in the atmosphere within a building may prove fatal.

Many of these gases are obnoxious, and, if not fatal when

breathed by human beings, are nauseating and undesirable and may be a contributing element to lesser diseases of man.

The basic function of a trap on a drainage system is to prevent these objectionable gases from entering the plumbing system. Mechanics often are of the opinion that its purpose is much more extended and often expect the trap to do much more than it was designed to do.

Because of extreme conditions caused by simultaneous fixture use and overtaxed waste conditions, the plumbing system is subjected to minus and plus pressures which affect the liquid content of the trap. (This phase of the plumbing industry will be discussed in the next chapter under "Trap Seal Loss.") Traps consisting of movable internal mechanisms which form their seals were produced to overcome these difficulties. The principle of these devices is to form a mechanical barrier against the passage of sewer air. These various types of traps fail because of the effect the dissolved and suspended materials contained in the waste have upon them. The movable parts corrode readily and, in this condition, fail to operate. Most state authorities prohibit their installation for these reasons.

Traps also have been designed with internal metal partitions, which are subject to acid conditions of the waste. The objective of this design was a device that would be compact and have a rather neat appearance. These forms of traps also have failed and should be discarded.

Today, sanitary authorities depend on, and design plumbing systems which use traps that have a water seal. Elaborate installation systems are essential to maintain a constant pressure of one atmosphere (14.72 lb. or 760 mm.) to make these traps effective; but even though the cost tends to mount with this type of installation, it appears as though not much ever can be done about it. Spending to promote efficient trade practices or to maintain a high standard of health is true economy, and the more foolproof an installation can be made, so that the consumer need not suffer from it, the greater its benefits.

In the preceding paragraphs many new trade terms have been used which might confuse the uninformed reader. The trap has been defined, but its acceptable design and construction has not been explained clearly. The most practical form of trap is constructed in the form of the letter **P**. Hence its name, **P**-trap. Plumbers of the older school referred to it as a gooseneck and it is often so designated in modern practice. The reason for this nomenclature is that the trap is curved in the same manner as the neck of a goose. There are other forms of water-sealed traps: the **S**-trap, shaped like the letter **S**, and the ¾ **S**-trap. These traps will be discussed in detail later in this chapter.

The **P**-trap, Fig. 131, must be installed as close to the fixture as is possible. This practice overcomes the tendency for the inlet side of the trap to become fouled. Each time that the fixture is discharged, a quantity of the liquid waste is arrested and retained in the dip of the trap. The liquid content is termed the trap seal, and may be defined as the column of water retained between the overflow and the dip of the trap, which separates the inlet and outlet arms.

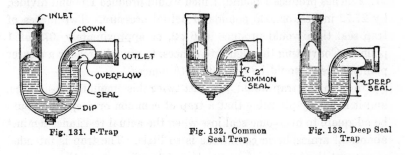

Fig. 131. P-Trap   Fig. 132. Common   Fig. 133. Deep Seal
                               Seal Trap   Trap

There are two forms of water-sealed traps. These are known as the **common seal**, Fig. 132, and the **deep seal**, Fig. 133. The common seal trap has a depth of 2 inches between the overflow and the dip. The deep seal retains twice this column of water, or 4 inches of liquid content.

The trap of common seal is used on plumbing fixtures subjected to normal conditions, and should be adequate. The deep seal trap also may be used under normal conditions, but it is intended for abnormal situations, such as extreme heat, increased or decreased atmospheric conditions, and circumstances where complete ventilation cannot be obtained.

It is not necessary to enter into a complete description of trap seal loss, and ventilation to overcome it, but one essential should be pointed out in detail to clear any mistaken ideas at this time. Many mechanics are of the opinion that a trap of deep seal eliminates the problem of seal loss because of the increased resistance the additional 2 inches of water it retains offers against increased or decreased atmospheric conditions. This distorted belief probably has a traditional aspect—a "hand-me-down" conception from the time of unscientific plumbing installations, when a sink and hopper closet constituted the sanitary facilities of the average residence. With very little serious thought and with simple mathematical calculation this false belief can be dispelled.

A trap of common seal has a liquid depth of 2 inches and will offer resistance against abnormal conditions only to the amount of pressure 2 inches of water will develop. It is a scientific fact that a column of water 2.31 feet or 27.72 inches in height exerts pressure of 1 pound per square inch at its base under normal conditions. If 27.72 inches produce 1 pound, 1 inch would produce 1 pound divided by 27.72 inches, or .036 pounds of actual pressure. Two inches of trap seal then would produce .036 x 2, or approximately .072 of 1 pound. Converting this figure to ounces, to give the reader a clearer point of view, would be equal to 1.15 ounces.

A deep seal trap would offer just twice this amount of resistance, and it is hardly plausible that a trap of common or deep seal would be adequate to overcome seal loss when the actual resistance against abnormal atmospheric conditions is so little. The trap is intended to prevent the passage of sewer air, not to offer adequate protection against minus or plus pressures. Without the necessary systems of ventilation, its seal content would have to be of a depth to resist a variation in pressure of one atmosphere (approximately 33 feet.)

A deep seal trap does offer some advantage over the trap of common seal in that its resealing quality is greater. The term "reseal" is applied to the scientific principle that water at rest tends to seek a level and maintain it. This principle may be applied to traps of deep seal, because, after their liquid content has been disturbed, the water tends to level itself sufficiently to seal the trap partially. Successive fixture discharges, however, will eventually unseal the trap. There is one solution to the problem of seal loss, namely, the practical application of scientific principle in the form of efficient ventilation systems. No trap has yet been designed which offsets this phase of the plumber's work.

Traps, called **anti=siphon** traps, were designed to increase re-sealing quality. The additional reseal is obtained by enlarging the volume of water in the trap by building a bowl of increased diameter into the outlet leg of the trap. The principle associated with its operation is the draining down of the additional volume contained in the bowl into the dip of the trap and, again, sealing it. These traps prove to be effective when installed under normal conditions and when the installation is comparatively new. Continued dis-charge of greasy wastes tends to reduce the diameter of the bowl, and no benefits can be expected of the trap when this occurs. Plumb-ing systems so constructed cannot be depended upon to give satisfac-tory service over a long period. The installation must be positive and remain so during its life to be effective.

All traps because of their construction are subject to stoppage and must be provided with a cleanout, or be so designed that they can be disassembled with little effort.

Materials Used in Construction. Traps are constructed of spun and cast brass and galvanized and tarred cast iron. The weight and gauge of the material must be in accord with the A.S.T.M. (American Society for Testing Materials) standards.

Size of Traps. The size of the trap required for a specific fixture has been established by laboratory test. Section 56 of Recommended Minimum Requirements is offered as a basis for this requirement.

SEC. 56. TRAPS, KIND, AND MINIMUM SIZE.—Every trap shall be self-cleaning. Traps for bathtubs, lavatories, sinks, and other similar fixtures shall be of lead, brass, cast iron, or of malleable iron galvanized or porcelain enameled inside. Galvanized or porcelain-enameled traps shall be extra heavy and shall have a full-bore smooth-interior waterway, with threads tapped out of solid metal

The nominal size (nominal inside diameter) of trap and waste branch for a given fixture shall not be less than that shown in the table. (See page 152.)

TYPES OF PERMISSIBLE TRAPS

There are two varieties of traps which may be used in connection with plumbing fixtures at this time: the **P**-trap, sometimes called ½ **S**, for fixtures suspended from the walls or supported on pedestals, and the drum trap, for fixtures which are set on the floor. Both of these traps have natural water seals and have been found to be most practical.

P=Trap. The **P**-trap may be obtained in sizes from 1¼ to 6

Trap Sizes for Various Fixtures

Kind of Fixture	Size (In Inches), Trap and Branch	Kind of Fixture	Size (In Inches), Trap and Branch
Bathtubs[1]	1½	Sinks, hotel or public	2
Bath, shower, stall[1]	2	Sinks, large hotel or public	2
Bath, sitz	1½	Sinks, small, pantry or bar	1¼
Bath, foot	1½	Sinks, dishwasher	1½
		Sinks, slop, with trap combined	3
Bidets	1½	Sinks, slop sink, ordinary[1]	2
Combination fixture	1½	Urinals, lip	1½
Drinking fountains	1¼	Urinals, troughs	2
Fountain cuspidors	1¼	Urinals, pedestal	3
		Urinals, stall	2
Floor drains	2	Washbasin[1]	1¼
Laundry trays	1½	Water-closet	3
Sinks, kitchen, residence	1½		

[1]The present tendency is toward an increase in the size of trap and waste pipe for this fixture, in order to reduce the time required for emptying it.

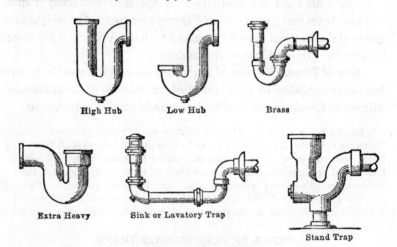

High Hub   Low Hub   Brass

Extra Heavy   Sink or Lavatory Trap

Stand Trap

Fig. 134. Various Types of P-Traps

inches in diameter and usually is constructed of nickel or chrome-plated brass, galvanized malleable and cast iron and other metal alloys. Each manufacturer has changed the style to some extent so that there are many acceptable varieties. Fig. 134 shows P-traps of approved design.

The P-trap is desirable for use in connection with lavatories, sinks, urinals, drinking fountains, and, in some instances, shower baths, and installations which do not require the wasting of large volumes of water. The flow of water through a **P**-trap is somewhat slower than through other standard types.

Fig. 135 shows how a **P**-trap may be used in connection with wall-hung fixtures. It must be installed as close to the fixture as is practical, and care should be exercised not to require too long a vertical leg between the trap and the fixture proper. This practice eliminates high velocities, which often are responsible for pronounced

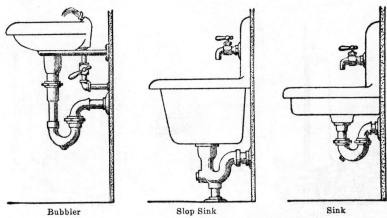

| Bubbler | Slop Sink | Sink |

Fig. 135. Traps on Wall-hung Fixtures

trap seal loss, and lessens the possibility of fouling. The dip portion of the trap should be as short as is practical, so the flow of water will not be retarded and make of the trap a veritable cesspool. The horizontal leg connection to the waste system must also be short for ventilation purposes.

Fig. 136 indicates how the **P**-trap may be used in connection with fixtures set into the floor. In this installation the trap is concealed between the building joists and serves the fixture very well.

Drum Trap. The drum trap is also a water-sealed device and derives its name from its large diameter. These traps are designed in two styles—the 4 x 5-inch type, Fig. 137, and the 4 x 8-inch drum trap, Fig. 138. The drum trap has many uses and its advantage over the **P**-trap lies in the fact that a greater volume of water may be passed through it in a shorter interval. Its resealing quality is greater than that of a **P**-trap, and, because trap seal loss is more prevalent in fixtures discharging greater volumes of water, it is more practical for this kind of an installation.

The drum trap does have some disadvantages. The trap is large and cumbersome, and when it is installed on fixtures where the trap is exposed to view, the appearance of the installation is affected. Attempts have been made to produce a modified drum trap for wall-hung fixtures, but these efforts have not been entirely successful. Drum traps are also objectionable because their cover or cleanout mechanism is above the water seal, and unless the mechanic uses a

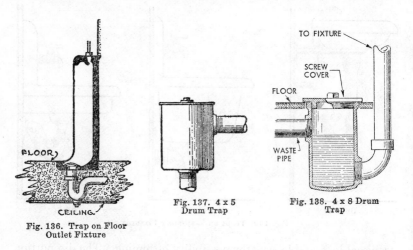

Fig. 136. Trap on Floor
Outlet Fixture

Fig. 137. 4 x 5
Drum Trap

Fig. 138. 4 x 8 Drum
Trap

lubricant and a fiber washer in the joint between the cover and the body of the trap, it may leak and form a sewer air by-pass. Up to the present time, however, the drum trap is the best that trap designers have to offer and sanitary authorities are almost obliged to accept them.

Drum traps of 4 x 5-inch and 4 x 8-inch size are used on bath-tubs, foot baths, sitz baths and, in some instances, in connection with urinals. A modified form of drum trap may be used for shower bath service. Drum traps also serve as the terminal for soda fountains, bar waste, and any type of drip funnel or indirect waste terminal. The same precautions apply to their installation as those indicated in the use of **P**-traps. Fig. 139 shows how a drum trap is installed on fixtures whose waste outlets are close to the floor. Drum traps generally have a seal of more than 2 inches, and are resealing in nature.

Anti=Siphon Traps. Fig. 140 illustrates anti-siphon **P**-traps which may be used on the plumbing installation. These traps are

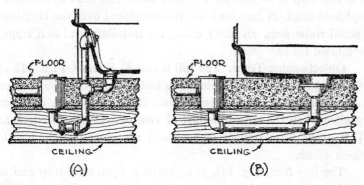

Fig. 139. (A) Drum Trap on Bathtub Outlet; (B) Drum Trap on Shower Installation

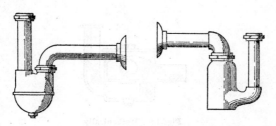

Fig. 140. Anti-Siphon Traps

to be dealt with in the same manner as are ordinary **P**-traps as far as ventilation needs are concerned.

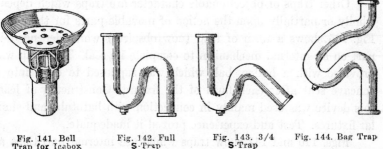

Fig. 141. Bell Trap for Icebox   Fig. 142. Full S-Trap   Fig. 143. 3/4 S-Trap   Fig. 144. Bag Trap

Bell Traps. Bell traps, Fig. 141, are intended only for certain kinds of fixture use. Their installation is common on indirect or local wastes, such as ice boxes and similar fixtures. The purpose of a bell trap is to prevent the passage of odorous gases. The seal in a bell trap is formed by a raised metal rim that is cast into a depressed bowl. A furrowed bell or cap, placed over the rim, forms a small water seal. In many cities, the installation of bell traps is prohibited for new work.

Objectionable Traps. The full **S** and ¾ **S** traps, Figs. 142 and 143, should not be used in plumbing installations because they embody obstacles to proper ventilation. These traps were commonly installed when the crown method of venting was permitted. **S** and ¾ **S** traps form perfect siphons and are objectionable for that reason alone.

The *bag trap,* Fig. 144, is an extreme form of **S** trap and one that is seldom found except in old plumbing installations. Where

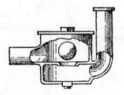

Fig. 145. Mechanically
Sealed Trap

it is encountered in repair work, it should be replaced with a trap of approved design.

Other traps of objectionable character are traps which depend wholly or partially upon the action of movable parts for their seal. Fig. 145 shows a form of trap (now obsolete) which depended in part on an internal mechanism to complete its seal. This trap was provided with a hollow ball which was supposed to drop into a concave seat after each flush of the fixture. Constructed of lead, this device was used mainly in connection with bathtubs and similar fixtures. Test and experience proved it inadequate.

Figs. 146 and 147 show traps which use internal partitions **as** the means of providing water seal. Although some traps of this type are permitted in certain areas, their efficiency is doubtful. With respect to practice, the double trapping of fixtures and crown venting are today prohibited by most cities.

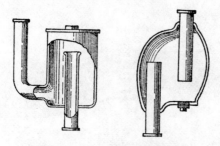

Fig. 146 Internal Partition Traps

In conclusion, it should be strongly emphasized that a trap is not a device designed to withstand pressure variations, and to be

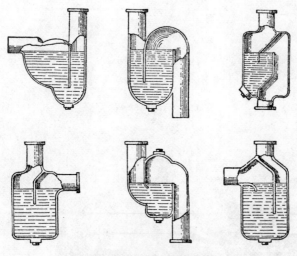

Fig 147 Light Metal Partition Traps

effective and maintain its water seal it must be ventilated adequately.

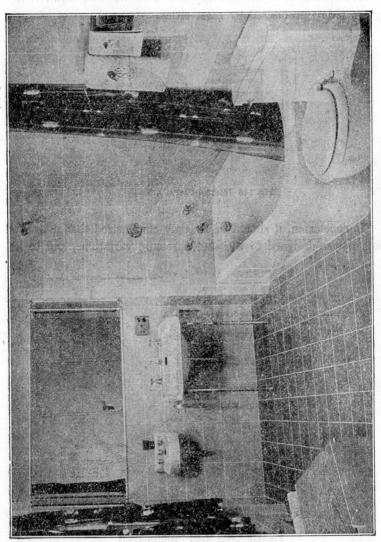

A MODEL BATHROOM INSTALLATION FOR A RESIDENCE

Courtesy of Kohler Co., Kohler, Wis.

CHAPTER XII

VENTILATION

Ventilation of the plumbing system is that portion of the drain-age installation designed to maintain atmospheric pressure within it and prevent at least three major difficulties, namely, trap seal loss, retardation of flow, and material deterioration. This phase of plumbing science is a most important one and yet, up to a few years ago, it was not seriously regarded. The average mechanic of thirty years ago was of the opinion that all that was essential to make a plumbing installation function efficiently was to install a waste pipe of sufficient size to discharge the waste materials from plumbing fixtures in a reasonable time, and to provide the waste with some sort of trap to prevent sewer air from entering the building. Plumbing during that period was not considered a scientific industry. The sanitary facilities in a building were few, so no serious results developed from the objectionable practices of those days. The consumer as well as the mechanic regarded a plumbing system as a means of waste disposal and did not feel that health standards could be affected by improper installations Up to twenty-five years ago, the average mechanic did not know, or at least could not give an intelligent reason why, fixture traps had to be ventilated, even though he had been ventilating traps for many years; nor did he associate scientific principles with any phase of his work. Plumbing during that period followed traditional patterns. Installations were made in the same manner year after year. There were exceptions however. Leaders in the industry were working constantly to improve practice, to show the layman the relation of health to sanitary conditions of living, and to educate mechanics of the trade in the physical principles associated with their work. The erection of large buildings and increased congestion in growing cities also played a very important part in bringing plumbing practice to a scientific level. Over a period of many years orthodox practices have changed radically and today plumbing is regarded as work of professional caliber.

In order to ventilate a drainage system properly, the mechanic must possess a thorough knowledge of the principles governing the atmosphere. He must know of what elements atmosphere is composed and the effects gases it contains have on piping materials. He must be able to correlate the principles of the siphon, pressure, and vacuum with the vent pipe installation, so that when the system is put into operation it will function indefinitely.

Recent experiments have proved that proper ventilation of the plumbing system is of the foremost importance. A waste pipe may be of smaller diameter, and more fixtures can be served on a given size of pipe, provided that a proper condition of atmosphere is maintained within the plumbing system during its operation. These factors have tended to increase the number of vents as well as the diameter of pipe required to serve for ventilating purposes, and they have decreased waste pipe size by permitting a greater number of fixtures to be installed on the waste.

There is still much experimental research needed to strike a scientific medium, but many states have adopted what findings and data have been given to them and are using them to good advantage. Plumbing systems, as a whole, will undoubtedly pass through an important period in the next few years, and it is rather certain that more ventilation will be the demand.

The Atmosphere. Because of the importance of the atmosphere and its properties in ventilation and ventilation problems, it may be well to devote a few paragraphs to the subject. Frequent reference to atmosphere will be made in this chapter and unless the reader has a knowledge of it, the value of the chapter may be obscure.

Surrounding the earth's surface is a volume of a mixture of gases, often referred to as a blanket, or ocean, of atmosphere. It contains approximately 21 per cent of oxygen, 78 per cent nitrogen, .94 per cent argon, .003 per cent carbon dioxide, as well as neon krypton, ozone and other gases of lesser importance. Although scientific data indicates a depth of at least 600 miles, the exact depth of the atmosphere is still unknown. However, in terms of weight, approximately one-half the atmosphere is below 18,000 feet, as the atmosphere is proportionately denser in its lower regions. The atmosphere has a density of about 1.29 grams per liter (1000 c.c.) under S.T.P. (Standard Temperature and Pressure) conditions; and a column of atmosphere 1 inch square and as high as the atmosphere exerts a pressure on the earth's surface of 14.72 pounds. The conclusion derived from these scientific facts is that every inch of the earth's surface, or all objects on it, be they liquid, solid, or gas in form, are under pressure of 14.72 pounds per square inch at sea level. Any level above or under the level of the sea would be subjected to lesser or greater pressure as the total volume of air above it is lesser or greater accordingly.

One of the common properties of the gases which compose the atmosphere is compressibility. Air can be compressed and, in this condition, develops pressure greater than atmospheric. This fact can be proved by the automobile tire under pressure, to name an object with which everyone is familiar. Air also can be withdrawn from a space or container, and this condition is termed a vacuum or partial vacuum, depending on the volume of air removed. A partial vacuum, therefore, would indicate a pressure less than that of one atmosphere.

The power of some of the gases to combine with other elements to form chemical compounds is also of importance, because these compounds may be of an acid nature and thus affect the piping material of the plumbing system. Some information pertinent to this matter has been made available through investigation conducted by the National Bureau of Standards in recent years.

In the discussion of trap seal loss and retarded flow, a pressure of less than one atmosphere (14.72 pounds) will be referred to as a minus pressure. A pressure greater than one atmosphere will be called a plus pressure.

Trap Seal Loss. One of the most common and objectionable difficulties occurring in a drainage system is trap seal loss. This failure can be attributed directly to inadequate ventilation of the trap and the subsequent minus and plus pressures which occur. There are at least five ways in which a trap seal may be lost.

1. Siphonage
   (a) Direct self-siphonage
   (b) Indirect or momentum siphonage
2. Back pressure
3. Evaporation
4. Capillary attraction
5. Wind effect

Siphonage. Loss of the trap seal by siphonage is the result of a minus pressure in the drainage system. The seal content of the trap, by the phenomenon of siphonage, is forced into the waste piping of the drainage system through exertion of atmospheric pressure on the fixture side of the trap seal. The principle on which a siphon functions can be demonstrated by simple laboratory experiment. In the preceding chapter it was stated that every column of water 2.31 feet (or 27.72 inches) in height would indicate a pressure at its base of 1 pound. Pressure may be defined as the force required to move a substance, be it liquid or gas.

If a trap of common seal were exposed to the atmosphere, it would be under pressure of 14.72 pounds per square inch on its inlet and outlet orifices. The seal of the trap under these standard conditions would remain in a static or neutral state. It would have no tendency to move because of the equality of pressure on both sides of the seal.

In the laboratory, this same principle may be demonstrated by using a jar that is partially filled with water and placing over its rim a tube which is bent to form two legs of unequal length, the shorter of which is immersed in the water in the jar. The long leg of the tube extends to a level below that of the immersed leg. This arrangement is identical with that of the trap on the plumbing system. The water in the jar represents the trap seal, and the immersed end of the tube may be identified as the trap and waste-pipe plumbing system.

In the laboratory experiment, the water in the jar is static, and is subjected to atmospheric pressure on its surface as well as through the tube. When a minus pressure is created in the long leg, through withdrawal of part or all of the atmosphere it contains, the water in the jar moves upward in the tube and discharges at the long extremity until the jar is emptied. The action which takes place is the result of unequal atmospheric conditions. The water is moved upward in the tube by the pressure of the atmosphere, which is greater than the pressure which has been partially removed from the tube.

Siphonage of the trap seal occurs in two forms—self-siphonage, or direct siphonage, as it is sometimes called, and indirect siphonage, or siphonage by momentum.

Self-siphonage or direct siphonage is commonly found in unventilated traps which serve oval-bottomed fixtures, such as a lavatory, or a small slop sink. It is the result of unequal atmospheric conditions caused by the rapid flow of water through the trap. The fixture having a rounded oval bottom discharges its content abruptly and does not offer the small amount of trickling waste needed to reseal the trap. Figs. 148, 149, and 150 give examples of this type of fixture, with its connected, unventilated waste. The seal content of the trap in Fig. 148 is in a neutral position. Both the inlet and outlet sides of the trap are exposed to the atmosphere and are under identical pressures.

Suppose the lavatory were filled with water and the waste plug suddenly would be removed, as shown in Fig. 149. The water, then, would rush through the trap and into the vertical waste pipe. It is a scientific fact that water is practically incompressible, and

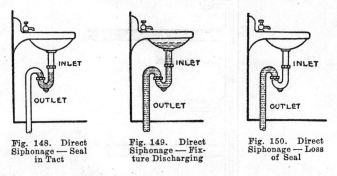

Fig. 148. Direct Siphonage — Seal in Tact

Fig. 149. Direct Siphonage — Fixture Discharging

Fig. 150. Direct Siphonage — Loss of Seal

when it is forced into a vessel containing air, it compresses the extremely elastic gases; or, if the vessel is open to the atmosphere, it replaces them entirely. This is precisely what occurs in the installation. The atmosphere in the vertical leg is replaced by the liquid discharge of the fixture, and a minus pressure in the waste results. The pressure of the atmosphere on the inlet side of the trap, Fig. 150, continues to force the water from it until the seal is broken. If the trap or fixture does not have an adequate re-sealing quality, the trap seal remains in this condition and allows the objectionable gases of the drainage system to enter the room in which the fixture is located.

Traps which serve fixtures of the flat bottom variety, such as sinks, bathtubs, shower baths, and laundry trays are subject to the same difficulty, but the seal content is never entirely removed. The last trickle of water is sufficient to replenish the depleted trap seal.

Fixture manufacturers are also interested in correcting self-siphonage, and in designing the brass accessories of plumbing fixtures accordingly. The waste area of the fixture plug, as well as the diameter of the inlet arm of the trap, usually is reduced. The reasoning in this practice is sound. If a quantity of water of insufficient volume to completely fill the outlet passageway is passed into a trap, there will be no danger of eliminating the atmosphere contained in the trap, and it is quite likely that the seal will remain intact. This precaution is a good one, but it does not solve the problem entirely.

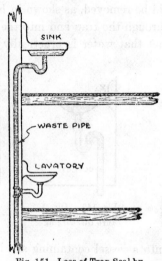

Fig. 151. Loss of Trap Seal by
Indirect Siphonage

A plumbing system is expected to serve indefinitely, and it is only natural that its waste pipe in time becomes fouled with grease and other materials. In this case the area of the waste is eventually diminished and becomes equal in size to the already reduced trap inlet. Self-siphonage under these circumstances is inevitable.

Siphonage by Momentum. Loss of the trap seal by siphonage caused in an indirect manner or by the momentum of water, as it passes a fixture trap outlet, is a difficulty commonly experienced in plumbing systems. This form of trap seal loss is encountered in small as well as in large plumbing installations and is the result of a minus pressure in the waste piping caused by discharge of water from a fixture installed on a line which serves a fixture placed at a lower elevation.

Fig. 151 shows a lavatory on the first floor and a sink on the second floor of a building, both installations having a common waste. When the fixtures are not in use, the waste line and the room where the fixtures are located are under atmospheric pressure and the trap seals of both fixtures remain stationary because of the equality. When the sink is discharged into the waste, the volume of water rushes past the trap outlet of the lavatory and tends to withdraw the atmosphere the trap leg contains. A minus pressure occurs in the trap outlet, and the seal is forced from the trap by the atmospheric pressure contained in the room. There is no possibility of re-seal under these circumstances, and once the trap seal of the lavatory is lost, the gases contained in the drainage system may enter the room through the fixture trap. This type of installation should never be used. Fig. 151 is used merely as illustration of the condition described.

Back=Pressure. Back-pressure, which is caused by a plus pressure, is responsible for trap seal loss often experienced in large plumbing installations. It is a serious form of trap seal loss, for not only does it allow sewer gases to enter the building, but if a person is using the fixture at the time back-pressure occurs he may be injured, or he may receive a thorough and unpleasant ducking. Back-pressure, as the term implies, practically blows the water out of the fixture into the room, and, where sufficient pressure is produced, the content of the fixture trap often strikes the ceiling of the room. The fixtures in which this occurs most commonly are those located at the base of soil stacks, or where a soil pipe changes its direction abruptly.

The flow of water in a soil pipe varies. A single fixture may produce only a small trickle which spirals down the sides of the soil pipe. Larger flows tend to drop and form a slug, because the compressed atmosphere, offering resistance, is unable to slip past the flow of water and exhaust itself at the roof terminal. The pressure becomes greater as the area into which the air is squeezed is reduced, and soon the resistance the trap seal offers is overcome.

Fig. 152 illustrates how back-pressure occurs. The soil pipe is connected to the house drain in the ordinary manner. The house drain in this instance is partially submerged because of the large number of fixtures it serves. A basement water closet is connected to the house drain close to the base of the soil pipe which serves a number of toiletrooms on the upper floors of the building. Fixtures have been discharged, and the downward flow compresses the air above atmospheric pressure in the partially submerged house drain. The back-pressure created has blown the trap seal from the basement closet into the room. The only way in which this condition can be corrected is by ventilating the base of the soil pipe, and this practice generally is employed on all vertical lines more than 3 inches in diameter and three floors in height.

Evaporation. Evaporation of the trap seal is one of the lesser forms of trap seal loss and is a phenomenon of nature. The atmosphere

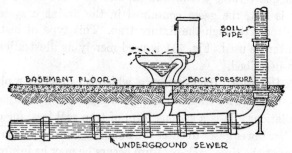

Fig. 152. Loss of Trap Seal as a Result of Back-Pressure

absorbs moisture and the amount varies inversely to the temperature. An atmosphere of low temperature may be saturated with but a few hundredths grams of water per cubic foot of air and, when in this condition, will evaporate no more. Air of extreme temperature has a high saturation point and will continue to take on water until the maximum amount of moisture it can suspend has been reached. A trap seal located in a room where the air is not saturated with water serves as a source of supply and, gradually, it is assimilated by the atmosphere and sewer gases are allowed to pass through the unsealed trap.

Under ordinary conditions it would require many weeks to evaporate a trap seal, and frequent use of the fixture would eliminate this difficulty entirely. Ventilation of the trap is not a solution to this problem; neither does it affect the trap content as one might suppose, because the air circulating within the plumbing system usually is in a saturated condition. The use of a deep seal trap is recommended as a means of prolonging the interval of total loss of the trap seal in the hope that the fixture will be used before complete evaporation of the trap's liquid content has occurred.

Capillary Attraction. Loss of a trap seal by capillarity sometimes, though seldom, occurs. It is caused by suspension of a foreign object such as a rag, string, or lint into the trap seal extending over the outlet arm of the trap, as shown in Fig. 153. The string in this instance forms an absorbing siphon. It soaks up water until it drips from the end reaching into the outlet arm of the trap. Once it reaches this stage the water flows from it rather rapidly, and the seal of the trap is soon displaced.

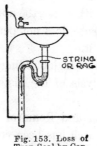

Fig. 153. Loss of Trap Seal by Capillary Attraction

Wind Effects. Wind of high velocity passing over the top of the soil pipe roof terminal affects the trap seal. A downdraft occurring in the plumbing system tends to ripple the liquid content of the trap and spill a quantity of it over its outlet leg into the system. This is not a serious problem because it is quite improbable that the entire seal will be removed. Some precaution can be taken to terminate the soil stack away from valleys, gables, or abrupt projections of the roof where the wind may strike and be directed into or across the soil pipe roof terminal.

Retarded Flow in Drainage System. Retarded flow in a drainage system may be the result of improper atmospheric conditions, because of insufficient ventilation, or incorrect installation of fittings.

Air, because of the elasticity of the gases it contains, may be compressed into pressures far in excess of atmospheric. The flow of water in a soil pipe tends to compress the volume of air against which it flows, and pressures greater than atmospheric are bound to prevail unless the system is properly ventilated. Increased pressure causes retarded flow in the vertical stack and also affects the discharge capacity of its branches. Under these circumstances the drainage system is decidedly unbalanced. It contains pressure greater than that of the atmosphere and it cannot function properly.

There is also the possibility that a partial vacuum may be developed in the drainage system which affects its discharge capacity. This is the result of atmospheric pressure on the flow side of the waste creating resistance to the volume of water because a minus pressure developed on the opposite side. It indicates a lack of proper relief ventilation or partial closure of the soil pipe terminal, or it may be the result of too long a vent pipe. The condition may be likened to a tube filled with water, the upper end of which has been sealed. Atmospheric pressure does not permit the water contained in the tube to drain from it.

These difficulties occur on high building installations and must be prevented by careful analysis of the installation before it is constructed.

Material Deterioration and Removal of Objectionable Gases. In recent years much research has been conducted for the purpose of discovering the various causes of deterioration of pipe metals. The wastes of the plumbing system contain many chemical elements which, in combining, create compounds of an acid nature that may be detrimental to the piping material of the plumbing system. Hydrogen, which is an important element in all acids, is found in large quantities in the drainage system. It may be found in a free state, but usually it is combined with other chemical elements. It is objectionable in either form and should be eliminated by adequate ventilation. The serious effects of the acid-forming elements are minimized by this means. Much could be written in explaining the processes of deterioration of piping materials. The subject involves complete knowledge of the fundamentals of chemistry and would require too much space for inclusion in this volume.

Some experiments have been made to determine what volume of objectionable and odorous gases can be diffused through trap seals. The liquid content of a trap absorbs the gases contained in the drainage system, and, when it becomes saturated, allows them to pass through the seal into the room where the fixture is installed. The time required to complete this process, however, is quite long, and the fixture is likely to be used before any reaction is evident. Therefore the volume of gas diffused is negligible and adequate ventilation overcomes possible dangers.

Ventilation Methods. There are many forms of ventilation which can be applied to the plumbing installation. Choice is determined largely by the manner in which the plumbing fixtures are to be located and grouped. As a rule, the completed vent pipe system is a combination of several methods.

At one time all fixture traps were individually ventilated, but through gradual experiment and the introduction of forms of soil and waste pipe relief vents, it was found that individual trap ventilation was costly and not entirely necessary. It is a fact, however, that in a plumbing installation in which the vent pipe system is of proper size, and wherein every trap is ventilated with an individual vent, the danger of trap seal loss is practically eliminated. Nevertheless, as long as practical tests have indicated that these conditions are not absolutely essential to a safe system of drainage, various methods of ventilation may be adapted for the sake of economy.

There are several kinds of vent pipe systems, each of which has a definite function in the completed plumbing system. The various types may be grouped under two principal classifications. The vent pipes used to ventilate the soil and waste pipes are the main soil and waste vents. The main vent and various other forms of relief vents are classified according to the purpose they serve, and are referred to as "relief" and "yoke" vents. These methods of ventilation serve the fixture trap only in an indirect way. Their primary purpose is to maintain atmospheric pressure in the waste pipe system.

The ventilation methods whose primary purpose is to protect trap seals against back-pressure and siphonage are individual, or back vents, as they are sometimes called, unit vents, circuit or loop vents, wet vents and looped vents.

Materials Used. The vent pipe system may be constructed of cast iron, galvanized wrought iron, galvanized open hearth iron, galvanized steel, brass, copper, or lead. Fittings should conform to the type of pipe used, except that cast-iron steam pattern or malleable-iron fittings may be used with threaded pipe.

Grades. The vent pipe must be graded slightly so that no water may accumulate in it. No definite amount of pitch is required. It is advisable that the pipe be graded to many points of the waste pipe system to assure rapid elimination of any condensation which might occur in it.

Supports. The vent pipe system generally is concealed in building partitions, and is supported by them. The runs of pipe usually are short and require no specified method of suspension.

When the building partitions are constructed of wood, it is advisable that the 2 x 4 studs be drilled so the construction is not weakened too much. It is advisable that the hole through which the vent is to pass be of larger diameter than the pipe, so that settling of the building does not place strain on the drainage system.

When the vent pipes are placed in partitions constructed of tile, brick, or similar materials, they must be temporarily suspended from the ceiling or blocked up from the floor on wood or metal supports until the mason has completed that portion of the partition on which the vent pipe will rest.

Long runs of exposed horizontal waste pipes may be suspended from band iron or other type of ring hangers substantially anchored in the building construction.

Required Sizes of Vents. Sizing of the vent pipe installation offers the same puzzling problems as those encountered in sizing the soil and waste pipe. The conditions under which every plumbing system operates vary, and it would be impossible to set up a sizing method for every installation. The factors which must be taken into consideration in establishing a method which, in a general way, would apply to all vent pipe systems are many. The volume and velocity of waste flow is one phase which cannot be ascertained definitely, because no one individual or group of individuals can control this element. A large volume of water flowing through the waste pipe system naturally requires a greater quantity of air moving at higher velocity to maintain atmospheric pressure. Velocity of flow in the waste pipe, tremendously increased in taller types of buildings, has the same effect. Long runs of vent pipe reduce the flow and volume of air because of the friction which occurs between the air in motion and the interior surface of the pipe.

Small discharges using large diameter waste pipes, in small residence work, vary inversely to the large type of building, and vent pipe installations may be grossly exaggerated to the demand.

All these factors must be studied carefully and may be partially solved mathematically, although it is essential that the sizing method be established by experience and actual installation tests.

The most logical method of vent pipe sizing for most forms of vent pipes is the unit system, as established by the Subcommittee on Plumbing and presented in Recommended Minimum Requirements, Section 103. Table 10, presented here, is taken from this section. It gives the number of units which can be served by a vent pipe of given diameter and indicates the length of vent pipe required to serve units up to 5,400. It should be remembered that the table gives

TABLE 10. Maximum Permissible Length of Vents (in Feet) for Soil and Waste Stacks*

Diameters of Soil or Waste Stack (Inches)	Number of Fixture Units	Diameter of Vent (Inches)									
		1¼	1½	2	2½	3	4	5	6	8	10
1¼	1	45									
1½	Up to 8	35	60								
2	Up to 18	30	50	90							
2½	Up to 36	25	45	75	105						
3	12		34	120	180	212					
3	18		18	70	180	212					
3	24		12	50	130	212					
3	36		8	35	93	212					
3	48		7	32	80	212					
3	72		6	25	65	212					
4	24			25	110	200	300	340			
4	48			16	65	115	300	340			
4	96			12	45	84	300	340			
4	144			9	36	72	300	340			
4	192			8	30	64	282	340			
4	264			7	20	56	245	340			
4	384			5	18	47	206	340			
5	72				40	65	250	390	440		
5	144				30	47	180	390	440		
5	288				20	32	124	390	440		
5	432				16	24	94	320	440		
5	720				10	16	70	225	440		
5	1,020				8	13	58	180	440		
6	144					27	108	340	510		
6	288					15	70	220	510	630	
6	576					10	43	150	425	630	
6	864					7	33	125	320	630	
6	1,296					6	25	92	240	630	
6	2,070					4	21	75	186	630	
8	320						42	144	400	750	900
8	640						30	86	260	750	900
8	960						22	60	190	750	900
8	1,600						16	40	120	525	900
8	2,500						12	28	90	370	900
8	4,160						7	22	62	252	840
8	5,400						5	17	52	212	705

*Local ordinance should be consulted.

average requirements; therefore the sizes of vents selected will, in most cases, be larger than seems necessary. This fact is illustrated in the example which follows, where a vent of 96-unit capacity is used for 63 units.

To familiarize the reader with the use of this table the following example is offered: What size of main vent would be required for a group of fixtures consisting of 6 water closets, 4 lavatories, 3 urinals, and 2 showers installed on the second floor of a building which is 100 feet in height? To determine the unit value of this group of fixtures, the same procedure used in sizing waste pipe is followed.

Fixtures	Units
6 Water closets	36
4 Lavatories	4
3 Urinals	15
2 Showers	8
Total	63

Assuming that each floor of the building is about 10 feet high, the vent pipe would have a length of 80 feet from the ceiling of the second floor (20 feet above first floor) to the top of the building (100—20=80). Horizontal runs must be added to this figure. Referring to Table 10, we find that on a 4-inch soil stack (the size required) 96 fixture units may be ventilated with a 3-inch pipe, provided it is not more than 84 feet long. Hence a 3-inch main vent would be required.

This example is given merely to acquaint the reader with the method of Table 10. It is not a practical example of what might occur. The vent pipe system would include installations of fixture groups on more than one floor, and must be a combination of back, unit and circuit vents terminated into a main vent, which probably would be connected with the main soil and waste vent before it passed through the roof. The vent pipes would be graduated from a small diameter pipe, for single fixtures, to a large diameter pipe, as more fixture traps were added.

In the following paragraphs, the plan of ventilating single fixtures and fixture groups will be discussed as well as the method of determining the size of their respective vents.

Main Soil and Waste Vent. The main soil and waste vent is that portion of the soil-pipe stack above the highest installed fixture branch extending through the roof. The main soil vent is the source through which air is admitted to the plumbing system. It serves also as a means of eliminating objectionable odors. Mechanics of the past did not consider this portion of the soil pipe of great importance, apparently because of the fact that it conveyed no liquid waste. Defective fittings and pipe often were utilized in its construction, and, in most instances, testing was not considered necessary. There is no portion of the drainage system so unimportant as to excuse indifferent procedures. Whether the installation conveys water, air, or waste, the

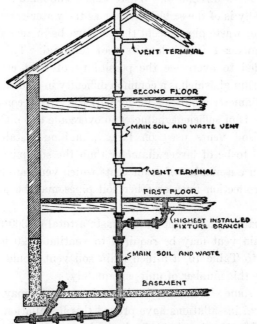

Fig. 154. Main Soil and Waste Vent Installation Where Flat Venting Is Permitted

same careful workmanship is necessary. The drainage system is a combination of waste, vent, and water, and the efficiency of one phase of it is dependent on the other. Careless workmanship in the vent pipe is bound to interfere with proper function of the waste pipe, and the consumer is the one who is finally affected economically as well as physically.

The main soil and waste vent usually is the terminal for the many main vents the plumbing installation requires. The main soil and waste vent must be run as direct as possible. It is objectionable to use short radius fittings in its construction, because short turns and offsets reduce the flow of air through it materially and this difficulty affects the entire plumbing installation and may result in trap seal loss even though each fixture is individually re-vented. Long horizontal runs must also be avoided wherever possible.

Fig. 154 indicates the portion of the soil pipe defined as the main soil and waste vent, and shows tees which serve as the terminal for the main vent.

Sizing the Main Soil and Waste Vent. The main soil and waste vent generally is of the same diameter as the water-carrying portion of the soil or waste pipe, except that it must be increased to at least 4-inch diameter 1 foot below its roof terminal. This practice is recommended to overcome the possibility of frost closure of the soil-pipe terminal, which is a common difficulty in cold climates. The increased diameter requires a longer period to freeze completely shut and usually is of sufficient diameter to overcome this difficulty.

Main soil vents used on large plumbing installations may be required to be of larger diameter than the soil pipe which they serve, because a number of main vents which ventilate fixtures connected to collection lines or other soil pipes may be joined to the soil vent.

For example, a soil pipe may waste a total of 300 fixture units and its main vent may be required to ventilate 420 fixture units. According to Table 10, a 5-inch main soil vent would be required to ventilate this number of units adequately.

Under some circumstances, the main soil vent may be reduced, and practical installations have proved this installation to be satisfactory. This practice should be limited to small residence installations. Some states having local control permit a 2-inch main soil vent to serve a soil-pipe stack, provided there is one 4-inch soil-pipe stack in the building. For each full-sized stack one reduced stack is permitted. These practices are limited to installations which consist of a water closet, lavatory, and bathtub underfloor work installed not higher than the second floor of the building. All reduced stacks must be 4-inch, however, as they pass through the roof.

Main Vent. The main vent is that portion of the vent pipe system which serves as a terminal for the smaller, tributary forms of individual and group fixture trap ventilation. It may be construed as a collecting vent line.

The main vent usually is located within a few feet of the soil-pipe stack. This is not a set policy, however, and its location

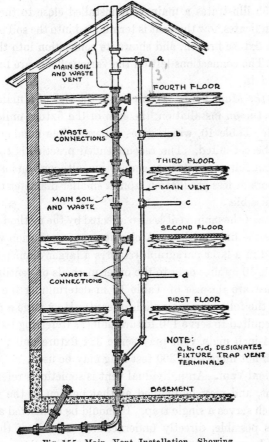

MAIN SOIL AND WASTE VENT

FOURTH FLOOR

a

WASTE CONNECTIONS

b

THIRD FLOOR

MAIN VENT

MAIN SOIL AND WASTE

c

SECOND FLOOR

WASTE CONNECTIONS

d

FIRST FLOOR

NOTE:
a, b, c, d, DESIGNATES FIXTURE TRAP VENT TERMINALS

BASEMENT

Fig. 155. Main Vent Installation, Showing Fixture Trap Vent Terminals

depends a great deal on the building construction. It is vertical in design ordinarily, and is installed at the same time as the soil-pipe stack. Openings are left at the correct height and proper floors to accommodate the forms of fixture trap ventilation.

The main vent begins at the base of the soil-pipe stack, where its purpose is to relieve any back-pressure which might occur at this location. It terminates in the soil-pipe stack, at least 3 feet above the highest installed fixture branch.

The main vent must be direct and free from offsets to permit an unobstructed movement of air. It must be well supported on each floor.

Fig. 155 illustrates a main vent installed close to the soil-pipe stack. It indicates how the vent is terminated into the soil pipe above the highest fixture branch, and shows its connection into the base of the stack. The connections left on the various floors are fixture trap vent terminals.

Sizing the Main Vent. To ascertain the size of a main vent required to serve an installation, the sum of the fixture units must be determined. Table 10, which gives a permissible number of units, may then be consulted. The recommended practice is to continue the main vent full size to its base connection, although on installations of three floors or less graduation from a smaller diameter to a larger one is permissible.

The size of the main vent is also affected by the method employed to ventilate the fixture traps. A circuit-vent installation which will be discussed in a later paragraph requires a larger diameter of main vent. Table 10 applies only to the other methods of ventilation.

To illustrate the use of Table 10 in determining the size of a main vent, the following example is offered: How large a main vent would be required to serve 140 fixture units? Referring to Table 10, we find a 4-inch main vent may serve 384 fixture units. Hence a 4-inch main not more than 300 feet long may be used.

Individual Vent. An individual vent is sometimes referred to as a back vent, and may be defined as that portion of the vent pipe system which serves a single trap. It should be connected as close to the trap as possible, directly underneath and back of the fixture, and it must be reconnected into the main vent above the overflow line of the fixture it serves.

The individual vent is by far the most practical method of ventilating a fixture trap. Danger of trap seal loss when this type of vent is employed is negligible. The plumbing system is relieved of minus and plus pressure at every fixture trap, and the entire waste system is benefited. There has been much discussion among sanitary authorities as to whether or not separate re-venting of each fixture trap is necessary. The general opinion seems to be that when fixtures are closely grouped this practice is unnecessary, and that a single re-vent would serve the purpose adequately. Tests conducted by the Bureau of Standards prove the point. Stack venting is permissible on one-story jobs, and on the second floor of two-story jobs. It must be remembered, however, that after an installation is put in operation, many things can occur which alter the situation. One of the uncontrollable elements is the danger of fouling. Where group ventilation is used it is evident that some part of the vent-pipe line must be used as a waste pipe, and because of the materials suspended in the water-carried waste, reduction of the diameter of the vent is inevitable. Complete stoppage of it is not unimaginable. Some authorities offer the evidence of installations which, though group vented, have functioned with apparent effectiveness for many years. Admittedly, some of these installations seem to perform as they should; others have failed dismally and at least partial seal loss occurred. It appears that conditions within the building have noticeable effect, and determine to some extent the amount of ventilation required. It is evident that the plumber must use his own judgment in the matter and be guided by the ample amount of information available to him.

Sizing of Individual Vents. The diameter of pipe which may be used to ventilate fixture traps individually has been established by practical experience as well as by laboratory test. It is illogical to use a pipe of less than $1\frac{1}{4}$ inches diameter for this purpose, not because the flow of air in it is insufficient, but because it becomes stopped up quickly, and in this condition it is worthless. The table shown below offers a basis for determining the size of a back vent for a specific type of plumbing fixture. It is more applicable, however, to the

Type of Fixture	Minimum Size of Vent, Inches	Unit Value
Lavatory	$1\frac{1}{4}$	1
Drinking fountain	$1\frac{1}{4}$	$\frac{1}{2}$
Sink	$1\frac{1}{2}$	2
Shower	2	2
Bathtub	$1\frac{1}{2}$	2
Laundry tub	$1\frac{1}{2}$	2
Slop sink	2	3
Water closet	3	6

graduation of the back vent as fixtures are added to the main portion of it.

The number of fixtures permissible on a pipe of given size may be ascertained from the following table.

Size of Pipe, Inches	Number of Units Permissible	Size of Pipe, Inches	Number of Units Permissible
1¼	1	2½	36
1½	8	3	72
2	18	4	384

This information, as tabulated above, is suggested as a method of arriving at a size for the individual vent pipe, and may be deviated to conform to local plumbing regulations without serious results.

Individual Ventilation of Plumbing Fixtures. Fig. 156 illustrates the advisable method of individually ventilating fixtures of the wall-hung or pedestal variety, such as sinks, lavatories, drinking fountains, wall-hung slop sinks and laundry tubs. As may be noted, the trap discharges into the side opening of a 90-degree drainage tee. The safe length of arm between trap and vent, using a sanitary tee, a long turn **T-Y,** or a combination **Y** and ⅛ bend, is as shown in the table on opposite page.

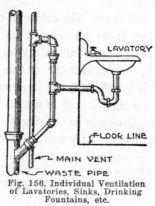

Fig. 156. Individual Ventilation
of Lavatories, Sinks, Drinking
Fountains, etc.

Thousands of tests conducted by the Bureau of Standards have eliminated guesswork. Tests using both a 1¼-inch and a 1⅛-inch pull-out plug showed that the fall of the arm and the type of fitting into which it discharged had a definite bearing on trap seal loss.

The top opening of the 90-degree drainage tee is used as a connection for the individual vent, which must be extended above the overflow line of the fixture so that in case the fixture waste pipe is stopped up the vent cannot serve as a waste for the fixture. There is a decided advantage in connecting the vent pipe to the tee as indicated in Fig. 156. Accumulation of dust, rust, or foreign material in the vent automatically drops into the waste line and is carried away by the discharge of the fixture, thus offering a clear vent.

Fig. 157 illustrates the method by which fixtures such as bathtubs, urinals, shower baths, foot tubs, and sitz baths, which set on the floor, may be ventilated.

The fixture trap discharges into the top opening of a drainage tee or **Y** fitting, the side opening of which is used as a connection for the individual vent. This form of back vent is not as satisfactory

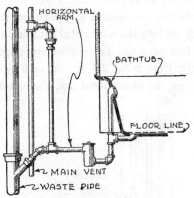

Fig. 157. Individual Ventilation of Bathtubs, Showers, and Urinals

Size of Fixture Drain (In.)	Permissible Length (Ft.)			
	Sanitary Tee		Long Turn **T-Y** or Combination **Y** and ⅛ Bend	
	¼" Slope	½" Slope	¼" Slope	½" Slope
1¼	4'0"	2'6"	1'6"	1'0"
1½	4'6"	3'0"	4'0"	2'0"
2	5'0"	4'0"	4'6"	4'0"
3	6'0"	6'0"	6'0"	6'0"
4	8'0"	8'0"	8'0"	8'0"

as the method previously described. The discharge of the fixtures into the waste may tax its capacity and the waste materials are liable to back into the horizontal arm of the vent pipe. Whenever possible, it is advisable to rise vertically from the fitting and then extend the vent horizontally. This cannot always be done, because of the floor construction, and the situation is one of compromise, using the best method possible under the circumstances. Whenever the installation may be designed to permit the venting of floor-type fixtures by the wall-hung fixture method of ventilation it is unquestionably desirable to do so.

Backing of the waste may be avoided by installing a clear water fixture on the vent-pipe line to serve as a means of washing the line each time the fixture is discharged. This form of ventilation will be discussed in detail later.

Fig. 158 illustrates a method of ventilating such fixtures as water closets, clinic sinks, slop sinks and similar fixtures. The individual vent is connected to the vertical portion of the closet waste as close to the fixture as is practical. This form of venting is contrary to good practice as, should stoppage occur, the matter discharged will back up into the horizontal arm and remain there, thus interfering with free venting. To prevent stoppage, the vent can be taken off vertically from a No. 2½ fitting, or a ½-**Y** can be used on the vertical between the closet bend and the closet bowl.

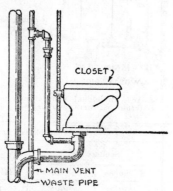

Fig. 158. Individual Ventilation of Water Closets, Clinic Sinks, Slop Sinks, etc.

Because the fixture it ventilates is as a general rule connected directly to the soil pipe, the vent, as it is located, eliminates the danger of seal loss by the momentum of the water waste rushing down the soil-pipe installation. Its value in this respect, however, has been questioned because of the design of the fixture trap.

Fig. 159 illustrates how a battery of fixtures may be ventilated by the individual vent method.   The installation consists of two wall-hung fixtures, each individually vented.   The size of the vent pipe

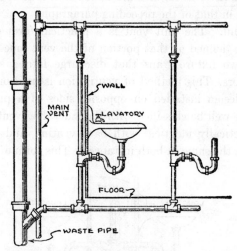

Fig. 159. Fixture Groups May Be Ventilated by the Individual Vent Method

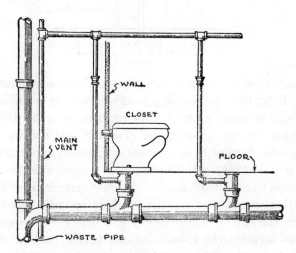

Fig. 160. Individual Venting of Traps in a Battery Installation

increases in diameter as the fixtures next to the last one are added to the collecting vent line.   More specific examples of this installation will be presented later.

Fig. 160 illustrates water closet traps in group installation. Each trap is ventilated individually. The same procedure of increasing the vent pipe diameter as fixtures are added to it is followed in this installation as in that of the preceding paragraph.

Unit Vent. The unit vent is a practical form of ventilation that may be defined as that portion of the vent pipe system which ventilates two fixture traps that discharge into a sanitary cross with deflectors. This method of ventilation is used on two fixtures of similar design installed on opposite sides of a partition. Unit venting may well be classified with the individual vent. The design of it is practically identical with back-venting, and the principles involved are the same in both instances. This form of ventilation is

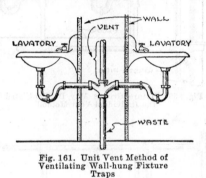

Fig. 161. Unit Vent Method of
Ventilating Wall-hung Fixture
Traps

commonly applied to fixture traps serving apartment and hotel toilet-rooms. Because of the economy involved, sanitary facilities usually are located in rooms containing the roughing in of the bathroom on both sides of a common partition. The unit vent and waste can be applied when fixtures have the same roughing-in measurements.

Sizing the Unit Vent. The diameter of pipe required to unit vent two fixture traps is determined in the same manner as for individual ventilation, except that the size of the pipe must be sufficient to ventilate the sum of the two fixture unit values. For example, suppose two lavatory traps are to be unit vented, each lavatory consisting of one unit. The sum of their unit values would then be two. Since only one unit is permissible on 1¼-inch pipe, a 1½-inch unit vent would be required for this installation. The same method is employed for other fixtures and the size of the vent naturally increases as the unit value of each fixture becomes greater.

Unit Ventilation of Plumbing Fixtures. Fig. 161 illustrates how wall-hung fixtures may be ventilated by the unit vent method. The fixture traps discharge into a sanitary cross which has deflectors. The top opening of the cross is used for the unit vent connection, which is completed in the same manner as an individual vent. A short pattern cross, Fig. 162, should be used on installations of this kind for the reason that one of long radius offers a perfect siphon and the trap may lose part of its seal under these conditions. The situation is one in which the extended dip of the long radius cross exceeds the inside diameter of the waste and provides a vertical outlet leg to complete a siphon.

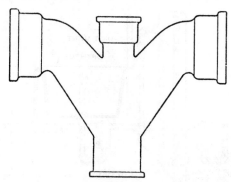

Fig. 162. Double Combination **Y** and ⅛ Bend with Deflectors

There has been some objection to the use of a short pattern tee because of the fact that one fixture may discharge its waste into the trap of another. This difficulty is experienced on larger fixtures, such as water closets, but the small fixture discharging into a large diameter waste does not present the same problem.

Bathtubs, floor drains, showers, and similar fixtures must be ventilated in the same manner. It is rather inconvenient, because of space limitations, to employ this method of ventilation on these fixture traps, hence it is not common practice. Fig. 163 illustrates how these fixtures may be accommodated with a unit vent.

Water closets and fixtures of similar design may also be unit vented, as illustrated by Fig. 164. On fixtures of this type it is advisable to use a fitting of long radius because of the volume of discharge of one fixture affecting the operation of the other. It is not common practice to unit-vent this kind of installation except in the case of the highest fixtures on the soil-pipe stack. In this event the main soil and waste vent serves in the capacity of the unit vent. The unit vent is also employed on water closets installed on patented fittings which are specially constructed for this purpose, and in which the objectionable reversal of flow does not occur. New types are available for all new installations.

Circuit or Loop Ventilation. Modern buildings demand more plumbing facilities, and toiletrooms are being provided with batteries of fixtures which may be ventilated by the circuit-vent method. A circuit vent may be defined as that portion of the drainage system

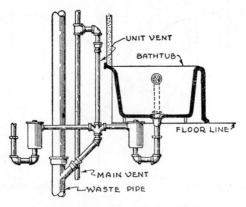

Fig. 163. Unit Vent Used in Bathtub Installation

which ventilates two or more fixture traps that discharge into a horizontal soil or waste branch extended at slight grade.

Circuit vents a few years ago were considered impractical and because of the lack of relief vents at proper intervals of the branch a marked disturbance of the trap seal was noted each time two or more fixtures were simultaneously discharged. At the present time, however, this condition has been somewhat changed. Relief ventilation has become an important phase of circuit venting, and the fixture traps thus served apparently suffer no objectionable effects. Large plumbing installations are now circuit vented, and few plumbing designers object to this method. Circuit venting offers economy because the installation is simplified by its use.

Sizing the Circuit Vent. The size of circuit vent required to

serve fixtures other than water closets adequately may be determined
by the unit system. For example, suppose a battery of six lavatories
is to be circuit vented, each lavatory having a unit value of one.
The sum of all fixtures would be six units. By referring to the table
on the number of units permissible, page 178, it is found that eight
units of fixtures may be ventilated with 1½-inch pipe. Hence a
1½-inch circuit vent may be used on this fixture group. The same
procedure may be applied to similar fixtures.

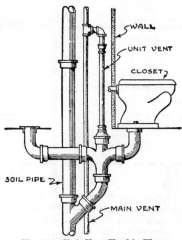

Fig. 164. Unit Vent Used in Water
Closet Installation

To ascertain the circuit vent pipe size for a battery of water
closets the following information is offered:

Number of Closets, Series Installation	Size of Circuit Vent Inches
2	2
3 to 6	3
7 or more	4

Not more than eight water closets or other type of fixture may be
permitted on any one loop or circuit. Suppose a battery of six water
closets is to be circuit vented. By referring to the above information
a 3-inch circuit vent would be required. Should eight water closets
be provided with a circuit vent, then a 4-inch pipe would be required.
If fifteen closets were to be circuit vented then two 4-inch loops or

circuits connected to a 4-inch main would be required. These figures with proper relief venting have proved adequate in practical tests.

Circuit Ventilation of Plumbing Fixture Trap. Fig. 165 illustrates how a battery of three water closets may be circuit ventilated. The circuit vent is taken off of the soil branch between the second and third closets. This is a precautionary procedure. In case the soil line becomes partially stopped or is overtaxed, the third fixture tends to scour the vent of any fecal waste which might obstruct it. The vent

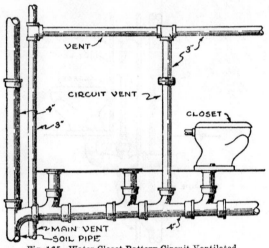

Fig. 165. Water Closet Battery Circuit Ventilated

should be connected into the top of the soil line to minimize this common difficulty.

Fig. 166 illustrates a battery of three lavatories that is circuit ventilated. The same precautions are needed on this installation as are required for water closets. The waste branch consists of long radius drainage tees installed on their sides. The vent pipe is connected vertically between the last two fixtures. This installation, although not unreliable, does not represent good plumbing practice. Its use is prohibited in large cities.

Relief Ventilation. A relief vent is that portion of the vent pipe installation which primarily eliminates minus and plus pressures in the drainage system. It ventilates the soil and waste pipe and connecting branches rather than the fixture trap. There is no definite rule as to where relief vents are to be installed, except in the case of circuit ventilation. It is largely left to the judgment of the mechanic, who should place relief vents wherever he believes back-pressure is likely to occur. As a matter of fact, it is necessary that this be done or the installation will cause no end of trouble. Experience, and knowledge of the scientific principles involved in a drainage system, should make it possible for him to ascertain where this difficulty might occur, and the vent pipe system may be planned accordingly.

Relief vents are commonly used in connection with waste branches which are circuit vented. Simultaneous use of fixtures on this kind of installation often taxes the branch to its capacity, and in this case the circulation of air produced by the circuit vent is affected.

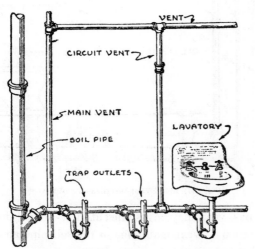

Fig. 166. Lavatory Battery Circuit Ventilated

The discharge of the branch also contacts the waste from fixtures on higher floor levels passing down the vertical soil pipe, and a compression of air, which may affect the trap seals of the circuit-vented branch and cause retardation of the waste flow, would result at this junction. Here, a relief vent must be installed to eliminate plus pressure and allow the fixtures to function properly.

Back-pressure is also likely to occur at the base of a soil-pipe stack, especially on plumbing installations in tall buildings. A relief vent can also be provided at this point. It often consists of the connection of the main vent into the base of the soil pipe below the lowest installed fixture branch.

Relief vents may be installed at intervals on the soil pipe where changes of direction are made. It is often necessary in tall buildings to offset the soil pipe frequently, and vertical flows of water contacting these changes produce increased pressures which affect the flow of water in the stack and may blow out trap seals in close proximity to it.

On long vertical soil pipes a relief vent, often referred to as a

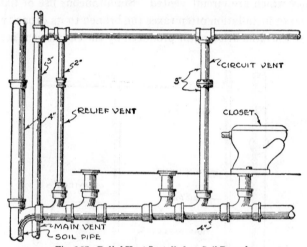

Fig. 167. Relief Vent Installed on Soil Branch

yoke or by-pass vent, may be installed at three- to five-floor intervals. This installation of vent maintains in the soil pipe an atmospheric condition that is necessary to proper function.

Sizing the Relief Vent. The relief vent which serves an installation of a circuit-vented branch must be at least one half the diameter of the soil pipe, and is never less than 1½-inch diameter pipe. If the soil pipe is of 4-inch diameter, the relief vent must be at least 2 inches in diameter.

Relief vents used at changes of direction and at the base of the soil pipe should be of the same diameter as the main vent, except that no relief vent should be less than 2 inches in diameter.

Yoke and by-pass vents installed between the main vent and the soil pipe at five-floor intervals must be of the same diameter as the

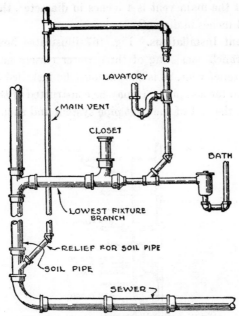

Fig. 168.   Reconnection of Main Vent to Base
of Soil Pipe

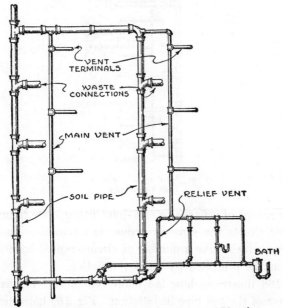

Fig. 169.   Relief Vent Applied to Changes of Direction
on the Soil Pipe

main vent. **If** the main vent is 4 inches in diameter, **the yoke vent** must also be 4 inches in diameter.

Relief Vent Installations. Fig. 167 illustrates how a **circuit**-vented soil branch consisting of three water closets must be relief vented. The relief vent, as is shown, must be installed between **the** first fixture and the soil pipe. It may be constructed of the same ma-**terial** used for the rest of the vent pipe system and should **be joined**

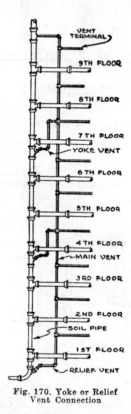

VENT TERMINAL

9TH FLOOR

8TH FLOOR

7TH FLOOR

YOKE VENT

6TH FLOOR

5TH FLOOR

4TH FLOOR

MAIN VENT

3RD FLOOR

2ND FLOOR

SOIL PIPE

1ST FLOOR

RELIEF VENT

Fig. 170. Yoke or Relief
Vent Connection

solidly by means of a flange or calk-joint fitting. On fixtures other than water closets the same procedure is advisable. In soil-pipe installations which serve a number of circuit-vented branches, each branch must be protected by relief ventilation.

Fig. 168 illustrates how the main vent may serve as **a relief** for the base of the soil pipe installation. Fig. 169 indicates **a relief** vent applied to changes of direction on the soil pipe.

A yoke or relief vent connection is illustrated in Fig. 170. Care must be exercised to allow a sufficient amount of space between the soil and main vent so this connection can be made. The fittings usually are of large diameter and require space to make them up. Often it is necessary to construct these connections with swing joints to make a rigid connection as the illustration indicates.

Wet Ventilation. A wet vent is a method of ventilation used rather extensively for small groups of bathroom fixtures. It may be defined as that portion of the vent pipe system through which liquid wastes flow. The practice of wet ventilating plumbing fixtures is one that is discussed often. The problem always offers an affirmative and negative argument. Some believe wet ventilation to be an efficient method of supplying the fixture trap with adequate atmospheric pressure to assure no loss of seal. Their strongest contention is the fact that a fixture placed on a dry vent line tends to scour the line at each discharge of the fixture. Laboratory tests on new and clean fixture wastes have proved this form of ventilation to be sufficient.

Those who oppose this argument contend that the portion of the vent pipe which is used as a waste for another fixture becomes fouled very rapidly and that its diameter is reduced tremendously. Complete stoppage often occurs and trap seal loss may result.

Both arguments have merit. Wet ventilation, if it is not properly designed and the fixture to be wasted on it is not carefully selected, does present stoppage problems and effect a reduced diameter of the vent line. Situations of this kind are common. When the waste and vent installation is clear there appears to be adequate ventilation, and very seldom do difficulties occur. Federal authority in Recommended Minimum Requirements permits this type of ventilation, and their test proved it to be sufficient. Many states have used it for years and apparently have experienced no great or continuous amount of trouble with it. It does have a place in the ventilation of small fixture groups, or it may be used to advantage on the circuit-vent installation provided a fixture is wasted into it which has none but clear water discharge, such as a wash basin or drinking fountain. The scouring effect of the fixture wastes tends to increase the efficiency of the installation. It is entirely impractical, however, to waste a sink or similar fixture into the vent line because of the variety of waste matter it discharges.

Sizing the Wet Vent. The wet vent size is determined in the same manner as is that of the individual vent. The unit system is the basis of its size. For example, suppose a wet vent is to be employed to ventilate a small fixture group consisting of a closet, lavatory, and bathtub. The unit values of these fixtures are: water closet, 6; lavatory, 1; and bathtub, 2. The sum of these values is 9 units. By referring to the table which lists the number of units permissible,

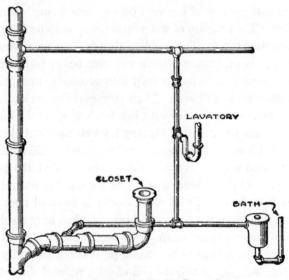

Fig. 171. Wet Vent Used in Connection with Bathroom Group of Fixtures

page 178, only eight units may be ventilated by pipe of 1½-inch diameter. Hence a 2-inch pipe must be used to ventilate this installation.

Wet Vent Installations. Fig. 171 illustrates a wet vent which may be used in connection with a small bathroom group of fixtures consisting of a closet, lavatory, and bathtub. In this instance the lavatory is installed on the increased diameter vent pipe which serves both the water closet and bathtub.

Other fixtures may be wet vented in the same manner, especially those of the floor variety, such as shower baths, urinals, slop sinks, and floor drains. It is not permissible, however, to ventilate wall-hung fixtures by this method.

Fig. 172 shows how a lavatory or drinking fountain may be wasted into a circuit-vent installation. The circuit vent usually is of large diameter and does not become fouled readily. The lavatory serves as a wash for the vent line and unquestionably adds to its efficiency.

Looped Vent. The looped vent is a method of ventilation used on fixtures which are located in the room away from partitions that

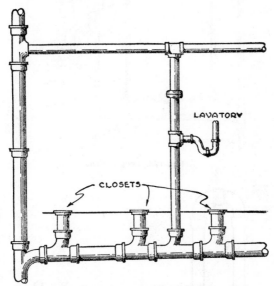

Fig. 172. Lavatory Wasted into a Circuit Vent Installation

might be utilized to conceal the waste and vent. These installations are common in barber shops, beauty salons, and hospital surgical rooms.

Looped ventilation is not deemed practical and it should be resorted to only when other methods of ventilation are not applicable. It is, however, the best that can be accomplished under certain circumstances and its use should not be too restricted.

Sizing the Looped Vent. Looped vents generally are of the same design as individual vents and are sized in the same manner.

Looped Vent Installation. Fig. 173 illustrates how a lavatory located in the center of a room may be loop-vented. The fixture in the illustration is back-vented. The vent pipe is extended vertically

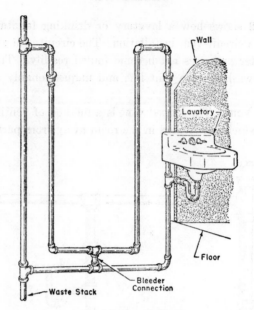

Fig. 173. Looped Vent

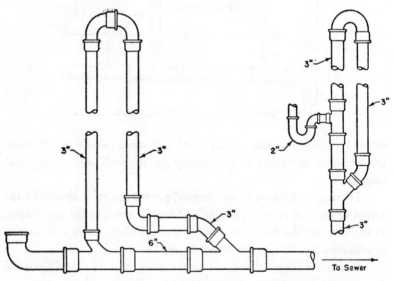

Fig. 174. Two Methods of Utility Venting

above the fixture overflow line and is then looped or returned downward to a point below the floor. The vent is then run horizontally to the partition and reconnected into the main vent a few inches above the top rim of the fixture it serves. A full size bleeder or drip connection must be made between the waste pipe and the lowest point of the vent line to permit any accumulation of water to drain from the trapped looped vent.

Local Vent. A local vent has no connection with the plumbing system. It may be defined as a conduit or pipe shaft used to convey the foul odors from a fixture or room. It is used to some extent in connection with water closets, and is connected to the fixture at a point below the seat. It terminates at the roof.

Rooms which have no natural ventilation generally are provided with a local vent. Pipe constructed of sheet metal is used for this purpose.

Utility Vent. The utility vent is sometimes used for underground public restrooms. Because of the park lawn or cement walk directly above, a vent pipe extending 10 feet above ground to the open air would be unsightly and would constitute a hazard. Therefore a utility vent usually is installed. Fig. 174 shows two methods of utility venting that may be adopted.

MECHANICAL AERATORS IN ACTION AT ACTIVATED SLUDGE SEWAGE TREATMENT PLANT, SYCAMORE, ILLINOIS

Courtesy of Chicago Pump Company, Chicago

CHAPTER XIII

SOIL, WASTE, AND VENT PIPE PRINCIPLES

In the preceding chapters soil, waste, and vent pipe installation practices were discussed as separate units of the drainage system. It may be difficult for the reader to correlate the several units of a complete plumbing system, because of the many complex problems which arise.

The drainage system, as a whole, consists of many forms of waste and vent correlated into one specific unit. Problems in graduating the sizes of the waste and vent become more difficult, and many important details are liable to be overlooked, unless the reader is given an opportunity to become familiar with the complete plumbing installation and see how to apply the information which has been presented.

In order that the reader may acquire better understanding of the principles involved, several plumbing layouts are presented in this chapter. Each of these layouts serves to explain basic principles of soil waste and vent pipe installations, as these principles are commonly applied.

Fig. 175 illustrates a drain, soil waste, and vent pipe arrangement that is permitted in some localities. The installation consists of a water closet, lavatory, sink, and bathtub on the first floor, and a laundry tub in the basement.

The house drain should be constructed of 3- or 4-inch cast-iron soil pipe and be extended to a septic tank or street sewer. A cleanout should be provided on the inside of the basement wall, and at the end of each horizontal run. Cleanouts should be extended at least an inch above the floor line so as to discourage the inexperienced from removing the cleanout plug in order to use the cleanout as a floor drain.

All basements and public washrooms should be provided with a floor drain to which water is supplied for protection of the trap seal. This safeguard is established by connecting a fixture drain on the house side of the trap. If the drain is not more than 5 feet away, venting will not be required. In order that siphonage may be prevented, the floor drain should be vented.

The piping layout shown in Fig. 175 represents the minimum requirement for a waste and vent pipe installation. An arrangement of this kind would not be permitted in localities where sanitary control is strict. The simplest installation is that designed for a bathroom, the bathtub and lavatory waste of which is served by a pipe not less than 1½ inches in diameter.

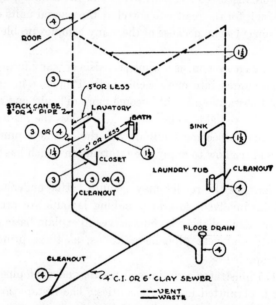

Fig. 175. Drain, Soil Waste, and Vent Pipe Arrangement
for a One-Story Residence

Tests have proved that separate re-venting of these fixtures is not necessary, provided the distance between the trap and the soil pipe, which in this instance is the point of vent, does not exceed 5 feet, as indicated.

The waste stack which serves the laundry tub and sink is built of 1½-inch diameter pipe. The branch of the drain is extended 4 inches above the floor level, where a 4 x 2-inch tapped cross is provided for the laundry tub connection. The sink waste is discharged into a 4 x 2-inch reducer provided for this purpose in the top opening of the tee.

The vent for the sink is 1½ inches in diameter, as the illustration indicates, and is connected to the main soil pipe above the attic floor of the residence.

Fig. 176 presents the layout of a drain, soil waste, and vent pipe installation designed to satisfy maximum requirements of nearly all localities. This layout is recommended over that presented in Fig. 175.

In this installation all fixtures are trapped, and each trap is vented except the floor drain, which is supplied with water from the

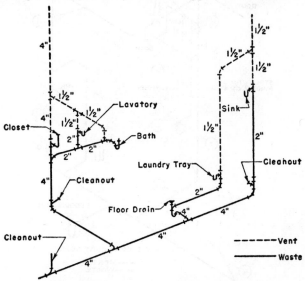

Fig. 176. Drain, Soil Waste, and Vent Pipe Layout
Recommended over That Illustrated by Fig. 175

laundry tub. This arrangement observes good practice, since any floor drain may lose its seal by evaporation unless it is supplied with water.

A 4-inch line is installed from the floor drain intersection to the wall to accommodate a 4-inch cleanout, which device is required to be as large as the pipe it serves. The pipe size is reduced above the 4-inch **Y**, and a 2-inch soil line is extended up to the sink. A 1½-inch galvanized iron vent extends through the roof. The laundry vent intersects at a point 3½ feet above the sink floor line. The water closet, lavatory, and bath are served by a 4-inch soil pipe and 2-inch soil laterals.

Fig. 177 is a layout of plumbing consisting of a closet, lavatory, and sink on the first floor, and a closet, lavatory, and bathtub on the second floor of a residence. A laundry tub is located in the basement. The purpose of this figure is to illustrate the principles of individual ventilation.

The house drain and soil pipe installation which serve the fixtures

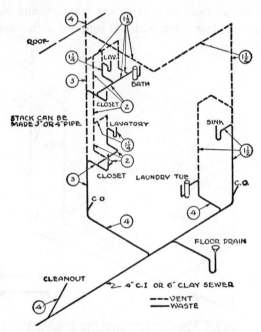

Fig. 177. Plumbing Layout for a Two-Story Residence, Showing Individual Ventilation of Fixtures

in Fig. 177 are like those of Figs. 175 and 176 except that a connection for the second-floor toiletroom must be included.

The lavatory in the first-floor toiletroom is wasted with a 1¼-inch pipe into the closet bend. The water closet is individually vented with a 2-inch pipe extended to a point in the soil pipe 3 feet above the closet connection. The lavatory trap is ventilated with 1¼-inch pipe.

The waste pipe for the lavatory and bathtub of the second-floor toiletroom is constructed of 1½-inch pipe and discharges into the closet bend. Both fixture traps are individually ventilated with

1¼-inch and 1½-inch diameter pipe respectively. The vents terminate into the side opening of a tee provided for this purpose in the main vent. The laundry tub and sink waste and vent are identical with that of Fig. 175.

Fig. 178 represents the plumbing installation for a duplex residence. It consists of a closet, lavatory, and two laundry tubs in the basement, and toiletrooms on the first and second floor, in which

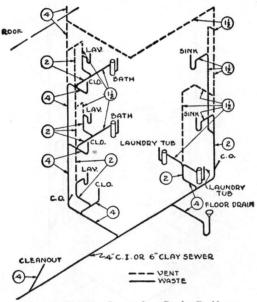

Fig. 178. Plumbing Layout for a Duplex Residence

a closet, lavatory, and bathtub are installed. A sink is provided in each kitchen.

The house drain, with the exception of variations in stack and waste locations is the same as in the preceding illustrations. The soil pipe is of 4-inch diameter and has toiletroom branch connections on three levels. This pipe is installed in the partition directly in back of the water closet.

The basement toilet is vented with a 2-inch wet vent into which the basin has been wasted. The vent is extended from the basement toiletroom vertically to a point 3 feet above the second-floor toiletroom branch and serves as a main vent.

The bathtub waste of the first-floor toiletroom is constructed of 1½-inch pipe and is increased at the lavatory branch to 2-inch pipe and then discharged into the closet bend. The bathroom group is wet vented through the lavatory waste, which has been increased to 2-inch pipe, and is reconnected into the main vent as the illustration indicates.

The second-floor bathroom is identical with the first except that a 1½-inch lavatory waste is ample.

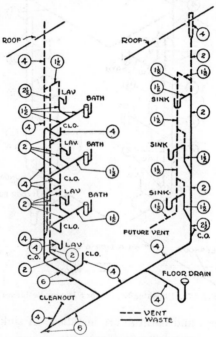

Fig. 179. Soil and Waste Stack Installation, Three-Story Apartment Building

The laundry tubs are connected to a unit waste of 2-inch diameter and are vented by the unit method with a 1½-inch pipe connected into the waste vent of the sink stack.

Both sinks discharge into a common waste pipe which has been increased to 2-inches at the first-floor sink connection. Each is individually ventilated.

This installation may be individually ventilated to conform with regulations of local authorities. Each fixture in this case, except the second-floor closet and sink, would require an individual revent.

Fig. 179 is a typical installation of a soil and waste stack in a three-story apartment building. The principles involved in ventilating the fixtures are those of Fig. 178, in which wet ventilation of the bathroom groups is employed.

The main vent of the soil-pipe installation has been increased to 2½-inch diameter, because the sum of the unit values it serves is in excess of those permitted on a 2-inch pipe. It is reconnected full size

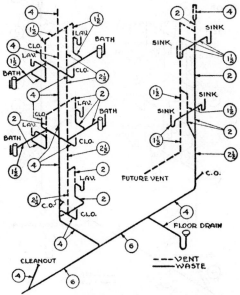

Fig. 180. Typical Plumbing Layout for a Two-Story
Apartment Building

into the base of the soil pipe to relieve any condition of back-pressure which might occur at this point.

The sink waste stack is also increased to accommodate the number of fixture units it serves. Each fixture is individually vented with 1½-inch pipe and is reconnected into a main vent, which is extended through the roof. The roof terminal has been increased to 4-inch diameter.

Fig. 180 illustrates how fixtures discharging into sanitary crosses may be wasted and vented. This installation consists of a soil pipe which serves 4 complete bathrooms and a sink waste on which 4 sinks are installed. It is typical of layouts in two-story apartment buildings.

The soil pipe is constructed of 4-inch diameter pipe and is provided with two 4-inch sanitary crosses for the water closet connections. The first-floor bath waste has been increased to 2 inches in diameter, because ventilation of the water closet is essential. The first-floor toiletrooms are ventilated by the wet vent method through

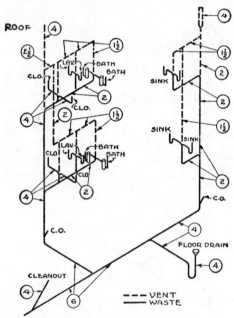

Fig. 181. Unit Waste and Vent Applied to Bathroom Fixture Group in Apartment House Installation

the lavatory waste, which also has been increased to 2 inches. The second-floor toiletroom underfloor and vent pipe arrangement is identical with those of the previous installations.

The main vent has been increased to 2½-inch diameter because of the number of fixture units it serves. It is reconnected into the base of the soil pipe. The sinks are discharged into a unit waste pipe and ventilated in the same manner. The waste line has been graduated in size in accordance with the number of fixture units it serves.

Fig. 181 illustrates how the unit waste and vent may be applied to bathroom fixture groups located on opposite sides of a partition. This design of installation is typical of hotel and apartment house layouts.

The soil-pipe stack accommodates four complete bathrooms, each consisting of a closet, lavatory, and bathtub. The closets are connected to a 4-inch sanitary cross, the top opening of which serves as a vent connection. Very often a fitting of special design is used

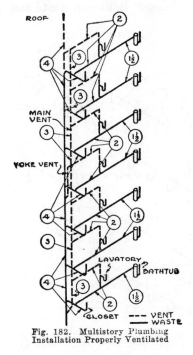

Fig. 182. Multistory Plumbing
Installation Properly Ventilated

for this purpose. The fitting is provided with a waste pipe opening to which the underfloor work may be connected. The closets traps are ventilated with a 2-inch unit vent.

The lavatory and bathtub wastes discharge into a 2-inch diameter common waste pipe. Both groups of fixtures are connected into the waste by means of a 90-degree drainage cross, the top opening of which serves as a unit vent. The lavatory and bathtub traps are ventilated with 1½-inch diameter pipe. The vents from the first floor are extended with 2-inch pipe to a point 3 feet above the topmost branch of the soil pipe, where the main vent is increased to 2½-

inch diameter to accommodate the second-floor bathroom. The second-floor fixture connections are identical with those of the lower floor. This type of installation is very practical and effective.

The sinks are installed on an independent 2-inch diameter vertical waste pipe installed on a common dividing partition. The fixtures are unit vented, as the illustration indicates.

Fig. 182 is an installation which is common to multistory apartment and hotel buildings. Bathroom waste and vent pipe connec-

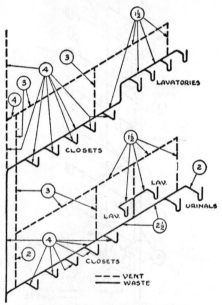

Fig. 183. Waste and Vent Pipe Installation
for a Two-Story Industrial Building

tions are typical of those of a small residence. The fixture traps are ventilated by the wet vent method.

The main vent is built of 3-inch pipe because of the number of fixture units it serves. It is provided with a 3-inch yoke or bypass vent between the fourth and fifth floors.

Fig. 183 represents the waste and vent pipe installation for a small toiletroom on each floor of a two-story industrial building. The lower floor men's toiletroom consists of five water closets, two lavatories, and two urinals, all of which are located on a common partition. The upper floor is a ladies' toiletroom containing five water closets and four lavatories. The fixtures of this toiletroom also are located on one partition.

The water closets of the first floor are connected to a 4-inch branch, which is circuit vented with a 3-inch diameter pipe extended to a point in the soil pipe above the highest fixture branch, where it has been increased to 4-inch diameter. The urinals and lavatories are connected to a 2½-inch common waste line. The lavatories are

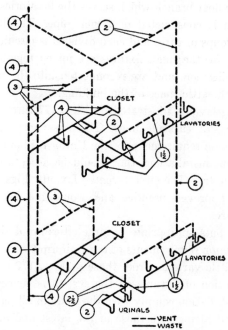

Fig. 184. Toiletroom Plumbing Installation
for Large Building

unit vented, and the urinals are provided with a circuit vent as the illustration indicates.

The second-floor toiletroom is typical of the first, except that it has four lavatories instead of two. The lavatory branch is ventilated by means of a 1½-inch circuit vent.

Fig. 184 is another installation which, with slight variations, is common to buildings of larger design. The figure represents a women's toiletroom (second floor) and a men's toiletroom (first floor) in which the fixtures are located on opposite partitions.

The toilet groups are ventilated by means of a circuit vent and a relief vent similar to the closet groups in Fig. 183.

The branch for the lavatories and urinals of the first floor installation is taken from the soil branch between the first and second closet, and extended with 2½-inch pipe to serve the urinal group.

The urinals are ventilated with a 1½-inch pipe, increased to 2-inch where it joins the vent pipe of the lavatory group. The lavatories are circuit vented.

The upper floor branch which serves the lavatories in the women's toiletroom is constructed of 2-inch pipe. The lavatories are divided into groups of three and each division is circuit vented.

The vents for the small fixtures are joined together above the second-floor toiletroom and are reconnected into the main soil and waste vent in the attic space of the building.

The piping layouts illustrated by Figs. 177 through 184 represent installations of least adequacy. These basic arrangements comply with *minimum* requirements of the Uniform Plumbing Code but do not represent installations that would be acceptable in all sections of the country. The circuit-vented layouts, Figs. 183 and 184, deviate from approved practice and would not be permitted in metropolitan areas.

Piping layouts appropriate to large cities are designed to comply with *maximum* requirements of the Uniform Code. These installations show little variance from those of Figs. 177 through 184 so far as the location of fixtures on the system is concerned, but the methods of installation comply with the Code in all particulars, and the fixtures and fittings are those of the latest and most approved design.

CHAPTER XIV

INSPECTION AND TEST

Testing the Plumbing System. The plumbing system must be subjected to a rigorous test once the roughing in of the soil, waste and vent pipe system has been completed. This practice assures the consumer a safe plumbing system and, to a large degree, certifies the plumber's installation. To permit a drainage system to be enclosed in building partitions without the formality of testing it, is too great a risk for the plumber to assume. Without complete testing, faulty workmanship (sometimes unavoidable) and defective materials will often become apparent only after the plumbing system is put in operation. The plumber, regardless of who is to blame for error or poor workmanship, is the person responsible, and he may be held liable for the damage.

Inspection of the Plumbing Installation. Not only should the drainage system be tested, but inspection by some person of authority who is familiar with plumbing practices is essential. The individual best suited for this responsibility is one who has had previous plumbing installation experience and who possesses complete knowledge of the physics principles involved in plumbing systems. Often the responsibility of inspection is placed in the hands of building inspectors—men whose knowledge of building crafts is general rather than specific. These men, when inspecting a plumbing installation, look for leaks and other obvious faults and do not recognize mechanical defects. The consumer is not properly served by inspection of this character.

Some states have established plumbing divisions, which are directed by competent plumbers who are responsible for sanitation standards. Field men who are responsible for designated areas are employed, and it is their duty to make inspections and offer installation advice in their territory.

Larger cities often employ a staff of full-time plumbing supervisors, and it appears that this practice is a most efficient one. It is hoped that in the near future a universal plumbing code specifying minimum standard practices will be established—each state adopting a policy of inspection to be carried out under local supervision. Wisconsin has enjoyed a policy of this kind for nearly twenty-five years and the beneficial results of this program cannot be measured. All mechanics under the Wisconsin plan serve bona fide apprenticeships, are registered and indentured with state commissions, receive technical training by compulsory school attendance and are required to pass a rigid state examination. Once the apprentice passes this test he is licensed by the state to operate within its boundaries, and he may suffer revocation of his license to practice for any serious violation of the code.

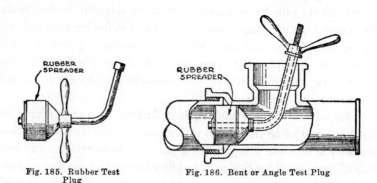

Fig. 185. Rubber Test
Plug
Fig. 186. Bent or Angle Test Plug

PLUMBING SYSTEM TESTS

Cast=Iron Drain Water Test. All openings of the house drain must be extended above the level of the basement floor before the drain can be tested with water. Connections which serve as floor drains, and similar devices, must be sealed thoroughly with a rubber test plug, as illustrated in Fig. 185.

The test plug consists of a capped stem of ½-inch pipe, equipped with a running thread. The pipe is provided with two flanges, the lower one of which is stationary. The top flange is moved up or down by a large wing nut and cast-iron body. A heavy rubber spreader placed between the flanges completes the test plug. The rubber portion of the plug may be inserted into the inside diameter of the pipe and expanded by turning the wing nut to the right. Where more than a few pounds of pressure exist, it often becomes necessary to wire or block the plug to prevent it from being blown out of the opening. The test tee must be sealed in similar fashion, except that a bent or angle test plug is used, as shown in Fig. 186.

The water is admitted into the drain through one of its extended connections until the entire house drain is filled with water to the level of the floor. The drain is then inspected for leaks and, should any be found, the joint must be recalked to make it watertight. Defective material must be replaced.

Vitrified Clay Pipe Water Test. The test of a vitrified clay house drain is applied in the same manner as that of a cast-iron drain. Rubber test plugs may be used, but care must be taken not to overexpand them, because clay pipe is liable to break from the pressure.

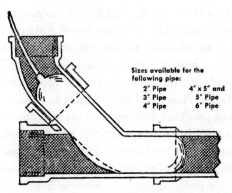

Sizes available for the following pipe:

2″ Pipe   4″ x 5″ and
3″ Pipe   5″ Pipe
4″ Pipe   6″ Pipe

Fig. 187. Crude Rubber Test Plug
Courtesy of Theis Bros., Minneapolis, Minn.

A test plug manufactured of natural crude rubber, reinforced with nylon, is shown in Fig. 187. The plug comes with a gage, and is available in a variety of sizes, with desired lengths of hose included. It is put in the pipe, inflated to the necessary pressure (the usual vitrified clay line test pressure is 5 to 10 pounds), and deflated and withdrawn upon completion of the test.

If a stronger test is required, a pour of quick-hardening compound, composed of equal proportions of plaster of paris and portland cement can be used. This method of sealing is effective in closing closet openings for tests up to 50 pounds. A piece of shaped cardboard is pushed into the pipe, prior to pouring the plaster and cement which is watered to the consistency of mortar.

Testing the House Drain with Air. The house drain may be tested with air pressure. When this test is used, the drain must be completely sealed. The air can be admitted through a test plug provided with a valve and gauge arrangement. About five pounds of air should be applied. Leaks are difficult to detect when the air pressure test is used. A small quantity of odorous solution, such as oil of peppermint, ether, or similar substances can be used to aid in locating the defect. Soapsuds may also be applied to the joints as a means of finding leaks. The air test is not used frequently and is resorted to only when a water test would be impractical.

Soil, Waste, and Vent Pipe Water Test. The test of a soil, waste, and vent pipe installation is accomplished in much the same manner as is the test of a house drain. All fixture openings must be sealed substantially with test plugs or by means of a short nipple and cap. The water may be admitted to the system through a tester which has been provided with a valve and a hose connection. After the entire pipe installation has been filled to the roof terminal, it may be inspected for leaks. The usual procedure is to check the installation from the top floor, working downward on the soil pipe to the basement. Leaking joints must be made watertight, and any defective material must be replaced before final inspection has been made. Tall building installations are usually tested by sections, because of the tremendous pressure produced by their extreme height. It is never advisable to test more than five floors of plumbing at one time.

Soil, Waste, and Vent Air Test. The soil, waste, and vent pipe installation may be tested with air, whenever the use of water would be impractical. The entire system is sealed and filled with air, which is compressed by means of a pump to a pressure of about five pounds. This test may be used in cold weather. The air test is not considered as thorough as the water test, and the difficulty encountered in finding leaks often makes the test a troublesome one.

Smoke Test. The smoke test is generally applied to plumbing systems in old buildings. A smoke test requires the services of at least three men. One man is assigned to the house, another operates the smoke producing machine, and the third is stationed on the roof. The man who is to check the installation for leaks must use care not to come in contact with the smoke before the test is conducted. He must depend on his sense of smell to a large extent to detect leaks, and once his clothes absorb the odor of the smoke, his task becomes more difficult.

All the plumbing fixtures must be sealed with water and the stack terminal stopped with a wet rag. The smoke machine which consists of a hollow drum provided with a cover and a rotating air pump may be filled with oily rags, or any substance which produces dense smoke.

The machine may be attached to the plumbing system through some medium outside of the building walls. The ignited rags are smothered by sealing the hollow drum, and the smoke from the smoldering materials is forced into the plumbing system by the air pump provided for this purpose.

The smoke test is not an efficient or safe test and should be used only when better methods are not practical.

HERE'S TO WATER

An uncontaminated and plentiful supply of water is vital to health and promotes community welfare.

Courtesy of American Radiator & Standard Sanitary Corporation, Pittsburgh, Pa.

CHAPTER XV

WATER SUPPLY

An adequate supply of palatable, safe water is an essential prerequisite for mankind's continued progress. The seemingly inexhaustible supply of water, which we take so much for granted, actually involves complex storage, purification, distribution and sewage systems, particularly in urban areas. It is the plumber, sanitation engineer, and members of related fields who bear the responsibility of providing sufficient safe water for mankind's needs, without depleting or polluting the natural sources from which this water is obtained.

Commonly Used Heat and Water Terminology. There are many common scientific terms which the plumber will come across and use in his daily work. As an aid to better understanding of the following chapters, some of these terms, often used in water supply and heat system work, are defined here.

Pressure is simply force exerted on an area, this relationship being expressed in various units, such as pounds per square inch, pounds per square foot, or grams per square centimeter. A column of water one foot high exerts a pressure of 0.43 psi (pounds per square inch) regardless of the diameter of the column.

The pressure exerted by water at rest is called *static pressure*, an illustration being the pressure exerted by the water at the base of a stack or service pipe when the water is not in motion.

The pressure forcing a stream of water, gas, or steam through an opening is called the *service pressure*. This pressure is reduced in practice, because of the friction normally encountered by liquids and gases flowing through pipes, valves, etc.

Normal pressure refers to the pressure range measured over a period of twenty-four hours.

Any pressure over and above the manufacturer's rating, or that is generally shown to overtax equipment, is called *excess pressure*, while *critical pressure* refers to the maximum or minimum pressures at which proper functioning can be maintained.

Any liquid or gas flowing through a tube creates friction as it contacts the walls of the pipe. This friction makes the rate of flow slower, the speed loss being measurable in feet or inches by its pounds pressure on the height of the column above the outlet. (For a specific application, see the section entitled Sizing The Water Service, in Chapter XVIII, THE HOUSE WATER SUPPLY.) The loss in rate of flow is often referred to as *head loss by friction* or *friction head loss.*

If water flowing at a high speed is suddenly stopped or slowed down, it is very possible that *water hammering* will result. This is caused by rebounding pressure waves, initially generated by the water's impact with the stopping factor (often suddenly-closed valves, or air-filled pipes and pipe sections). The force is suddenly directed backward and waves of varying pressure are started, which can set up vibrations in the pipe and its supports, making a hammering noise and sometimes causing breakage or bursting.

In the plumbing profession, heat is commonly measured in terms of the *British Thermal Unit* (*Btu*). The Btu is the amount of heat required to raise the temperature of one pound of water 1° Fahrenheit. (In Europe, the *calorie* is the common unit of heat measurement, this being the heat required to raise the temperature of 1 gram of water 1° centigrade.)

Heat can change substances from solid to liquid, or liquid to vapor. During this transformation there is a point at which heat is still being applied to the substance, but the temperature of the substance remains static. This heat, not registered in temperature rise but absorbed during the changing process, is called *latent heat.* For example: if a piece of ice is heated, its temperature will steadily rise until it reaches the freezing-point, at which point it starts to melt. But then, during the whole melting period, it remains at the freezing point temperature, although heat is still being applied to the ice.

A primary aim of plumbing is the continued supply of suitable water, with efficient disposal of contaminated water after usage. Many complex processes and mechanisms become understandable when viewed as part of this endless supply cycle. Understanding the "why" of things is a great aid to understanding the "how" of things. Nature, and man through nature, replenishes the water sources from which we obtain our water. Our share in this amazing cycle requires constant care to avoid contamination of the water, particularly at its sources.

Purification of Municipal Water Supply. Water taken from any natural source—ground, lake, or river—contains many objectionable elements. These may be noxious gases, such as carbon dioxide (CO_2), or hydrogen sulphide (H_2S); or they may be bacteria, mud, suspended vegetable matter, and mineral elements such as calcium, iron, or magnesium.

The noxious gases are removed by aerating the water. Some of the mineral elements contained, such as certain forms of iron, also are removed by this means. Suspended materials require a coagulation and settling process, while bacteria are eliminated by the addition of chemicals and by sand filtration. A chemical element that tends to render the water hard is replaced by an element that is favorable. This is done either by the city or by the individual consumer, who uses commercially prepared water-conditioning compounds for the purpose.

Aeration of the Water Supply. In this process, which consists of spraying the water into the atmosphere through jets, or passing it over rough surfaces, the gases are expelled into the atmosphere and formed into other chemical compounds by contact with oxygen. The iron associated with the oxygen forms a ferric compound known as an oxide and, in this form, can be precipitated readily.

Coagulation and Precipitation. This process is one in which the larger suspended materials, such as decayed vegetable matter, mud, and bacteria, are removed by coagulation and precipitation. The coagulants used are ferrous sulphate ($FeSO_4$) and lime (Ca), ferric chloride ($FeCl_3$), aluminum sulphate, $Al_2(SO_4)_3$, and others. The addition of these chemical compounds tends to form a gelatinous substance known as floc, which precipitates quickly. The coagulants form compounds of a harmless nature and are precipitated with the flocculent material. The precipitate is gathered into large sumps by drag equipment of the rotating or endless belt variety and is disposed of in the most practical manner.

Sand Filtration. After the water has been treated with a coagulant it is filtered through layers of sand and gravel placed in concrete basins. The water is passed over the sand beds, which form a fine screen and retain the finer particles held in suspension as well as the smaller organisms and bacteria deposited on the surface of the sand. The filtered water is collected in conduits and conveyed to storage basins of adequate size. After a short time a gelatinous scum forms on the top of the filter sand, reducing the efficiency of the filter. The sand itself may become partially clogged as well. The flow of water under these circumstances is reversed and the objectionable materials collected on the top of the filter sand are removed by water that is sprayed through jets and collected in a skimming trough surrounding the filter. This process is an elaborate one and requires a large expenditure for mechanical equipment.

Disinfection. Disinfection is accomplished by injecting chlorine gas into the water supply. Though all domestic water is chlorinated before delivery to the user, not always does this treatment destroy the pathogenic germs of water-borne diseases. Laboratory tests have shown that the virus causing poliomyelitis can remain in chlorinated water for four or five days. To be safe during an epidemic, one should boil all water intended for human consumption before using it.

Calcium and Magnesium Removal. The removal of calcium and magnesium salts on a large scale has not proved entirely practical. The process will be discussed in the chapter dealing with private water correction.

Public Water Distribution. The method of water distribution usually is determined by the source from which the supply of water is secured. Where the supply is taken from lakes or rivers, the direct method of distribution is used.

Direct Pressure Distribution. The direct pressure method of water distribution obtains its supply of water through a large diameter intake installed on the lake basin and extended into deep water. The intake in some instances is provided with a shaft, which is known as a "crib," to make it accessible. The inlet end of the intake is, as a rule, provided with a metal hood and is extended vertically from the bottom.

It is the usual practice to take the water from the lake to a receiving well by gravity, from which point it passes through the filtration plant by the same method or by means of pumping equipment. The filtered water is pumped from the storage reservoir by centrifugal or piston pumps into the city water mains under a degree of pressure sufficient to serve specific needs.

The pumping equipment used in the direct pressure method of water distribution operates continuously in order to maintain a constant pressure in the city water mains. The mains are constructed of cast-iron pipe and usually are laid in the street—on the north side of streets running east and west, and on the east side of streets running north and south. They are located about ten feet from the curb.

Water mains usually are cross connected with one another so that a constant movement of water is maintained. Thus the possibility of the water becoming stagnant is averted. Sections of the main are controlled with valves that are made accessible through stop boxes extended to street grade. By this means, parts of the system can be controlled in the event of a breakdown in the piping material.

The contour of the soil varies in many cities and it becomes necessary to use pressure equalizing stations in some locations. These stations consist of an overhead storage tank and booster pump to provide additional pressure to serve these high areas. The direct pressure system is a very practical one and may be employed where water is in boundless supply.

Indirect Pressure Distribution. Where the water supply is taken from a drilled well, distribution is usually obtained by indirect pressure. For this, a turbine pump is used. A pump of this classification is constructed with the motor mounted on top of the stand pipe, which extends into the well below the water table. The number of pumps employed is determined by the amount of water to be lifted, the depth of the well, and the elevation of the storage reservoir, which in most cases is a large hemispherical bottom tank.

These turbines or pumps are similar to a centrifugal pump that is operated in a horizontal position. One is placed above the other, each pump representing a *stage*. Three pumps form a three-stage turbine, four pumps, a four-stage turbine, and so on. Sometimes as many as twenty-five stages are employed in a deep-well installation.

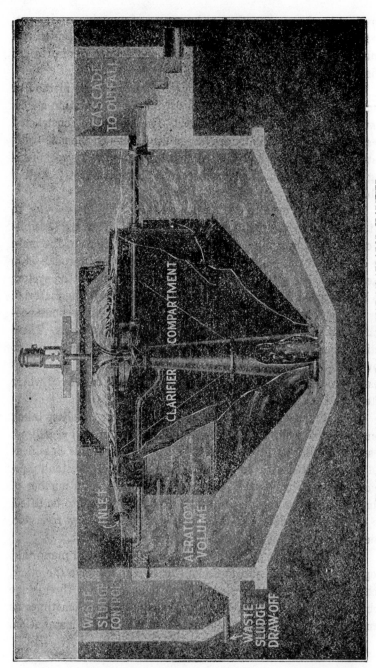

CROSS-SECTIONAL VIEW OF A COMBINATION AERATOR CLARIFIER

Courtesy of Chicago Pump Company, Chicago

_____

CHAPTER XVI

MATERIALS USED FOR WATER DISTRIBUTION

The mineral content of water is the factor which determines the type of pipe to be used for water distribution in any given case. For hard water, which has a large calcium and magnesium content, pipe of any kind is suitable. But for soft water, which has a damaging effect upon materials such as galvanized or lead pipe, the utmost care must be exercised in selection of the pipe. The effect of soft water upon lead pipe may result in cases of mild lead poisoning, or it may produce a harmful effect upon the teeth.

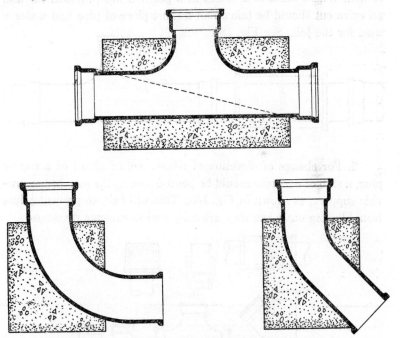

Fig. 188. Above, Water Main Tee. Below (*left*), Water Main Elbow; (*right*) Water Main Curve

Cast=Iron Pipe and Fittings. Most municipalities and water services use heavy cast-iron pipe, known as *corporation pipe*, for construction of city mains, and for general purposes of water distribution. As pipe of this kind comes in sizes ranging **from 2 inches** to 36 inches in diameter it well suits the purpose.

Although the plumber is seldom called upon to install corporation pipe, he frequently is required to make an extension or put in a branch line from the main. In order that he may do a good job, therefore, there are two rules that he should keep in mind, viz:

1. When a tee for a branch line is to be installed, the length of pipe to be removed from the main to accommodate the tee shall equal the distance from the top of the inside of the hub (or bell) to the bottom of the inside of the hub of the run of the tee. See dotted line in sketch showing a *water main tee*, Fig. 188.

If the cut is to be made within 2 feet of the hub, the short piece should be melted out, then the tee installed and leveled up **and** calked. Where the hub is found in a position not practical for use, an extra cut should be taken and a short piece of pipe and a sleeve used for the job. See Fig. 189.

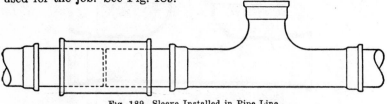

Fig. 189. Sleeve Installed in Pipe Line

2. For change of direction of elbow, tee, or ahead of a cap or plug, a slab of concrete should be poured around the member to provide support, as shown in Fig. 188. This will help to prevent joints from blowing out when they are subjected to extreme pressures.

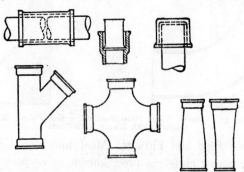

Fig. 190. Water Main Fittings Used on Special Installations. *Left to right* (top). Sleeve, Plug, and Cap; (bottom) Y-Branch, Cross, Reducer, and Increaser

The fittings used on cast-iron corporation pipe are mainly *tees*, *elbows*, and *curves*, see Fig. 188. Tees are used to form branch connections to the main. Elbows are used to make right-angled turns or changes of direction. Curves, which come in short and long patterns, are used to make changes of direction less than right angles to the vertical or horizontal run. Fig. 190 presents some fittings that are used on special installations, namely, *sleeves, plugs, caps, crosses,* **Y** *branches, reducers,* and *increasers*.

Asbestos-Cement Pipe.   Pipe of this designation is gaining favor in industrialized countries throughout the world. This is because it does not rust or corrode.

Asbestos-cement pipe is manufactured in sizes ranging from 3 to 36 inches in diameter for pressures of 50 to 200 pounds. It is composed of asbestos fiber and cement and formed under high pressure into a dense, tough material of good strength and durability. This pipe offers excellent resistance to destructive elements, and its smooth inner surface promotes ease of flow. It is not subject to electrolytic action or to tuberculation, which condition is the internal corrosion that increases friction and impedes flow.

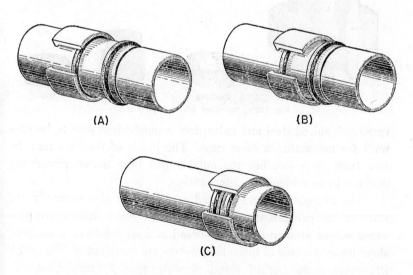

(A)   (B)

(C)

Fig. 191.   Asbestos-Cement Sleeve Coupling, with Sleeve Cut to Show
How Rubber Rings Are Compressed between Sleeve and Pipe.   *A.*
Rings and Sleeve at Start of Operation.   *B.* Sleeve Pulled over One
Ring.   *C.* Sleeve Centered over Joint
Courtesy of Johns-Manville Sales Corporation, New York, N.Y.

Asbestos-cement pipe is quickly installed since it is light in weight. Maintenance cost is low, as it does not require regular cleaning out or replacement. The joint used on this pipe consists of an asbestos-cement sleeve, Fig. 191, and two rubber rings called *simplex couplings*. When the couplings are installed, the rubber rings are compressed tightly between pipe and sleeve to form a flexible yet tight connection which holds joint leakage, water loss, breakage and expensive repairs at the minimum.

Galvanized Steel and Galvanized Wrought=Iron Pipe. Pipe made from these materials is affected by the character of the soil in which it is laid. The acidity and alkalinity of certain soils will

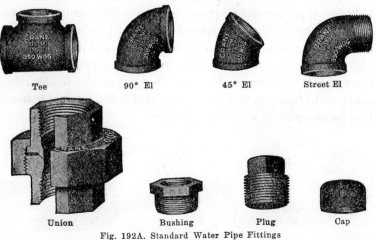

Tee   90° El   45° El   Street El

Union   Bushing   Plug   Cap

Fig. 192A. Standard Water Pipe Fittings

cause galvanized steel and galvanized wrought-iron pipe to become unfit for use within a short time. The inside of the pipe may be free from corrosion, but the outside will bear marks resembling pockmarks as evidence of deterioration.

The fittings used on steel and wrought-iron pipe generally are those of the galvanized malleable variety. These fittings are provided with a standard tapered thread and do not have a recessed shoulder as do those of the calked water-pipe installation. Fig. 192A illustrates a number of standard water-pipe fittings. Cast-iron screw fittings are used for temporary water and steam lines, as these fittings can be easily broken by a hammer blow when tearing down or breaking into a line.

Brass Pipe and Fittings. Brass is probably the material best suited to water service. It is used extensively in industrial plants, where the use to which it is subjected is at times severe. In fact, it is often standard equipment for the reason that corrosive waters have little effect upon it, and because it can be worked with greater ease than can steel. Owing to their high cost, however, brass pins and fittings usually are not used in home plumbing systems.

The fittings used on brass pipe are the same in design as those used on malleable iron. In some cases malleable iron fittings are used on brass pipe installations, though generally this is considered poor practice because of the chemical action likely to occur between the two materials.

Copper Pipe and Fittings. Copper pipe gained immediate popularity with the introduction of swedge and solder fittings. The fact that copper is affected by soil conditions, however, has caused many cities to abandon its use for service water supply.

Fig. 192B.   *Left*, Soldered Fitting; *Right*, Swedge Fitting

The use of copper pipe for home plumbing, under the house, and in between walls is becoming increasingly common. This is because copper does not require the cutting of large holes in plates and studding, as its outside diameter is much smaller than that of screw pipe. Moreover, threading is eliminated.

Fig. 192B shows the swedge and solder joint types of fittings used in joining copper pipe. The swedge or compression type is not practical, however, unless it is exposed so that breaks can be seen and repair readily accomplished when trouble occurs.

Lead Pipe. Lead pipe enjoyed wide use in the past, when lead was plentiful and its quality was good. When the supply began to diminish, and as quality suffered, the use of lead as a pipe material declined so that only a few cities (those built on filled earth) continued its use for water service. Since the flexibility of lead is a feature of accommodation as settling takes place, its use greatly reduces breakage at the meter.

As to whether or not to use lead pipe, local ordinance should be consulted. Questions of weight and quality are matters to be determined by the prevailing condition or need.

Block Tin. Block tin is seldom used for water supply because of its cost and also because of the difficulties encountered in making tight joints when this material is used. Since the metal is extremely soft, there is, as well, the danger that a man inexperienced in use of the soldering iron may permit the heat from the iron to destroy sections of the pipe.

Aluminum Water Pipe. This material has appeared on the market in quantities only sufficient to augment, in some sections, the limited supply of galvanized and copper pipe. Two reasons why aluminum pipe is not used extensively are that its cost remains high and that chemicals contained in the water may attack the metal.

Magnesium Pipe. Though magnesium is comparatively a new element, the fact that it can be claimed from sea water at low cost tends to promote its use. The magnesium tube or anode, used in most water heaters, counteracts electrolytic action and prevents precipitation of lime upon the inner walls of the tank.

CHAPTER XVII

JOINTS ON WATER SUPPLY SYSTEMS

The joints of the water supply system must be made with precision for the reason that the piping comprising it is commonly subjected to high pressures. A defective joint will not only permit a large quantity of water to escape in a few minutes but it also can be the instrument of serious property damage.

A great part of the water supply system is concealed in building partitions and often the system is installed under adverse conditions or is subjected to abuse through carelessness on the part of workmen in other trades. A leak may occur as the result of jarring of the pipe installation, or because a heavy object was dropped on the connection, thereby breaking it. The plumber cannot be held responsible for mishap of such kind, but he can do a great deal to avoid trouble by supporting his work thoroughly and by inspecting it regularly during the course of construction of the building.

Even though it is substantially made, the joint on a water supply system is the weakest part thereof. The most commonly used joint is the screwed connection, which consists of a male and a female end screwed together. This part is standardized to such extent that a joint made in Chicago and a fitting made in San Francisco can be connected to pipe manufactured in Alabama, using tools produced in Toledo with which to run the thread. Such standardization, of course, is possible because of nation-wide adoption of the American Standard pipe thread specifications. (See Table 11.)

In order to secure a good screw joint, certain things are to be kept in mind. First, both male and female threads should be wiped clean of all matter tending to obstruct normal engagement. Second, if pipes and fittings have been subjected to abuse in shipping, or to extremes of temperature, a tap should be screwed in the female (inside) thread and a die run over the male thread to assure smooth run of the threading and a complete final fit.

The third thing to keep in mind is that use of compound (pipe dope) to compensate for faulty workmanship is to be avoided. This

TABLE 11. American Standard Pipe Threads

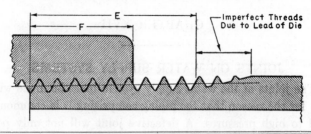

Nominal Pipe Size Inches	E Length of Effective Thread	F Normal Engagement by Hand between Male and Female Thread	Number of Threads Per Inch
⅛	0.2638	0.180	27
¼	0.4018	0.200	18
⅜	0.4078	0.240	18
½	0.5337	0.320	14
¾	0.5457	0.339	14
1	0.6828	0.400	11½
1¼	0.7068	0.420	11½
1½	0.7235	0.420	11½
2	0.7565	0.436	11½
2½	1.1375	0.682	8
3	1.2000	0.766	8
3½	1.2500	0.821	8
4	1.3000	0.844	8

substance should be used only as a lubricant tending to reduce friction in making the screwed joint. Never use lubricant on the female thread, as by so doing the excess amount of this material will be squeezed into the pipe, where not only will it cause contamination of the water contained, but most likely will damage the valve seats and working parts.

Always begin to screw a joint by hand, and make sure that the fitting turns readily for the length of approximately three threads. This practice will serve to avoid the possibility of cross threading. With the fitting in place, a few turns of the wrench will insure a tight joint.

Be careful not to pull too hard on the joint, and do not use an oversize wrench. Many good valves and fittings have been ruined by running the pipe into the valve seat. Practice is necessary in order to acquire the knack of securing normal engagement between male and female threads (see sketch, Table 11), thus producing a tight joint.

Calk Joint on Cast=Iron Water Main. Cast-iron water main is joined by means of molten lead, which is substantially calked after it has been allowed to cool. The joint usually is made in a horizontal position, although vertical joints are not uncommon.

Horizontal Calk Joint. The first step in the process of making a horizontal calk joint on water main is to dry the hub thoroughly and wipe it clean of foreign substances. It is also advisable because of the thickness of the metal, to heat the hub and bell ends of the pipe in cold weather, using a plumber's furnace for the purpose or igniting a small amount of hemp which has been placed in the end of the pipe. Moisture and cold material are liable to cause the molten lead to sputter and may cause an imperfect run of the lead.

The pipe must be laid in place and aligned properly so the joint will be evenly spaced. A single ring of twisted or braided hemp must then be yarned into the bell to prevent the lead from running into the pipe. After the hemp has been inserted and packed thoroughly, a pour rope of asbestos, or a roll made of plastic clay is applied to the face of the bell and must be provided with an exceptionally large well to allow the molten lead to enter the joint as rapidly as possible. The joint must be made in one continuous pour and requires a large amount of lead, and it is well for the mechanic to have a surplus amount of metal in the pouring pot in case difficulty is experienced.

After the joint has cooled sufficiently the pour rope may be removed and the joint can be calked. Inside and outside calking irons and a heavy hammer are used for this purpose. The inside of the joint must always be calked first so the ridge left by the pouring rope can be calked into the bell. The outside edge is also calked to complete the joint. The surplus metal, accumulated as a result of pouring the well, must not be removed—it is well to calk it and in this way drive it into the joint. It is difficult to crack the bell of the water main, and the mechanic doing the calking need not restrain the blow on the calking iron. After the joint has been completed it should be inspected for defect and then be tested with water pressure.

Vertical Calk Joint. A vertical calk joint on water main is made in practically the same manner as a vertical joint on cast-iron soil pipe except that only a single ring of hemp is yarned into the bell and the molten lead is poured in one complete run. The joint must be substantially calked to be effective.

Swedge and Solder Joint on Copper Pipe. The swedge joint is commonly used on copper water service. It is made in connection with a specially designed fitting.

The first step in making a swedge joint is to straighten the ends of the copper pipe by filing them. After this has been done the make-up nut is slipped over the pipe and the end of the tube is expanded with a swedging tool until it seats perfectly on the convex face of the fitting. Pipe and fitting are now aligned and the nut tightened to complete the joint. Annealed copper may be hardened if too much pressure is exerted in tightening the make-up nut, thus producing a break. Solder joints on copper water pipe are identical to those of waste pipe, as explained in Chapter IV.

Wiped Joint on Lead Water Pipe. A wiped joint on lead water pipe is made in the same manner as for lead waste pipe. To overcome sweating of the joint, it is advisable to use a solder of greater

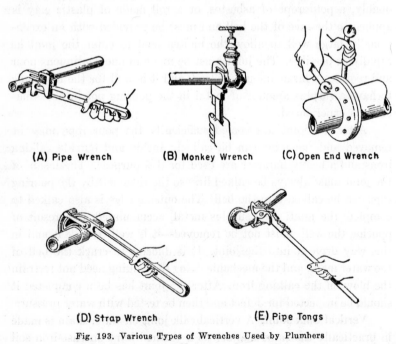

(A) Pipe Wrench   **(B) Monkey Wrench**   **(C) Open End Wrench**

(D) Strap Wrench   **(E) Pipe Tongs**

Fig. 193. Various Types of Wrenches Used by Plumbers

tin content than that contained in the solder used on lead waste pipe. Wiped joints on water pipe are required to withstand extreme pressures, therefore it is essential that they be made with the utmost care.

Wrenches Used on Joints. There is a proper wrench for every job. That being the case it is important to choose the right type and the right size of wrench for the particular job to be done. Five different types of wrenches are illustrated in Fig. 193, sketches *A* through *E*. The separate uses of each of these wrenches are explained in the paragraphs that follow.

Pipe Wrench. The pipe wrench, Fig. 193, at *A*, was designed to be used only on pipe and screwed-end fittings. It has the effect that the harder one pulls, the tighter it squeezes. Its efficiency does not match that of the monkey wrench on parallel-sided objects, and its squeezing action can be severely damaging. In using this wrench, men of little experience in this work often crush or otherwise injure the body of the valve.

Monkey Wrench. Because the monkey wrench (*B*) has smooth, square jaws, it is ideally suited for use on valves and fittings having hexagonal ends. On such parts the monkey wrench fits better than do other types of wrenches, and it does not have the crushing effect that the pipe wrench does. A monkey wrench should never be used on round parts or on pipe, as the grip of its jaws is not suited to shapes of this kind.

Open-End Wrench. Wrenches of this classification are well suited to the task of pulling up flange bolts. The open-end wrench (*C*) affords greater speed in performing this duty than do other types of wrenches. Be sure, however, to select a wrench of the proper size, as not to take this precaution is to risk injury should the tool slip. Another bad result of using a wrench of the wrong size is the effect produced upon the bolt heads, causing them to wear round.

Strap Wrench. Wrenches of this type (*D*) are used principally in working with plated or high-finished materials, since they afford polished surfaces greater protection from damage. Strap wrenches also are used in places too small to admit a pipe wrench.

Pipe Tongs. These wrenches are of the chain and lever variety. Although they are made for handling pipe ranging upward from ⅛-

inch diameter, pipe tongs (E) are commonly used for pipe measuring as much as 6 inches in diameter, and larger.

For the sake of emphasis it is urged again that the tool selected be of the proper size for the use intended. If an undersized wrench is used, difficulty will be experienced in pulling up the joint; if an oversized tool is chosen, there is the danger that the joint may be pulled up so tight that a fitting may be cracked, a valve seat be twisted out of shape, or a pipe be forced into the valve seat. Practice, experience, and care will soon enable the mechanic to acquire the "feel" necessary to handy manipulation of wrenches of every type and size.

The tools you work with are constantly being improved through the combined efforts of plumbers, designers, scientists, and manufacturers. Taking advantage of new developments can simplify and improve your work immeasurably.

A *pipe wrench* with an all-steel body has been developed which will hold its shape without distortion under extreme conditions. A wrench called the *hex wrench* has been designed for use on valves. For close work, where ordinary adjustable wrenches having wide jaws are impractical, a *speed wrench* is of great help. An improved *strap wrench* has been designed and manufactured, with no moving parts and greater holding strength.

Some other developments of interest to the plumber are a *portable bolt and pipe die set* which cuts pipe sizes ⅛ to 1 inch inclusive; a *portable tri-stand chain vise* that makes a serviceable, easily carried and set up work bench, with base room for oil cans, slots for tools, pipe rests, and three size pipe benders; a new type of *threader* for 4 to 6 inch pipe inclusive, an aid to threading larger pipe sizes on the job; a *three-wheel tubing cutter* which will cut tubing with only a short turn; a *sink installer*, with a head that holds all types and sizes of bolts, excellent for work in hard-to-get-at places; and a *sink strainer installer* consisting of a tool for holding the strainer body, while tightening the lock nut with a specially constructed wrench included in the set.

CHAPTER XVIII

THE HOUSE WATER SUPPLY

The house service is that part of the plumbing system installed underground between the water main in the street and the house water meter. It plays an important part in making a plumbing system sanitary.

Fixtures which do not have a sufficient flow of water become fouled and emit odors which are extremely disagreeable and may cause sickness. Difficulty of this kind usually can be traced to careless planning and poor workmanship on the water service installation.

Municipal authorities are interested in proper installation of the house service because of the added expense incurred in extra pumpage of water when a service develops serious fault.

There are a number of essential factors the plumber must observe when constructing a house service. It is important to size it adequately, and also to determine the pressure at the city main. (Pressure is determined by connecting a pressure gauge to the tapped main, or by consulting local authorities.)

Often the pressure of the main is low. Should this be the case, a service of additional diameter may be installed to correct this fault. This practice, however, would serve only homes and buildings of moderate height. In buildings of extreme height, pressure is the only solution to proper fixture sanitation and, should the city installation lack pressure, a pumping unit must be employed to obtain efficient results.

The piping of the house service must be installed in as direct and short a manner as is practical. Turns, offsets, and traps, should be avoided because these elements create friction and diminish the discharge capacity of the service, Fig. 194.

It is also important that the house service be placed in the ground at sufficient depth to protect it from freezing—four feet is considered adequate Should it be impossible to place the service deep enough in the ground to insure protection against frost, it may be insulated with a good quality of frostproof covering or hair-felt, or it may be boxed in wood. Failure on the part of the plumber to protect the service from freezing involves expense the owner must absorb and often results in the breaking up of city streets in order to make repairs.

Ground conditions are an important factor in service efficiency. Soils of an acid nature and surfacing such as cinders or ashes have a deteriorating effect on the piping. The service, when subjected to these conditions, must be protected with a covering of good earth or be painted and boxed to protect it against objectionable elements. It is good practice to run the service inside clay tile with cemented

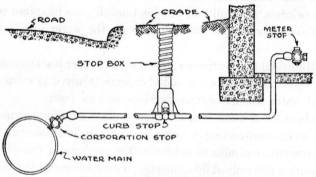

Fig. 194. House Service Installation

joints when the soil is of an acid nature. Too, there is danger of placing the service in soil where, because of some electrical leak, it acts as a ground, thereby effecting deterioration of the piping material.

Materials Used on a House Service. There are five kinds of pipe which may be used for the house service installation: Lead and copper are used on residences and small buildings, while cast iron is favored for larger buildings. Steel and wrought iron pipe may be used, but these materials do not withstand soil conditions quite so well. The installation of material is governed by state and municipal authority.

Joints of the House Service. The kind of joint used on the water service is determined by the kind of material used in its construction. The methods of making the joints on the water supply system were discussed in the previous chapter of this book.

SIZING THE WATER SERVICE

Up to the present time no definite information has been provided the plumber that would enable him to ascertain the size of a water service or water supply system accurately. Experience plays a large part in this phase of plumbing practice. Traditional factors, handed down from mechanics of the past, also have effect.

Trade experience can be accomplished only by making installations over a long period of time. Contacts with fellow workmen and schools having practical trade instructors do much to build trade experience. Analyses of past and present building installations that have given effective service are also valuable.

Tabulated material relative to sizing the service is limited. Some commendable work has been accomplished, but little that is specific enough to be of great value. Trade journals offer helpful information occasionally, but this, too, is general and should be carefully analyzed before it is applied.

There are many factors which affect the size of a water service, such as flushing devices, and the volume of water and pressure required to operate them. Flush valves, for example, require more pressure and volume than do closet tanks, and hospital, kitchen, and industrial equipment vary in water needs from standard fixtures. Shower bath installations require additional water. Filter devices, water softeners, and similar installations all affect rate of flow and discharge volume in a water system.

There are other factors to be considered before the size of service can be determined. The service must be ample to supply a given quantity of water per minute. It is impossible to use a pipe for the conveyance of water without encountering a loss in pressure resulting from friction within the pipe itself as well as through stops and fittings used in its construction. Friction in the plumbing system is the resistance produced by contact of flowing water with the interior surface of the pipe. It is also a resistance between the molecules of water. The only practical means by which friction may be overcome is by pressure, which may be defined as the force required to move the water within the piping system. The public water supply mains are under a pressure of sufficient force to serve an installation of moderate size. A water service connected to the main would be under the same pressure as the main as long as the water within it was at rest. Should some one suddenly draw water from it, however, a decided drop in pressure would occur at the outlet orifice. The variation in pressure between the main and the outlet end of the water service is referred to as pressure loss by friction. It is logical to assume that the more fittings, stops, and other devices placed on the distribution system the greater the pressure loss, and therefore the lower the discharge capacity of the water service.

Very little reliable information regarding pressure loss through stops and fittings was available until recently. Data largely consisted of laboratory conclusions based on short tests. Water usually contains mineral elements which can be removed by centrifugal force. When these elements are precipitated they adhere to the pipe interior. The size of the pipe is decreased materially under these circumstances, and its discharge capacity is tremendously reduced. There may be fluctuations of pressure in city mains caused by emergency draws, peak loads, or breakdown of equipment which also change the situation. However, these are beyond the control of the plumber and need not be considered in the design of the installation.

Table 12 indicates the approximate loss of pressure by friction in pipe valves and fittings of various design, and from it the reader can at least realize the necessity of giving this problem some consideration.

Additional pressure may be lost because of the height to which the water must flow; however, pressure loss resulting from elevation is a relatively simple problem. It is a scientific fact that a column of water 1 foot high exerts a pressure of .434 pounds per square inch at its base. This fact simply indicates the weight of a column of water 1 inch square by 1 foot long. The diameter of the pipe has no effect on the pressure at its base. A pressure gauge provided on plumbing installations always indicates the pressure per square inch. If a building is 50 feet in height, the loss because of head would be .434×50 or 21.7 pounds. This amount can be deducted from the original pressure to determine whether or not it is of sufficient force to elevate water to the required height.

To size the water service, the mechanic must positively establish two things. First the maximum demand of water to supply the

TABLE 12.—Lengths in Feet of Galvanized Iron Pipe to Give Same Pressure Loss as Fittings

Description of Fitting	Nominal Diameter of Pipe, Inches						
	3/8	1/2	3/4	1	1 1/4	1 1/2	2
Elbows:							
90° Pipe ends reamed....	.8	1.1	1.7	2.2	3.6	5.0	5.4
90° Pipe ends unreamed..	4.8	7.3	6.7	5.9	......	......	......
45° Pipe ends reamed....	.5	.7	1.0	1.2	2.1	2.7	2.8
45° Pipe ends unreamed..	4.6	6.5	6.1	4.7	......	......	......
Tees: (Pipe reamed)							
End to end..............	.2	.3	.3	.4	.8	1.1	1.5
End to side.............	1.5	1.6	2.5	3.2	5.2	7.1	8.4
Side Reduction Tees:							
(Flow end to side equivalent length same size as side outlet.)							
3/4 x 3/4 x.................	2.5	3.2	......	......	......	......	......
1 x 1 x....................	2.5	3.5	4.4	......	......	......	......
1 1/4 x 1 1/4 x.............	......	......	......	5.8	......	......	......
1 1/2 x 1 1/2 x.............	......	......	......	5.7	6.6	......	......
2 x 2 x....................	......	......	......	5.7	8.5	8.6	......
Valves:							
Gate.....................	......	.5	.8	.9	1.0	1.2	1.4
Globe....................	......	19–23	10–30	8–40	......	34–45	16–50
5/8" Corp. cock for lead...	2.3	6.7	28	90	......	......	......
3/4" Corp. cock..........	......	2.1	8.9	29	......	......	......
1" Corp. cock............	......	......	2.2	7.0	27	......	......
3/4" Curb cock for lead...	......	1.9	8.0	26	......	......	......
3/4" Curb cock for copper.	......	1.0	2.0	6.4	......	......	......
1" Curb cock for lead....	......	......	.8	2.6	9.8	......	......
3/4" Comp. stop & waste..	......	10	35	100	......	......	......
1" Comp. stop & waste..	......	3	12	38	148	......	......
Water Meters:							
5/8" with 1/2" connections..	......	6.7	28	90	......	......	......
5/8" with 3/4" connections..	......	4.8	20	64	......	......	......
3/4" with 3/4" connections..	......	3.4	14	45	......	......	......
1" with 1" connections..	......	2.2	9	30	115	......	......
1 1/4" with 1" connections..	......	1.1	4.4	14	54	......	......

Wisconsin State Plumbing Code.

Note. Table 12 was prepared from practical laboratory tests, and to give some idea as to the length of pipe which would equal the friction loss of a specific fitting, the reader has but to refer to it. For example, a 1-inch 90° elbow attached to an unreamed pipe would be equivalent to 5.9 feet of 1-inch pipe. A 3/4-inch compression stop and waste would be equal to 35 feet of 3/4-inch pipe, that is, as far as pressure loss caused by friction is concerned. There are methods of determining the size of water distribution systems by evaluating the friction loss of each fitting, but these methods are to some extent impractical and are applicable mainly to large buildings where each installation would require individual analysis.

total fixture need; second, the peak demand or the maximum load that a system would be subjected to because of simultaneous use of fixtures.

The maximum demand a water system might be subjected to is easy to calculate. It is quite logical that the maximum demand of water for certain fixture needs would be determined by the amount of water that could be discharged through the waste orifice of a fixture in one minute's time. This discharge has been established in units, and was discussed in Chapter VI. It is reasonable to assume that if a fixture can waste only a given amount of water and the water supply system may not be taxed over and above this volume of water in a given interval, without causing overflow of the fixture being used, the maximum demand of a water supply system would be the total sum of its fixture unit values, each unit representing 7½ gallons of water. As an example of procedure in determining maximum demand the following installation problem is offered.

Suppose a bathroom installation consisting of a basin and bath had, according to the unit system, a discharge capacity of 22½ gallons of water per minute. A water service of no larger size than is sufficient to supply this volume should be indicated. Should an installation consist of 20 wash basins, constituting 20 units or 150 gallons, a fixture service adequate to the demand will be required. A size of service large enough to supply maximum demand is impractical. Its diameter would be unreasonably large.

The factor of simultaneous fixture use enters into the problem of sizing the water service. Just how many fixtures will be used simultaneously is difficult to determine. It is reasonable to assume that the fewer the number of fixtures installed the greater the possibility of simultaneous use. Larger installations would have an opposite effect. Simultaneous use might be 100 per cent where two fixtures are installed to less than 10 per cent where 500 fixtures comprise the plumbing installation. The following table indicates the approximate percentage of simultaneous use of fixtures to establish probable demand.

No. of Fixtures	Per Cent of Simultaneous Use
1 to 5	50 to 100
5 to 50	25 to 50
50 or more	10 to 25

Large buildings offer complex problems, and each one is unique 'n some aspect. Friction loss, types of flushing devices, method of water distribution and many other factors must be taken into consideration.

The factor of probable demand or the amount of water a distribution system is called upon to produce in any period of a 24-hour day arises from the problem of simultaneous use of plumbing fixtures and is largely a guess on the part of the designer. He may be guided by the type and design of the building, the purpose for which it is to be used, the number of people who are to use it, and many other undetermined elements. Some statistical information is available, but it is still exceedingly vague and may only be used in correlation with practical trade experience. The probable demand is not likely to be more than 25 per cent of the maximum demand, especially for average size buildings of the residential type, and this figure may be used with safety to establish the size of a water service.

Method of Determining Demand. To show how the unit system may be applied in determining the maximum and probable demands of a water supply system, a five-fixture residence is given as an example. Using the unit values of the fixtures, as explained in Chapter VI, the following compilation results:

No. of Fixtures	Units	Gals. per Minute
1 Water closet	6	45
1 Lavatory	1	7½
1 Bathtub	2	15
1 Kitchen sink	2	15
1 Laundry tub	2	15
Total	13	97½

The total of 97½ gallons represents the maximum demand. As consumption of this volume of water is unlikely, the amount may be reduced materially. As previously indicated, the probable demand for this type of building would not be more than 25 per cent, so it is safe to use 25 per cent of 97½ gallons, or approximately 24 gallons of water, a minute.

There may be some question on the part of the reader **as to the** correctness of using a value of 6 units for a water closet, because a closet of the flush tank type discharges much less than 45 gallons per minute. As was previously indicated, the size of the water service as well as that of the distribution system is largely to be assumed. No method of calculation is entirely dependable. To justify the use of flush valve closets, no mention of lawn sprinklers has been made. The average residence is provided with two. Should the building be provided with tank closets instead of flush valve closets, then it is advisable to determine maximum demand on this basis.

A value of 2 units (15 gallons) of water may be used for flush tank water closets, but added to the maximum demand to offset this change must be added the discharge of two lawn sprinklers, which amounts approximately to 4 units or 30 gallons of water.

Once the probable demand has been established, it becomes essential that the plumber, in order to determine the actual diameter of pipe necessary to supply the required volume of water, must know something of the discharge capacities of the various diameters of pipe. The tables which follow this discussion are a compilation of tables contained in the bulletin, *Interior Water Supply Piping for Residential Buildings*, by Prof. Francis M. Dawson, formerly of the University of Wisconsin and now associated with the University of Iowa. The tables show the volume of water a pipe of predetermined length will discharge under a given pressure.

It is decidedly impracticable to use a water service of less than ¾-inch diameter, and should the plumber find, after referring to the tables, that a pipe of smaller diameter will provide the probable demand of the building, the ¾-inch service should be installed.

Calculating Flow (Discharge Capacity) of Pipe of Various Lengths

The method of determining flow, using the tables here presented, can best be illustrated by an example. It was established that a 24-gallon flow per minute would be required for an ordinary residence of five fixtures. Referring to the tables, it is seen that a 100-foot ¾-inch lead pipe under 60 pounds pressure at the main would supply 22 gallons per minute, which should be adequate to the demand.

Length of Pipe, Feet—⅜″ Galvanized Iron

(Actual Diameter 0.494″)

Pressure at the Main	Length of Pipe									
	20	40	60	80	100	120	140	160	180	200
	Gallons per Minute									
0	5	3	3	2	2	2	..	..	..	..
10	9	5	4	3	3	3	2	2	2	2
20	10	6	5	4	4	3	3	3	3	2
30	10	8	6	5	4	4	4	3	3	3
40	10	9	7	6	5	4	4	3	3	3
50	10	9	7	6	6	6	5	4	4	4
60	10	10	8	7	6	5	5	5	4	4
70	10	10	8	7	7	6	5	5	5	4

Length of Pipe, Feet—½-inch Galvanized Iron

(Actual Diameter 0.614″)

Pressure at the Main	Length of Pipe									
	20	40	60	80	100	120	140	160	180	200
	Gallons per Minute									
0	..	..	..	..	..	..	..	..	..	..
10	10	8	5	5	4	3	3	3	3	3
20	14	10	8	6	6	5	5	4	4	4
30	18	12	10	8	7	7	6	6	5	5
40	20	14	11	10	8	8	7	7	6	6
50	..	16	13	11	10	9	8	7	7	7
60	..	18	14	12	11	10	9	9	8	7
70	..	..	15	13	..	..	10	9	8	8
80	..	..	..	14	..	..	..	..	..	..

Length of Pipe, Feet—¾-inch Galvanized Iron

(Actual Diameter 0.82″)

Pressure at the Main	Length of Pipe									
	20	40	60	80	100	120	140	160	180	200
	Gallons per Minute									
10	22	14	12	10	8	8	..	6	6	6
20	30	22	18	14	12	12	..	10	10	8
30	38	26	22	18	16	14	..	12	12	10
40	..	30	24	21	19	17	..	16	15	13
50	..	34	28	24	21	19	..	17	16	15
60	..	38	31	26	23	21	..	19	18	17
70	..	..	34	29	25	23	..	21	19	18
80	..	..	36	30	27	24	..	22	21	20

Length of Pipe, Feet—1-inch Galvanized Iron

(Actual Diameter 1.035″)

Pressure at the Main	Length of Pipe									
	20	40	60	80	100	120	140	160	180	200
	Gallons per Minute									
10	40	28	22	18	16	15	14	13	12	11
20	55	40	32	27	24	22	20	19	18	16
30	70	50	40	34	30	27	25	23	22	20
40	80	58	45	40	35	32	29	27	25	24
50	..	65	57	45	40	36	33	31	29	27
60	..	70	58	50	44	40	36	34	32	30
70	..	76	63	54	47	43	40	37	34	32
80	..	65	57	50	45	42	39	37	35	33

Length of Pipe, Feet—1¼-inch Galvanized Iron

(Actual Diameter 1.359″)

Pressure at the Main	Length of Pipe									
	20	40	60	80	100	120	140	160	180	200
	Gallons per Minute									
10	80	55	45	37	35	30	27	25	26	24
20	110	80	65	55	50	45	41	38	36	34
30	..	100	80	70	60	56	51	47	45	42
40	..	..	95	80	72	65	60	56	52	50
50	..	..	107	92	82	74	68	63	60	55
60	..	..	..	102	90	81	75	70	65	62
70	..	..	..	..	97	88	82	74	69	67
80	..	..	..	..	105	95	87	79	74	72

Length of Pipe, Feet—1½-inch Galvanized Iron

(Actual Diameter 1.604″)

Pressure at the Main	Length of Pipe									
	20	40	60	80	100	120	140	160	180	200
	Gallons per Minute									
10	120	90	70	60	55	50	45	40	40	35
20	170	130	100	90	75	70	65	60	55	55
30	...	160	130	110	100	90	80	75	70	65
40	...	170	150	130	110	100	90	90	80	80
50	...	...	170	140	130	120	110	100	90	90
60	...	...	...	160	140	130	120	110	100	100
70	...	...	...	170	150	140	130	120	110	100
80	...	...	...	...	160	150	140	130	120	110

Length of Pipe, Feet—2-inch Galvanized Iron

(Actual Diameter 2.052″)

Pressure at the Main	Length of Pipe									
	20	40	60	80	100	120	140	160	180	200
	Gallons per Minute									
10	240	160	130	110	100	90	90	80	80	70
20	300	240	200	160	150	140	130	120	110	100
30	...	300	240	200	180	160	150	140	140	130
40	...	...	280	240	220	200	180	160	160	150
50	...	...	...	280	240	220	200	200	180	160
60	...	...	...	...	280	240	220	200	200	180
70	...	...	...	...	300	260	240	220	220	200
80	...	...	...	...	300	280	260	240	220	220

Length of Pipe, Feet—½-inch Copper or Lead Pipe

(Actual Diameter 0.527″)

Pressure at the Main	Length of Pipe									
	20	40	60	80	100	120	140	160	180	200
	Gallons per Minute									
10	8	5	4	3	3	2	2	2	2	2
20	12	8	6	5	5	4	4	3	3	3
30	15	10	8	7	6	5	5	4	4	4
40	17	12	9	8	7	6	6	5	5	4
50	..	14	10	9	8	7	6	6	5	5
60	..	15	12	10	9	8	7	7	6	6
70	..	15	13	11	10	9	8	7	7	6
80	..	..	14	12	10	10	8	8	7	7

Length of Pipe, Feet—⅝-inch Copper or Lead Pipe

(Actual Diameter 0.62″)

Pressure at the Main	Length of Pipe									
	20	40	60	80	100	120	140	160	180	200
	Gallons per Minute									
10	12	8	7	6	5	5	4	4	3	3
20	18	12	10	9	7	6	6	5	5	5
30	22	16	12	10	9	9	8	7	6	6
40	26	18	14	12	10	10	9	8	8	7
50	..	22	16	14	12	11	10	9	9	8
60	..	24	18	16	14	13	12	11	10	9
70	..	..	20	18	15	14	13	12	11	10
80	..	..	22	19	16	15	14	13	12	11

Length of Pipe, Feet—¾-inch Copper or Lead Pipe

(Actual Diameter 0.745″)

Pressure at the Main	Length of Pipe									
	20	40	60	80	100	120	140	160	180	200
	Gallons per Minute									
10	20	14	10	10	8	8	6	6	6	5
20	30	20	16	14	12	10	10	10	8	8
30	36	26	20	17	15	14	13	11	10	8
40	..	30	24	20	18	16	15	14	13	12
50	..	34	28	24	20	18	16	16	14	14
60	..	36	30	26	22	20	18	18	16	16
70	..	..	32	28	24	22	20	18	18	16
80	..	..	36	30	26	24	22	20	18	18

Length of Pipe, Feet—1-inch Copper or Lead Pipe

(Actual Diameter 0.995″)

Pressure at the Main	Length of Pipe									
	20	40	60	80	100	120	140	160	180	200
	Gallons per Minute									
10	50	30	24	20	18	16	14	14	12	12
20	70	45	36	30	26	24	22	20	18	18
30	80	55	45	38	34	30	28	26	24	22
40	..	65	55	45	40	36	32	30	28	26
50	..	75	60	50	45	40	36	34	32	30
60	..	80	65	55	50	45	40	38	36	34
70	..	..	70	60	55	50	45	40	38	36
80	..	..	80	65	60	50	50	45	40	40

Length of Pipe, Feet—1¼-inch Copper or Lead Pipe

(Actual Diameter 1.245″)

Pressure at the Main	Length of Pipe									
	20	40	60	80	100	120	140	160	180	200
	Gallons per Minute									
10	80	55	42	37	32	30	27	25	22	22
20	110	80	65	55	47	42	40	35	35	32
30	..	105	80	70	60	55	50	45	42	40
40	..	110	95	80	70	65	60	55	50	47
50	..	..	110	90	80	70	65	60	57	55
60	..	..	..	105	90	80	75	70	65	60
70	..	..	..	110	100	90	80	75	70	65
80	..	..	..	..	105	95	85	80	75	70

Length of Pipe, Feet—1½-inch Copper or Lead Pipe

(Actual Diameter 1.481")

Pressure at the Main	Length of Pipe									
	20	40	60	80	100	120	140	160	180	200
	Gallons per Minute									
10	130	90	70	60	50	45	40	40	35	35
20	170	130	100	90	75	70	65	60	55	50
30	...	170	130	110	100	90	80	75	70	65
40	...	...	155	130	115	105	95	88	80	77
50	...	...	170	150	130	120	108	100	90	88
60	...	...	...	165	145	130	120	110	105	98
70	...	...	...	175	160	142	130	122	113	106
80	...	...	...	...	170	155	140	130	122	115

Length of Pipe, Feet—2-inch Copper or Lead Pipe

(Actual Diameter 1.959")

Pressure at the Main	Length of Pipe									
	20	40	60	80	100	120	140	160	180	200
	Gallons per Minute									
10	280	180	150	145	110	100	90	85	80	70
20	...	280	220	190	165	160	140	125	120	110
30	...	...	280	240	210	180	170	160	150	140
40	...	...	...	280	240	220	200	190	175	160
50	...	...	...	320	280	250	230	210	200	190
60	...	...	...	...	300	280	260	240	220	200
70	...	...	...	...	...	300	280	260	240	230
80	...	...	...	...	...	...	...	300	260	240

INSTALLATION PROCEDURE

Before installing the water service it is necessary that city permits for the branch or tap into the city water main and for opening of a street be paid. Municipal employees usually make all taps and branches. It has been their experience that if this task is left to the plumber, very often the connection is not made properly. The tap is made with a special tapping device and is accomplished without shutting off the city water supply. Municipal water departments have also found it beneficial to supply the plumber with the brass goods of the water service, namely the corporation cock, curb cock, and the meter stop. This has a twofold purpose. First, it standardizes the brass material in the city street, and second, it assures administrative authorities a good quality of brass material. Each stop has a definite purpose and is of importance to efficient service.

The Corporation Stop. Fig. 195 shows a corporation stop which is inserted into the water main. The installation of this stop is required when the city main is tapped with the pressure on. The tapping device consists of a boring tool and tap and a check valve enclosed in a hollow casing. The machine is constructed in such a way that after the tap is made, the boring tool is replaced with the corporation cock. The machine is then sealed and the cock screwed into the main. Very little water is lost in this operation. Another important function of the corporation stop is that it serves as a control stop for the water service. It also serves as a shut-off for the service should the use of the building be discontinued.

Curb Stop. Fig. 196 represents a curb stop which generally is installed on the water service between the curb and the sidewalk line.

Fig. 195. Corporation
Stop

Fig. 196. Curb
Stop

It is accessible through a cast-iron stop box that is brought up to grade and equipped with a removable iron cover, as shown in Fig. 197. It has three purposes. First, it serves as a control stop for that portion of the service between the curb and the building; second, it serves as a shut-off for the building in the event that the basement becomes flooded; third, it serves as a control valve in case the building is not used in the winter time. The curb stop in this case can be closed and the water drained with safety.

The Meter Stop. The meter stop, Fig. 198, is placed on the water service on the street side of the house meter. It serves as a controlling stop for the building installation. Constant turning of the meter stop ruins it in a very short time. It is good practice and a law in some cities that the meter be equipped with a gate valve on the house side of the meter.

Care should be taken to protect the service where it passes through concrete floors or walls. This may be done with a metal sleeve or by wrapping the service with a material that offers adequate protection.

The Water Meter. The water meter, Fig. 199, is a device adopted by city authorities to measure in cubic feet or gallons the amount of water that passes through the water service. The meter is a very

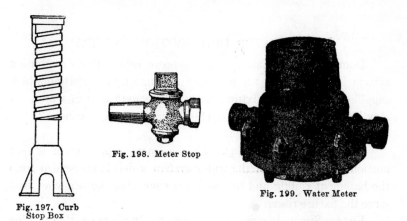

Fig. 197. Curb
Stop Box

Fig. 198. Meter Stop

Fig. 199. Water Meter

delicate instrument and should never be roughly treated by the plumber. The meter is generally read once every three months and the owner of the building is charged for the water consumption registered. Usually it is the property of the city and must never be removed unless permission is granted.

CHAPTER XIX

COLD WATER DISTRIBUTION SYSTEM

Domestic Cold Water. The domestic cold water distribution system is that part of the piping of a plumbing installation which supplies the plumbing fixtures with an adequate flow of water. It may be divided into two units: (1) basement supply mains; (2) fixture supply risers.

Basement Supply Mains. The basement supply main is that portion of the piping of the water system which is suspended from the basement ceiling and to which are connected the branches that serve the fixture risers.

Fixture Supply Risers. The fixture supply risers are that portion of the piping of the water system which furnishes the fixtures installed on the various floors with a flow of cold water.

Careful installation of the water distribution system is essential. Adequate size, sufficient grade, proper alignment and carefully installed connections are of the utmost importance.

There are mechanics who consider the water system of secondary importance. This is probably due to the fact that water under pressure will flow through a pipe even though it is installed carelessly. Workmanship of this kind on the part of the mechanic indicates poor judgment and lack of trade knowledge, as the flow of water is materially lessened when a condition of this kind prevails.

The resistance offered by the interior walls of the pipe as the water flows through them creates friction. Short offsets, long runs of pipe, and the use of unnecessary fittings, are responsible in part for increased friction.

Poor workmanship might also result in damage to property. Leaking joints, trapped piping, and water hammer are the defects in a water system that are the result of work that is done in a careless or indifferent manner. Installations having these defects are unsatisfactory and may render the fixture unsanitary.

Alignment and Grade of Water Pipe. Care should be taken

when installing basement supply mains that the runs of piping are neatly grouped and aligned. The exact spacing between pipes can only be determined by the plumber, as each job involves its own particular needs. However, for a neat appearing installation, the same spacing should be maintained throughout.

The plumber must be systematic in connecting the riser branches to the main. All connections to the branch should be made at right angles to the main, while those taken from the main employ 45-degree fittings, as shown in Fig. 200.

Basement supply mains must be graded carefully. It is advisable to pitch the water lines to one low point, and to equip them with a drain valve to make it possible to drain the entire water system

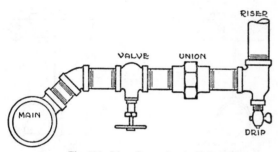

Fig. 200. Riser Connection to Main

quickly. This practice eliminates extra time and trouble and overcomes the danger of property damage and the expense of making repairs within the building should it have runs of trapped piping. If it is impossible to grade the piping to one point, all portions of the system that cannot be centrally drained should be provided with separate drain cocks.

Water piping does not require the amount of grade that waste lines do. Just off level, or up to ¼-inch pitch per foot, is sufficient to produce a well working installation.

Supports and Hangers. The water distribution system may be suspended from the basement ceiling in various ways. In large concrete buildings iron inserts are placed on the rough floor sheathing and slushed into the concrete slab. This method of pipe suspension, Fig. 201, is very satisfactory.

Hangers can also be fastened to concrete ceilings and brick walls where no provision has been made for inserts. This is accomplished by drilling a hole and inserting an expanding shield and lag bolt. The practice of using wooden plugs or similar material for fastening pipe hangers is impractical and should be avoided.

In buildings of wood construction the hangers can be screwed or spiked to the wood joists of the building. Eye lag bolts may also be used for this purpose, Fig. 202.

The fixture supply riser, too, must be well supported on each floor. The device generally used for this purpose is known as a pipe rest or clamp. It consists of two pieces of heavy band iron made to fit tightly around the vertical pipe. The two halves are drawn together by bolts that pass through drilled holes on each side of the

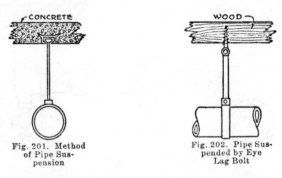

Fig. 201. Method of Pipe Suspension

Fig. 202. Pipe Suspended by Eye Lag Bolt

clamp. A vertical riser should never depend on support of its horizontal branches. Installations of this kind usually result in a breaking off of the branch at its connection into the riser. All horizontal runs of the fixture riser must be well suspended and given even grade to the fixture riser.

Use of Valves on Water Systems. The water system must be equipped with an adequate number of valves, placed at locations where they are of most use. Very often mechanics feel that the installation of valves is an unnecessary expense. This is not true, and a system which has an insufficient number of valves is usually inefficient.

Plumbing installations are subject to breakdowns because of their numerous working parts. The existence of a valve where breakdown has occurred makes it possible to control the water supply and avert serious damage. Valves also provide water control for portions of the installation and, in this way, localize the inconvenience caused by repair.

Gate Valve. The gate valve (Fig. 203) is one of the most common valves found on a water distribution system. It takes its name from the gate-like disc that moves across the path of flow. In such a valve, the flow takes a straight line as, with the valve open, the seat openings are practically the same size as the inside diameter of the pipe.

Gate valves are best suited to main supply lines and pump lines, for which operation is infrequent. Where the valves are operated frequently, the disc is held in a fully open or fully closed position, and there is no throttling to interrupt flow. It is not good practice to use a gate valve as a control valve on a lawn sprinkler, as eventually it will fail to close tight, thus leaving a wet area around the lowest sprinkler. For a lawn sprinkler system, a fully opening disc seat is recommended.

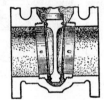 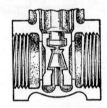

Fig. 203. *Left*, Wedge-Shaped Disc for Gate Valve. *Right*, Parallel Faces of Double Disc Are Forced Apart by Disc Spreader

These valves have two types of discs (see Fig. 203), of which the most commonly used is the *wedge-shaped* or *tapered disc*, illustrated at the left. This type of disc is recommended where the stem must be installed pointing downward. The other type, which is the *double disc* valve (*right* view) closes in the same manner as the wedge type except that its parallel faces drop in a vertical position and are forced apart by the disc spreader. Double disc valves are used chiefly in cold liquid and sewage disposal installations.

Globe Valves. These valves, which are actuated by a stem screw and handwheel, are recommended for installations that call for throttling. Because of the fact that the direction of flow is changed, the globe valve affords greater resistance to flow than does the gate valve. Moreover, the disc and seat in a globe valve can be reground easily, thus solving to great extent the problem of maintenance.

There are three types of globe valves, as shown in Fig. 204, the foremost of these being the *plug type* disc valve, shown at the *left*. This valve represents the latest design in valve engineering because of its long, tapered disc and matching seat. Its wide bearing surfaces give it good resistance to the cutting effect of scale, dirt, and other kinds of foreign matter commonly found in pipes. For this reason it satisfies the requirements of the most demanding flow control service.

The best known of the three types of globe valves illustrated is the *conventional* disc valve (*center*), a chief advantage of which is the ease with which it permits a pressure-tight bearing to be obtained between the disc and the seat. This valve is made in several seating styles of varying degrees of taper. It is recommended for cold and any temperature services.

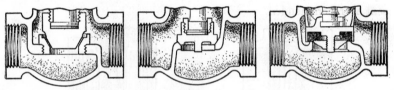

Fig. 204. Globe Valves Are of Three Types, i.e.: Plug Type (*left*), Conventional Type (*center*), Composition Type (*right*)

The third type of globe valve is the *composition* disc valve (*right*), which acts as a cap and consists of a metal disc holder, composition disc, and retainer nut. Various types of discs may be used to form its assembly, which feature allows this valve to be used for different services—oil, gasoline, steam, hot and cold water, and so on. This adaptability, together with the fact that the disc can be turned over or removed without removing the valve, accounts in good part for its popularity.

Angle Valves. Valves of this classification operate in the manner of the globe valve and are available in a similar range of disc and seat designs. They are used in making a 90-degree turn in a line, thus reducing the number of joints and saving much make-up time.

Check Valves. The function of check valves is to check or prevent reversal of flow in the line. They are used principally in industrial piping. In installing check valves it is important to make sure that flow is directed to the proper end.

Check valves are of two basic types, namely, the *swing-check* valve (*left*) and the *lift-check* valve (*right*), Fig. 205. When a swing-check valve is used, the flow moves through the body of the valve in a straight line. Somewhat similar in design to a gate valve, their low resistance to flow makes them suitable for use in control of low-to-moderate pressures of liquids and gases.

When a lift-check valve is used, the flow takes a winding course through a horizontal bridge wall on which the disc is seated. Valves of this type are used in conjunction with globe valves and are recommended for gas, water, steam, air, and general vapor services.

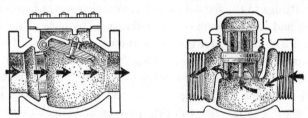

Fig. 205. *Left*, Swing-Check Valve; *right*, Lift-Check Valve

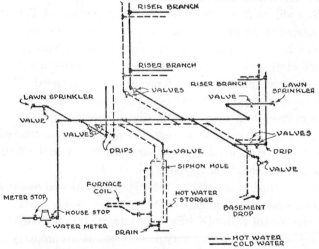

Fig. 206. Water Distribution System for Small Residence

Location of Drip Cocks. Fig. 206 presents a water distribution system for a small building. It may be taken as a model for one- or two-family residences provided the pressure at the city main is not less than 40 pounds. Flow sufficiently ample to provide for domestic and sanitary needs is assured occupants of buildings using a distribution system such as this.

The house service is constructed of ¾-inch diameter pipe with the usual stops. A 1-inch tap in the main is recommended. The water meter is of ¾-inch pipe size.

From the water meter a ¾-inch line is provided with a gate valve and drip and extends full size to the vertical hot-water storage tank in as direct a manner as possible without crossing diagonally over the basement ceiling. From the branch connection of the storage tank the main continues in a straight line to the rear of the basement, where a sill cock and sprinkler connection are installed.

A ¾-inch branch is taken from the main at right angles, using proper fittings to the fixture supply riser which serves bathrooms on the first and second floors of the building. The base of the riser is provided with a valve for control purposes.

Another ¾-inch branch is taken from the main at right angles to serve the laundry tub and kitchen sink. The sink riser is extended with ¾-inch pipe to the upper floors and is equipped with a valve at its base. The laundry tub drip may be of ½-inch diameter. A valve must also be used on this installation.

The sill cock which serves the front portion of the premises is connected to the main somewhere near the house meter.

The hot water portion of the distribution system is also constructed of ¾-inch pipe and is aligned with the cold-water piping. A spread between pipes of about 6 inches is recommended. The risers of the hot-water system are of ¾-inch pipe and must be provided with valves aligned with those used in the cold-water supply.

The distribution system may be suspended from the ceiling of the basement and should never be strapped lightly to the floor joists. A space of 6 inches is advisable.

The method of supporting the installation as well as the storage tank connections in the heating device is discussed in other chapters.

Insulation of Water Distribution Systems. Insulation of the water distribution system is a necessary precaution against property damage and inconvenience caused by condensation. Condensation occurs because the water within the system usually is at a low temperature and the warmer atmosphere which comes in contact with the pipe tends to release water held in suspension by the pipe. The nuisance can be overcome by covering the cold-water piping with approved antisweat covering. This material consists of layers of hair felt and tarred paper. The fittings of the distribution system may be wrapped with hair felt and an applied coat of asbestos. Cheesecloth is used to complete the job.

The hot-water piping should also be covered if for no other reason than that of economy. The heat loss which occurs when hot-water pipe is exposed to the atmosphere is tremendous, and over a period of time it may be costly. The covering used for this purpose should be of the air cell variety made of asbestos paper formed in such a manner that the pipe is encircled with a jacket filled with air. Air is a very poor conductor of heat, hence it becomes an efficient insulator.

CHAPTER XX

PUMPS AND LIFTS

The installation of a pump on a water supply system is very common, because the public water distribution systems of large cities usually carry a maximum average pressure of 50 pounds to the square inch, which is only sufficient to serve buildings less than five or six stories in height. Many of the buildings in central business areas are of greater height and necessitate the installation of additional pumping equipment.

The specifications of this kind of installation usually are prepared by technical engineering authorities who compute the probable demand as well as the pressure loss because of head and friction. This data is submitted to pump equipment manufacturers, who are familiar with the specifications of their product, and the technicians employed by them then fit a pump into the water supply system that will fulfill the requirements demanded of it.

This procedure does not give the plumber much opportunity to become familiar with the pump mechanism. Often he has but a vague idea of its operating principle. He must, however, make repairs on the equipment and he participates in its installation in the distribution system. It is advisable therefore that he have at least basic knowledge of the construction of a pump as well as of the physics principle involved in operation.

TYPES OF PUMPS

There are two kinds of pumps in common use, namely, the piston pump and the centrifugal pump. The piston pump is used frequently on the smaller water distribution systems for the elevation of water from private water wells or other sources. The centrifugal unit usually is associated with distribution systems of tall buildings.

Regardless of the design of the equipment, the scientific principle involved in the elevation of water remains the same. The pump is a simple mechanism that produces a vacuum within itself and permits

the lifting of water from the source of supply by atmospheric **pressure.**

If it were possible to produce a perfect vacuum on the **suction** side of the pump it would lift water to a height (theoretically) of approximately 34 feet. A perfect vacuum in the suction pipe, however, is impossible. Because of piston slippage, valve leaks, and resistance caused by vaporization and suction, the lift of a common pump is never greater than 27 feet

Water can be delivered to a much greater height with the use of a pump because of its reciprocating action. This is made possible by valves that are synchronized with the stroke of the pump's piston.

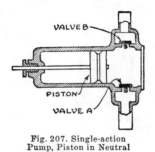

Fig. 207. Single-action
Pump, Piston in Neutral
Position

There are three types of piston pumps:
1. Single-action pump
2. Double-action pump
3. Duplex or twin piston pump

Single=action Pump. The working principle of a single-action pump is very simple. It consists of a cylinder constructed of cast iron, into which a piston is fitted accurately. The piston is made of cast iron or bronze and is equipped in most installations with leather cup packings, Fig. 207.

The suction side of the pump is equipped with a valve, *A*, which opens when the piston is drawn in the cylinder. Moving the piston in the cylinder removes atmospheric pressure and creates a partial vacuum. The atmospheric pressure on the surface of the water forces the water into the suction pipe, Fig. 208. The valve opens **and** the cylinder is filled with water.

The delivery side of the pump is also equipped with a valve, *B*.

After the suction or draw stroke of the piston has been completed and the cylinder is full of water, the return or power stroke drives the piston against the confined water. The suction valve closes and the confined water forces open the valve on the delivery side of the pump, allowing the water to pass through it, Fig. 209 When all the

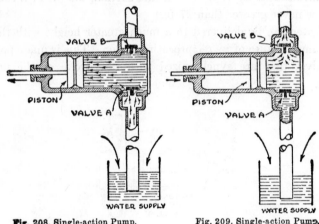

Fig. 208. Single-action Pump, Suction Stroke

Fig. 209. Single-action Pump, Delivery Stroke

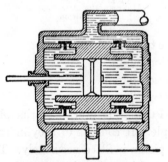

Fig. 210. Double-action Pump, Neutral Position

water has been expelled from the cylinder the suction stroke begins, again filling the cylinder.

Double=action Piston Pump. The double-action type of pump is similar to the single-action type in its operating principle. The difference is that the double-action pump contains four valves instead of two and delivers water on both the suction and delivery stroke of the piston, Fig. 210.

When the piston of a double-acting pump is put in motion, the valve on the suction side of the piston is opened. At the same time, the valve on the delivery side of the piston opens, thus allowing the confined water to be discharged from the cylinder, Fig. 211. Reversing the piston movement, the suction valve on the opposite side of the piston opens and allows water to re-enter the discharge side of the

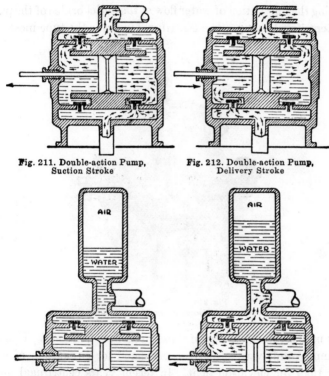

Fig. 211. Double-action Pump,
Suction Stroke

Fig. 212. Double-action Pump,
Delivery Stroke

Fig. 213. Air Chamber for Eliminating Pulsating Action of Pump

cylinder. Water confined on the opposite side of the piston is forced from the cylinder by the piston action, Fig. 212. Simultaneous opening and closing of the pump valves is effected by the action of springs, gears, or cam mechanisms.

Duplex Piston Pump. Duplex pumps deliver water on both strokes of two pistons operating in twin cylinders. Pumps of this variety usually contain eight valves, which are synchronized with the piston stroke.

Air Chamber. Fig. 213 represents an air chamber installed on piston pumps to minimize the pulsating discharge from the pump. Pulsation results in an uneven flow of water.

An air chamber is a vessel provided with an opening on its underside. It contains air under atmospheric pressure. When water is forced into it the pressure is increased in direct proportion to the amount of water admitted. The air in this case acts as a cushion, lessening the fluctuation of water flow at the outlet orifice of the pump.

Centrifugal Pump. The centrifugal pump is the type most com-

Fig. 214. Centrifugal Pump
Courtesy of Chicago Pump Company, Chicago

monly used for the elevation of water in modern buildings. This type of pumping unit was formerly constructed for circulation of water rather than for elevation of it. In recent years centrifugal pumps have been developed which will elevate water to great heights. The advantage of a centrifugal pump lies in the fact that its motive power can be supplied easily.

Construction of a Centrifugal Pump. A centrifugal pump consists of a housing made of cast iron. The inside of it is accurately machined, forming a chamber into which is built an impeller or water wheel, Fig. 214.

The impeller of the pump usually is built of brass or bronze to prevent corrosion. The impeller in some instances consists of paddles joined to a central shaft. The more efficient impellers are built of a solid piece of metal containing a number of small waterways or pockets. These cup-like recesses pick up the water on the inlet side of the pump and discharge it on the delivery side.

A pump of the centrifugal type permits a great amount of water to slip past the impeller. This cuts down its efficiency to some

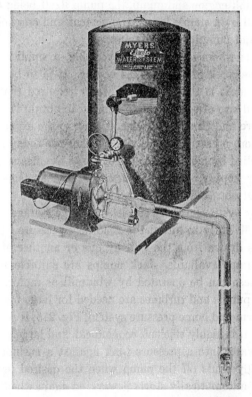

Fig. 215. Farm Home and Pressure Rural Water System
(HN Ejecto Shallow Well Pump)
Courtesy of F. E. Myers & Bro. Co., Ashland, Ohio

extent but is not particularly detrimental. The discharge capacity of a centrifugal pump is practically unlimited.

RURAL WATER SYSTEMS

Air Power. In rural areas, and on many farms, the old reliable windmill is still used to supply the power necessary to pump the water required for domestic use, and for stock needs. In the past, when prevailing winds failed, the farmer was compelled to resort to laborious hand pumping to obtain water. To his aid came the gasoline engine, which made it possible to operate the windmill under unfavorable wind conditions, a jack pump and gas-driven engine constituting the means of propulsion. If the winds failed, a belt could be used to link engine and pump, the pump itself being connected to the windmill by a simple lever arrangement and one bolt. Thus the need for hand pumping was eliminated.

In areas where electricity is available, a small electric motor performs the work of the gasoline engine. It is to be remembered, however, that it is the amount of water required that determines the type and size of equipment to be used to obtain it.

Pressure and Gravity Systems. Owing to the expansion of rural electrification, pressure and gravity systems are today replacing the power supplied by moving currents of air as the means of furnishing rural water supply. The efficiency of these systems rates high, whereas that obtained from air has a low rating.

Gravity systems usually consist of a large storage tank which is elevated sufficiently to afford the necessary pressure. The water to be stored is drawn from the well by one or another of the various types of pumps available. Jack pumps are suitable for the small farm, as these can be operated by windmill or motor. Centrifugal and rotary pumps and turbines are needed for large installations.

The farm and home pressure system, Fig. 215, is popular for the reason that it is highly efficient, economical, and largely trouble-free. It pumps water into a pressure tank against a cushion of air that automatically shuts off the pump when the desired pressure is attained, and automatically starts it working again when the pressure drops to the minimum needed for operation. For related material, see Rural Air Power Systems, Chapter XXI.

Submersible Pumps. The submersible pump is an efficient user of its motor energy because the pump and motor are installed in the well below the water. This type of installation also solves freezing, priming, and lubrication problems. Submersible pumps are utilized primarily for deep wells, but also find application in shallow wells where an extra large volume of water is desired.

COLD-WATER DISTRIBUTION SYSTEMS
IN TALL BUILDINGS

In tall buildings which cannot adequately be served by the normal water pressure of the city mains, mechanical means for water distribution are employed. There are two types of water distribution systems which may be used to serve this purpose, namely, the air pressure system and the overhead feed system.

AIR PRESSURE SYSTEM

The air (pneumatic) pressure system is an assembly of mechanical devices which raise water, using compressed air as the delivery agent. This method of water delivery is rapidly superseding other installations because it offers many advantages. The pumping unit is compact and may be installed in a limited amount of space. A small basement room usually is adequate for this purpose. The water is confined in air-tight equipment, which makes the installation a sanitary one—it is impossible for dust and other objectionable materials to come in contact with the water. In a slight way the oxygen contained in the compressed air, which passes through the water, serves as a purifying agent and tends to make water palatable.

The air pressure system offers economic advantage in that the pipe diameters may be smaller than those of other installations, and the cost of the original equipment is materially less.

The working parts of an air pressure system are few and operate efficiently. Only occasional maintenance, such as lubrication or minor repair, is required.

This method of water distribution may be adapted to a small rural residence as well as to the tallest urban building. In buildings of extreme height, air pressure systems serve zones of about ten floors. A thirty-story building may be provided with three distinct units, each providing water supply for a ten-floor interval.

Principle of Operation. The scientific principle on which a pneumatic pressure system operates is the fact that 1 pound of air pressure elevates water approximately 2.31 feet under atmospheric conditions of about 14.72 pounds.

The mechanical devices which are used in this method of water supply and the pipe connection of the separate units are as follows:

Assembly of Mechanical Devices. There are four mechanical devices used on an air pressure system:

1. A large storage tank, with all tappings located on the under-side.

2. A single or duplex centrifugal pump.

3. An air compressor.

4. An automatic pressure control switch.

Note: On small units a piston pump generally is used.

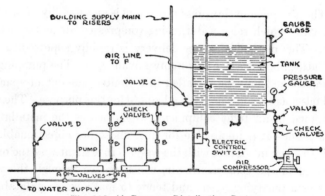

Fig. 216. Air-Pressure Distribution System

Figure 216 illustrates a complete pneumatic pressure distribution system. The centrifugal pump is set on a concrete footing. It must be insulated against noises produced by vibration. The inlet side of the pump is connected to the water supply main and is equipped with a valve, *A*. The outlet or delivery side of the pump is connected to an opening in the tank and is equipped with a valve. Two check valves, *B*, placed between the pump and tank, prevent the increased pressure within the system from exerting itself on the supply main. Two pumping units are used on most installations, especially those for large buildings, to insure a supply of water should one pump fail.

The building supply main is connected to the delivery side of the pump, between the check valves and the tank. A valve (*C*) may be placed on this connection.

It is advisable to install a valved by-pass between the city main and the building supply main. In case of pump failure, this valve (*D*) can be opened and a part of the building will be supplied with water.

The air compressor (*E*) is also connected to the underside of the tank. This line must be equipped with a valve and a check valve.

The electric control switch is located conveniently, and is connected to the tank with a pipe of small diameter. The varying pressures within the tank affect the mechanism of the control devices

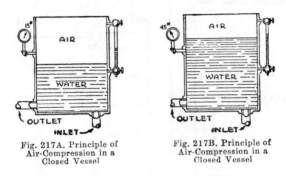

Fig. 217A. Principle of
Air-Compression in a
Closed Vessel

Fig. 217B. Principle of
Air-Compression in a
Closed Vessel

which start and stop the pumps, *F*. It is absolutely essential that all connections be made on the underside of the storage tank.

Operation of an Air Pressure System. The operating principle of an air pressure system is not difficult to understand. It is based on known facts of physics. The tank, being air-tight, serves as an air chamber. When water is forced into it the pressure of air is increased in proportion to the volume of water. Air is elastic or compressible, and water is inelastic or noncompressible. As the air is squeezed into less space, a pressure greater than that of the atmosphere is exerted on the surface of the water. *Example:* The tank is originally under atmospheric pressure of approximately 14.72 pounds at sea level. Atmospheric pressure under these conditions will not act as an elevating agent. This is because the same pressure exists on all sides of the confined liquid. A gauge installed on the vessel will indicate zero pounds pressure.

When water is forced into a closed vessel that is under atmospheric pressure to the extent of one half its cubical content, a gauge installed on the tank will indicate about 15 pounds. This represents 15 pounds of pressure above atmospheric, Fig. 217A, and it will **elevate water, theoretically, 34.6 feet.** Should more water be pumped into the tank to the extent of two-thirds its content, a pressure of approximately 30 pounds will be indicated on the gauge. If water is pumped into the tank to three quarters of its content, ap-

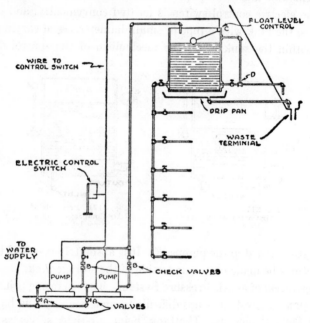

Fig. 218. Overhead Distribution System

proximately 45 pounds will be indicated on the gauge, Fig. 217B. Theoretically, this will elevate the water 103.9 feet.

Compressing only the original pressure contained in the tank is insufficient. The slightest draw on the distribution system will spend the increased pressure very quickly. There is also the possibility of the tank becoming water-logged. The air compressor is the unit installed to overcome water distribution difficulties. It usually is controlled manually.

When the system is put into operation the volume of air in the tank is increased by the air compressor to a pressure of one-half the minimum required. Admitting water to the tank increases the pressure proportionately by displacing a portion of the tank's air content. This permits a more extended draw on the distribution system.

A pressure range above the minimum pressure required is necessary. The pressure range usually is from 20 to 40 pounds. The air pressure switch governs this phase by starting and stopping the pumps. For example, suppose a pressure of 70 pounds is required to supply the water needs of a building. The centrifugal pump would operate until a predetermined maximum pressure of 100 pounds was attained. The pressure range would then be 30 pounds. This would permit the system to deliver water until the 30 pounds of pressure was spent. The installation would not require further use of the pumps until the system was again under the minimum pressure of 70 pounds.

The air compressor need be operated only when a part of the air contained in the tank has been eliminated.

Distribution Piping. The design of the distribution piping for the air pressure system of water supply is identical with that described in the preceding chapter.

OVERHEAD FEED SYSTEM

An overhead system of water supply is an installation which distributes water to the plumbing fixtures by gravity, Fig. 218. It is probably the oldest method of water distribution and dates back to the early Roman aqueducts. It is commonly used in municipalities where the source of water is taken from a drilled well, because a pneumatic pressure system for municipal use would be impractical.

Overhead feed systems of water distribution were the first practical installations used in buildings of extreme height. Although the overhead supply is becoming obsolete, many architects and building designers still favor it. Many large buildings are supplied with water by this method.

There are several disadvantages in an overhead system of water supply. The stored water is subject to contamination because the tank is of the open type. The installation requires more material and labor, thus increasing the cost, and the equipment is distributed over the entire building, utilizing valuable space.

The mechanical devices which make up an overhead feed system of water supply are as follows:

1. An open storage tank equipped with an overflow pipe and a condensation pan.

2. A twin set of centrifugal or piston pumps.

3. A float and electric control switch.

The pump usually is located in the basement of the building and is mounted on a well-insulated and substantial footing. The inlet side of the pump is connected to the city water supply and is equipped with a gate valve, *A*.

The outlet side of the pump is connected to the overhead storage tank just below its overflow rim and must be provided with two check valves and a gate valve, installed as close to the pump as is practical, *B*. To provide for shut off in an emergency, a valve should be placed close to the tank as well. The tank is placed above the highest installed fixture which requires water, usually the roof penthouse.

The supply of water to the tank is controlled by an electric float switch installed within the tank close to its overflow rim. When the water in the tank recedes and rises to established levels, the float switch starts and stops the electrically driven pumps, *C*.

Whenever a water system depends on a mechanical device, such as a float switch, the installation must be protected. Should the control device become defective, overflow of the tank would occur and damage would result. This possibility can be averted by connecting an overflow pipe of adequate size to the tank at a point below the top rim. The overflow should be drained onto the roof of the building and be run just as direct as possible. If this cannot be done, it may be connected indirectly to the storm sewer. Should the building be without a storm sewer, the overflow may terminate indirectly into the sanitary sewer.

The condensation which occurs on the outside surface of the tank is dripped into a condensation pan. The pan is provided with a drain pipe, which discharges separately onto the roof. Should the tank overflow pipe be taxed to its capacity, backflow into the pan would cause flooding of the building, therefore, cross connection of the two wastes is inadvisable. The indirect connection of both drains serves to prevent contaminated wastes contained in the storm or drainage system from backing up into the water tank.

The fixture riser supply main should be connected to an opening near the bottom of the tank. The main usually is suspended from the steel framework between the ceiling of the top floor and the roof of the building. The main is pitched from the tank to the risers and should be well suspended to avoid sagging of pipes. The risers are connected to the main at right angles. Inverted 45-degree connections are advisable. Each riser must possess a valve located close to the main.

In most instances the risers of the low pressure system (from the city service) and the high pressure system (from the overhead tank) are cross-connected. They are equipped with an accessible valve installed at the top floor that is fed by the low pressure system. This valve usually is allowed to remain closed.

There are a number of advantages contained in the cross-connection installation. High pressures on the lower floors are eliminated, and, should the pump unit fail, at least the lower floors of the building may be served by city pressure. Failure in city pressure would be overcome by using the pump unit as an auxiliary unit to provide water for the fixtures in the low pressure zone.

RURAL AIR POWER SYSTEMS

A supply of water may be furnished farm buildings by means of air power. The elevation of water by this method is somewhat obsolete, however. This is largely due to the perfection of electric pumping equipment.

The air system still has a place in rural sanitation. There are advantages to be gained by its use. Water may be taken from sources supplying both pure and impure water with the same pump unit without possibility of rendering the pure water unsafe. This type of system is economical in cases where water is taken from different sources. It is also considered sanitary due to the fact that it is a closed system.

The power unit consists of an iron tank of adequate size, an air compressor, an electric motor or gas engine, and a well cylinder or pump. Fig. 218 illustrates the method of connecting the equipment and the delivery piping.

The air compressor and tank are connected with a ¾-inch pipe. The pipe line is equipped with a check and a control valve, A, and enters the tank at a side tapping.

The compressor is mounted on a substantial and well-insulated base. Insulation of the base is necessary to eliminate noise caused by vibration. It is driven by electricity or gasoline.

The cylinder, B, located in the source of supply, and the tank are connected by a pipe usually ½- or ¾-inch in diameter. This line is equipped with two pressure gauges, one on each side of a pressure reducing valve, C.

Control of the compressor is governed automatically by the variating pressure in the tank, D.

Operating Principle of the Air Power System. The iron tank is

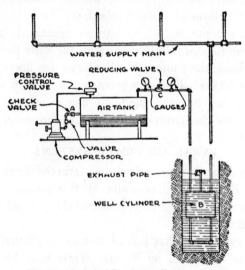

Fig. 218. Method of Connecting Equipment on
Air Power System

filled with compressed air. The compressor can be governed to build up any pressure range in the tank that may be desired. This pressure depends entirely on the height to which the water must be elevated.

For example, should 40 pounds of air pressure be required to elevate the water to the highest fixture in the building, the air in the tank could be compressed to 100 pounds. This would permit a reserve supply of 60 pounds. The air compressor would not start to operate until this surplus was used.

The pressure on the well cylinder is controlled by the hand-operated reducing valve installed on the air line. To make it possible for the operator to control the pressure in both units, a gauge must be installed on each side of the reducing valve. The air line is then run to the well cylinder.

A water line is brought back from the cylinder and connected with a valve in the well pit so that the cylinder can be removed. This method of water supply is automatic in operation and requires very little attention.

Construction and Working Principle of the Well Cylinder. The well cylinder is the most complicated of the entire air pressure system. It is constructed of a hollow iron body and a series of air and water valves operated by floats. It is submerged into the source of supply. Water enters the empty cylinder through a valve in the bottom. As the water rises in the cylinder it lifts the lower float, which action closes a valve within the cylinder. This furnishes water to a diaphragm operating the air-exhaust valve. The water continues to rise, as does the top float, which opens the air supply valve and closes the exhaust. As air enters the cylinder the confined water is discharged. After the water is discharged the floats drop back into place, closing the valves and allowing the cylinder to receive water. This action is performed with such speed that the pulsations of the cylinder valves are not noticed to any great extent.

THE UNCLUTTERED MODERN KITCHEN HAS A LIVING AREA AND A VIEW

Courtesy of Crane Company, Chicago

CHAPTER XXII

DOMESTIC HOT=WATER SUPPLY

The domestic hot-water system is that portion of the plumbing installation which produces and conducts heated water to the discharge fixture.

The writer recalls the time that heated water delivered to plumbing fixtures was a luxury and very few people even among those moderately well to do were provided with this convenience. Sanitation standards of the present day, however, require an installation of hot-water supply and, today, almost the humblest dwelling is provided with this convenience.

Domestic hot-water installations have given the plumber much difficulty and trouble. Many installations of the past and those now being constructed are giving only partially satisfactory results. This condition is due largely to a lack of knowledge on the mechanic's part of the hypothesis and physical principles involved.

Water, like gases and solids, is composed of small individual parts scientifically referred to as molecules. Each molecule of water consists of two atoms of hydrogen and one atom of oxygen in chemical union in definite weight proportion. The molecules are separate particles and, as such, have distinct properties.

When heat is applied to water, the molecular activity is intensified. Each particle expands in itself and tends to move in a direction opposite to the others. Heat convected into the water increases its volume, because of the expansion of each molecule. A cubic foot of water at 39.1°F. has a specific weight of 62.425. At 62°F. it has a weight of 62.36.

When water reaches the boiling point (212°F.) under atmospheric pressure, it changes character. The molecules have expanded to such an extent that they become lighter than atmospheric pressure. Disintegration occurs, and vapor, referred to as steam, rises from the surface of the water. As the temperature decreases, the molecular activity lessens. Thus the water contracts, attaining its normal density at 39.1 degrees. The boiling point of water is increased tremendously under pressure greater than atmospheric, which fact makes protection of the hot-water distribution system against explosion necessary.

The movement of water in a distribution system is the result of expansion and contraction of the molecules contained in the water. The molecular theory can be applied in its entirety to hot-water service.

STORAGE TANK AND WATER HEATING DEVICES

The hot-water tank serves the domestic hot-water system in a storage capacity, and care in planning is essential in making the pipe

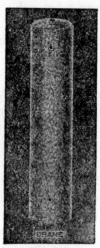

Fig. 219. Vertical
Hot-Water Tank

connections between it and the heating device. Much of the difficulty encountered in domestic hot-water service originates in this unit and, unless the piping arrangement for circulation of the heated water is properly installed, the efficiency of the entire hot-water system is affected.

Construction of the Hot=Water Tank. There are two types of tanks used for the storage of hot water: (1) The small hot-water tank (range boiler); (2) the large hot-water tank (storage tank).

The range boiler generally is constructed of galvanized sheet steel or copper, Fig. 219. It is built into a cylindrical shape having concave ends, and all of the seams are welded or riveted to assure strength. They are made in standard and extra heavy weights and are conveniently tapped for the heater connections. The range boiler varies in size from 12 to 24 inches in diameter and is not more than 6 feet long. It may be used in either a vertical or horizontal position.

Fig. 220 illustrates a storage tank which is constructed of heavy gauge sheet metal and generally is coated with a rust-proof paint instead of being galvanized as in the case of the range boiler. These tanks range in size from 24 to 54 inches in diameter and are not more than 15 feet long. They can be tapped for horizontal or vertical installations and are used on larger installations where hot water needs are greater.

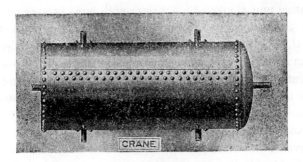

Fig. 220. Horizontal Storage Tank

The installation of tanks of standard gauge metal should be limited to working pressures of 85 pounds or less. For pressures over 85 pounds, the extra heavy tank should be used.

Sizing the Hot=Water Tank. To determine the proper size of a hot-water storage tank for a particular installation, three things must be considered. These are: (1) the design of the building; (2) the number of occupants; (3) heating capacity of the supply device.

Practical consumption tests have given an accurate estimate of the amount of water used per individual. It is safe to say that hot water consumption per person ranges from 2 to 10 gallons per hour. This depends entirely upon the type of building in which the system has been installed and the purpose for which the water is used.

A safe estimate of water consumption to allow per person according to the type of building is as follows:

	Gal. per Person per Hour
School buildings	2 to 3
Office buildings	4 to 5
Apartment buildings	8
Hotels	8 to 10
Factories	4 to 6
Residences	10

These figures are based on water consumption for hygienic and average uses only. In buildings used as creameries, laundries, canneries, etc., the size of the hot-water tank should be based on the water consumption of the various industrial devices together with the rated consumption per individual of the building occupants.

The size and design of the storage tank is augmented by the type of heating unit employed. A heater with large heating surface will reheat the water drawn from the storage tank in a shorter time interval, and, in this case, a smaller storage tank may suffice. However, in arriving at definite conclusions as to size of the storage unit and heating device practical experience and sound judgment prevail.

Determining the Working Load of a Hot=Water System. The working load of a building is a factor difficult to estimate. Experience has proved, however, that buildings of the school, office, or industrial type average about 25 per cent of the rated maximum consumption per individual. Apartments, residences, and similar buildings, average about 35 per cent of a working load.

Hotels average about 50 per cent of rated individual consumption as a working load for the domestic system of hot water.

For example, a hotel having 100 rooms averaging two persons to the room, would require a storage tank according to rated maximum individual consumption of about 2,000 gallons. Experience has shown that 50 per cent of the rated daily consumption per individual is a safe working load. Therefore 50 per cent of the 2,000 gallons would give the size of tank required for the installation. To this must be added any hot water used for purposes other than hygienic use.

Other buildings are governed in like manner, using the per cent of working load given above.

It is reasonable to assume that when 75 per cent of the water has been drawn from the storage tank, the remaining content will be cooled. The size of the heater may be based on this factor and must be of ample capacity to replace the drawn off water in a reasonable time interval.

For example, should the storage tank installed in a large building be required to serve an installation with 500 gallons of water in any one hour of a day, the heater would be required to replace this quantity in 1 hour. As a rule manufacturers' specifications indicate the efficiency of a heater, thus simplifying selection of a heater of ample capacity for the demand.

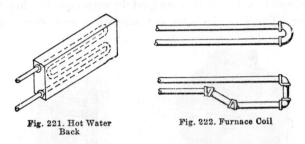

Fig. 221. Hot Water Back   Fig. 222. Furnace Coil

WATER HEATING DEVICES

The more common and standard types of heating devices are listed as follows:

1. Water back
2. Furnace coil
3. Coal heater
4. Gas heater
5. Instantaneous heater
6. Storage heater
7. Steam heater
8. Under water-line heater
9. Electric heater
10. Solar heater

Water Back. The water back is constructed of cast iron. It was possibly the first method of utilizing heat to furnish the domestic installation with hot water. It is installed in a range (which is used for cooking and heating purposes on one side of the fire box) and is provided with waterways or channels for the circulation of water. The water back is quite efficient, but has become obsolete with the innovation of more modern heating devices, Fig. 221.

Furnace Coil. Fig. 222 represents a furnace coil constructed of black steel pipe, usually of 1 inch diameter, made in the form of a large **U.** It is the most commonly installed heating device in use. This does not mean, however, that it is the most efficient type of heating unit. A furnace coil does not furnish the domestic system with a constant supply of hot water because of the irregularity of heat in the furnace due to the variableness of outdoor temperatures. The best results are obtained by placing the coil in the fire box directly above the live coal fire. It is poor practice to trap the furnace coil, because this is likely to result in stoppage.

Coal Heater. The coal heater shown in Fig. 223 is a very efficient type of heating device for buildings such as hotels, apartments, factories, or schools. It is constructed of cast iron and consists of a hollow water jacket, which completely surrounds the fire box of the heater. This provides large heating surface and permits a great

Fig. 223. Coal Heater

Fig. 224. Gas Heater

volume of water to be heated in a very short time. The heater is tapped in such a manner that it can be connected to the storage tank conveniently.

Gas Heater. The coil gas heater generally is used only as an auxiliary heater. Its installation, however, is common. This is due to the fact that the average home is equipped with a furnace coil. The heater is installed in conjunction with the coil to give the occupants a supply of hot water in warm weather, or as required.

The gas heater is constructed of a cast-iron body into which is fitted a ¾-inch copper coil. The heat is supplied by means of a gas burner placed below the coil. It must be equipped with a vent flue connected to a chimney so that the dangerous and objectionable gases and odors may be removed. This type of heater is efficient and inexpensive. See Fig. 224.

A gas heater is used in conjunction with a *solar heater*, Fig. 225, in places where chilly, sunless days occur, especially during the winter months. In many places, the combination of the two heaters has proved itself highly efficient even where freezing temperatures are not uncommon, as a nonfreezing type of solar heater is available. These heaters use a nonfreezing liquid fuel that is heated in the sun coil and circulated throughout the heater in another coil.

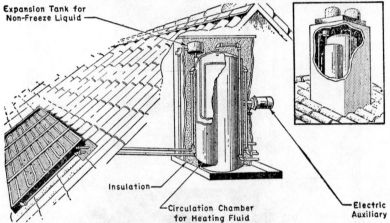

Fig. 225. Solar Water Heater. *Inset*, Storage Boiler Extending through Roof and Camouflaged as Chimney. When Conditions Do Not Permit Boiler to Be Placed below Gable
Courtesy of Day and Night Mfg. Co., Monrovia, Calif.

In installing a solar heater, make sure that the sun coil faces south and is set at an angle. The top coil should be at least a foot below the bottom of the storage tank.

Instantaneous Water Heater. This type of water heater is not as popular as the automatic because of its higher cost of operation and maintenance. It is used chiefly in barber shops and by hairdressers to provide an unfailing supply of hot water at short intervals. The heater consists of a cast-iron body in which many feet of copper coil are installed. See Fig. 226. It operates by means of gas and water valves which open and close by turn as the water is drawn. A vent flue connected to a chimney supplies the means of safely dispelling dangerous gases and offensive odors.

Automatic Storage Water Heater. Progress has been made in the manufacture of automatic storage heaters, whether operated by gas or electricity. Manufacturers are producing these heaters in quantities, and different types of automatic controls are employed.

The gas-operated type, Fig. 227, is engineered with special emphasis on low cost and economy of operation. The boiler is given an

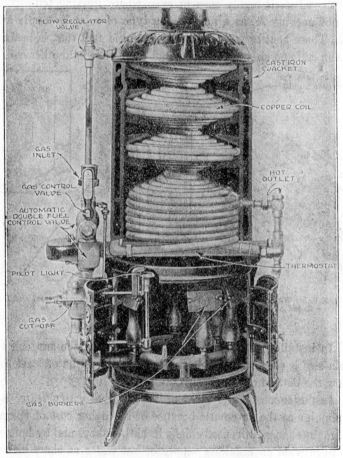

Fig. 226. Instantaneous Water Heater
Courtesy of Ruud Manufacturing Company, Pittsburgh, Pa.

extra heavy galvanizing to insure long service. It has a flue of the baffle type that is designed to make full use of the heat energy created by the Bunsen burner. It also has a magnesium rod, which device is a guarantee against rust and corrosion.

Although a thermostatic control is good insurance on any heater

designed for hotel and apartment house use, a control with a safety
pilot is required by law on heaters that use liquid petroleum gas
(L.P.G.). The control used by most manufacturers is the Grayson
(Fig. 228), but many companies manufacture their own.

Fig. 227. Automatic Storage Gas Water Heater
Courtesy of Mission Water Heater Co., Los Angeles, Calif.

Another type of heater is the *booster* type. It is an inexpensive
water-heating system that utilizes a magnesium rod as protection
against rust and corrosion.

An electric storage water heater is shown in Fig. 229. This heater
usually embodies two elements—one below and one above—both
being connected and controlled by a thermostat. The lower element
maintains a constant temperature, while that above speeds recovery

of heat by the water and cuts off as soon as the water in the upper part of the heater reaches the desired temperature. This heater, especially in localities that enjoy the benefit of low power rates, provides the cleanest and most trouble-free method of heating water at moderate cost.

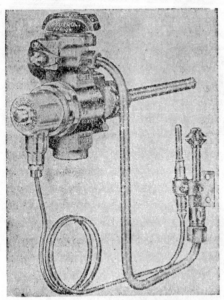

Fig. 228. Unitrol Junior Thermostatic Control for 15- to 40-Gallon Gas Water Heaters
Courtesy of Grayson Controls Div., Robertshaw-Fulton Controls Co., Linwood, Calif.

Fig. 229. Electric Hot Water Heater
Courtesy of Strauss Electric Appliance Co., Waukesha, Wis.

Under Water Heater. This type of heater, Fig. 230, is used on installations in residences where hot water or steam is used for heating. The heater is constructed of a cast-iron body, into which a copper coil is built. The inlet and outlet of the coil are connected to the boiler of the heating plant below its water line. This permits a circulation of hot water through the coil and furnishes the water of the domestic system. In some instances the coil is connected to the water system.

One precaution is necessary on this installation. The flow or return pipe must be equipped with a hand-operated gate valve between the heater and the tank. This is necessary in order to overcome the possibility of the auxiliary heater of the domestic system

heating the boiler during periods when the house heating plant is
not in service.

HOT-WATER CIRCULATION

The storage tank and heating device of a hot-water distribution
system are so assembled as to create a circulation of water within
them. The movement of the water is the result of molecular activity. The application of heat to a body of water causes it to expand
and become less dense, which gives it a natural tendency to rise. The
inequality of weights between the hot and cold water contained in
the tank results in circulation.

The storage assembly is the most important installation of the
entire distribution system. Improper connections render the entire
domestic hot-water system of little value.

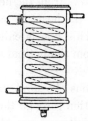

Fig. 230. Indirect Water Heater

The piping arrangement of a heater and storage tank are practically identical on every installation. Slight changes in connections
are necessary when more than one heating unit is used. The circulation of water, however, is not affected in any way regardless of the
number of units that may be connected to the tank.

Tank Connections. The hot-water storage tank may be set in a
vertical or horizontal position. This depends on the size of the installation it serves, the general construction of the building, or the
type of installation and heating device. It is advisable to set the
tank in a vertical position on small installations and in a horizontal
position on the larger ones.

It is necessary to place the tank at a point above that of the
heater when it is installed horizontally. When the tank is placed
in a vertical position, at least the greater portion of the tank must
be above the heating device. This practice allows the heated water
to rise and permits more rapid circulation.

The cold-water supply is delivered into the tank by means of a boiler tube. It extends to within 6 inches of the tank bottom. The inlet on the tank best suited for cold-water supply is located directly above the return tapping. The reason for extending the boiler tube to the bottom is to avoid the possibility of cooling the hot water which accumulates at the top of the tank. The cold-water tube must have a small hole within 6 inches from the top of the tank. This hole serves as a vacuum breaker and prevents siphonage. The supply line into the tank must be equipped with a control valve located as close to the hot-water tank as is practical.

The heating unit has two tappings, flow and return. The flow tapping of the heater is connected to an opening on the tank somewhere above its center point. This line is called the flow connection, because the heated water flows from the heater into the tank.

The return tapping of the heater is connected to a tapping on the bottom of the tank. This line is called the return, because it returns the colder water from the bottom of the tank to the heating appliance.

On all tank installations the flow and return lines should be connected so that a circulation through the entire tank results. The same procedure should be followed in making flow and return connections to the heating unit.

When more than one heating device is used, the flow and return lines must be connected so they will operate in conjunction with one another. The storage unit must be equipped with a drain valve at its lowest point.

The hot-water distribution piping is connected to a tapping on the top of the tank at a point near the flow inlet. This practice assures the house occupants a ready supply of hot water. It is not necessary to place a control valve on the hot-water supply to the building.

It is the practice in some instances to equip the storage tank with a blow-off valve to prevent serious difficulties should the tank become overheated. On installations of this type, temperature as well as pressure must be controlled.

The following illustrations, with a brief analysis of each, tend to simplify the general rules applied to tank and heating device connections offered in the preceding paragraphs.

Fig. 231 represents the connections of a furnace coil to a vertical range boiler, which installation may be found in almost any small residence.

The range boiler is mounted on a stand and receives its cold-water supply through a tube provided with a siphon hole and manually controlled gate valve.

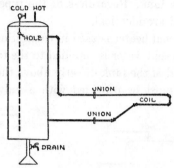

Fig. 231. Furnace Coil Connections
to Range Boiler

The return and flow connections of the coil are connected to the tank and provided with unions as the illustration indicates. The hot-water supply is connected to the hot-water distribution system. No valve is necessary on this piping. The cock in the bottom of the boiler serves as a means of eliminating sediment.

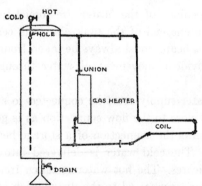

Fig. 232. Auxiliary Gas Heater Connection
to Range Boiler

A furnace coil is not a good medium for heating water unless an auxiliary gas heater is used in connection with it. Fig. 232 illustrates this type of installation.

The furnace coil is connected to the range boiler in the same manner as in Fig. 231. The gas heater is connected between the flow and return connections of the furnace coil. This practice is a common one but is not always permissible. The gas heater may have to be connected to the tank directly or its flow connection be extended to the hot-water supply pipe between the distribution main and the top of the tank. Regardless of the type of connection the principles involved are identical.

In Fig. 233 a coal heater is used in connection with a horizontal storage tank. The cold water is supplied to the tank through a tube installed on the end of the tank directly above the return connection to the heater. It must be provided with a valve.

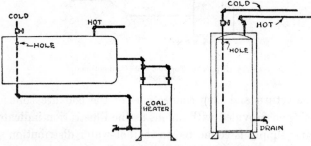

Fig. 233. Coal Heater Connection

Fig. 234. Storage Heater Connection to Storage Tank

The top openings of the heater are joined into one pipe and reconnected into one end of the tank above its center. The return connection to the heater must always be taken from the end opposite the flow to provide a movement of water through the entire installation.

The hot-water supply must be connected to an opening in the tank located as close to the flow connection as is possible.

Fig. 234 shows the connection of a storage heater to the water supply system. The cold water is delivered into the tank as the illustration indicates. The hot water is taken from the top of the insulated tank and connected to the distribution system.

Fig. 235 illustrates how a steam heater of the converter variety can be connected to a horizontal storage tank. This type of installation generally is used in industrial buildings. The heating unit is connected to the horizontal tank in the same manner as the coal

heater installation which was discussed previously. The steam
supply to the heater is controlled by a thermostat installed in the
storage tank.

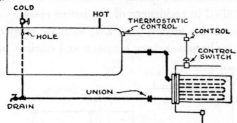

Fig. 235. Steam Heater Connection to Storage Tank

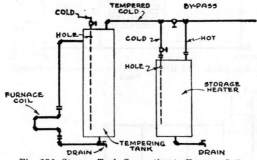

Fig. 236. Storage Tank Connection to Furnace Coil
and Tempering Tank

Fig. 236 shows a storage tank connected to a furnace coil. This
arrangement is unique in that the water is tempered before it passes

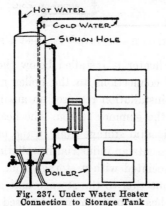

Fig. 237. Under Water Heater
Connection to Storage Tank

through the storage heater. There is economy in this type of installation, because a storage heater costs more to operate than does

this unit. The storage heater is connected into the hot-water supply pipe. A valve installed between the two connections serves as a by-pass device should the storage heater fail. This type of heater usually is installed in residences of the better class.

The under water heater, Fig. 237, with its connections to the storage tank, has become a very common and efficient installation in

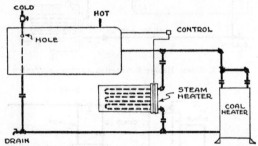

Fig. 238A. Two Heating Devices Connected to Storage Tank

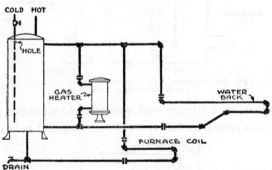

Fig. 238B. Three Heating Devices Connected to Storage Tank

recent years. The heater is installed below the water line of the heating boiler. Its connections to the boiler must be valved because this type of installation requires an auxiliary heater to provide hot water in the summer season. The heater is connected to the vertical or horizontal storage tank in the usual manner.

Fig. 238A represents an industrial installation consisting of two heating devices associated with a storage tank. The principles involved in their connection to the storage unit are typical of other installations. One heater or both may be used when connected in this manner.

Fig. 238B is an installation which may be found in older types of residences. It represents three heating devices connected to one storage unit. This type of installation is an uncommon one and is presented to demonstrate how the principles of circulation may be applied under almost any circumstances.

Pressure and Temperature Control. High temperature as well as pressure, when uncontrolled, may result in explosion of the storage unit. Water under pressure, heated above 212°F., if suddenly released results in vaporization. In this form it expands approximately 1,700 times its original volume and should the tank rupture

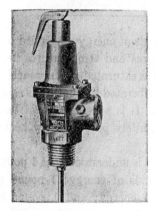
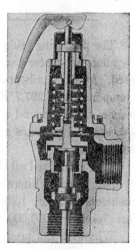

Fig. 239.   Temperature and Pressure Relief Valve
Courtesy of Watts Regulator Company, Lawrence, Mass.

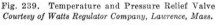

under these conditions the flashing off of the vapor may have sufficient force to send it through the roof. Temperature-pressure relief valves are available to prevent such occurrence. They should be provided on every installation as a safety precaution. The relief valve should be installed on the hot-water tank or on the hot- or cold-water supply as close to the tank as is practical.

Drip lines from relief valves should be extended down to the basement floor, over or near a floor drain, or discharge over a properly trapped and vented fixture with the end of the drip line above the flood-level rim of the fixture.

It should be understood that while excessive pressure can cause a tank to rupture, excessive temperature can cause an explosion. Until recently this was a controversial point; tests, lately undertaken, verify the statement. They show that at zero pressure the boiling point of water is 212°F.; if the pressure is raised to 10 psi, the boiling point goes up to 239.5°F, as shown in the table.

Pressure (P.S.I)	Degrees Temp. at Boiling	Energy Liberated (Ft. Lbs.)
0	212.0	. . . .
10	239.5	479,800
30	274.0	1,305,000
50	297.7	2,021,900
70	316.0	3,642,000
90	331.2	3,138,400

The preceding table shows the pounds of energy developed in a 30-gallon hot-water tank at the pressures and temperatures indicated. At 50 pounds pressure the water is saturated steam, reaching a temperature of 297.7°F. The energy liberated by rupture is equal to more than a pound of nitroglycerin. It is seen, then, that temperature and pressure protection is definitely part of every heater installation.

This is realized all the more when it is understood that 1 pound of black powder liberates 960,000 pounds of energy; 1 pound of smokeless powder, 1,260,000 pounds of energy; and 1 pound of nitroglycerin, 2,000,000 pounds of energy.

Whatever the need, there is a temperature and pressure relief valve suited to it. See Fig. 239.

Fundamentally, water-heating systems are of two types: a hot water space heating system, where water is confined within the system at relatively low pressure, and a hot water supply system, which is not a closed system and operates on much higher pressures, with possibly dangerously high temperatures. Since this equipment differs in construction, in purpose, and in operating conditions, the emergency safety devices used on each must also be different. Steam pressure relief protection is considered adequate for hot water space heating systems, whereas temperature and pressure relief protection is necessary for hot water supply systems.

Neither the plumber nor the home owner should tamper with the settings of relief valves. It is in this way that valves are often made ineffective and fail to provide relief when it is needed.

CHAPTER XXIII

HOT=WATER DISTRIBUTION SYSTEMS

The hot-water distribution system consists of the pipe installa-tions which convey the heated water from the storage unit to the plumbing fixtures. The pipe is installed in such a manner that a continuous movement or circulation of the water occurs.

There are three types of installations in common use today; namely, the upfeed and gravity return, the overhead supply system,

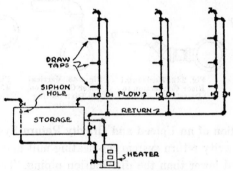

Fig. 240. Upfeed and Gravity Return System of
Hot-Water Distribution

and the pump circuit system. Each method has specific application to certain types of buildings.

In the following paragraphs the various systems will be dis-cussed.

UPFEED AND GRAVITY RETURN SYSTEM

The upfeed and gravity return system shown in Fig. 240 is commonly used in small residential and industrial installations. Its working efficiency depends on the accuracy of its construction.

Circulation of water is an action governed by laws of physics. In constructing a hot-water system the slightest mistake on the part of the installing mechanic may interfere with the operation of natural laws and thus cause unsatisfactory service. The proper style of valves and fittings and the correct sizing of pipe are essential to efficient function of the system.

The purpose of a gravity return and upfeed system is to permit a constant circulation of hot water within the piping arrangement. The circulation allows the building occupants to draw an immediate supply of heated water at any plumbing fixture. A circulating return offers an economy in that waste of water is eliminated.

The principle on which a gravity return system functions is provided in the unequal weights of two columns of heated water of uniform height. The inequality of weight is the result of a variation in temperature in the two columns. A mechanic can do much to help insure a temperature variation in a circulating system.

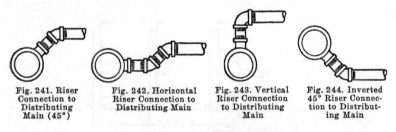

Fig. 241. Riser Connection to Distributing Main (45°)

Fig. 242. Horizontal Riser Connection to Distributing Main

Fig. 243. Vertical Riser Connection to Distributing Main

Fig. 244. Inverted 45° Riser Connection to Distributing Main

Construction of an Upfeed and Gravity Return System. On an upfeed and gravity return system the heating unit and storage tank must be placed lower than the distribution piping. They should be located so as to serve the piping system conveniently and efficiently.

The distribution main is connected to a tapping on the top of the storage tank close to the flow from the heater. It is suspended from the basement ceiling at a slight grade upward from the storage unit, and various risers are taken from it. On large installations the distributing main should be equipped with a valve. The intervals of suspension are determined by the diameter of the pipe.

The hot-water riser should be connected to the distributing main by means of a 45-degree connection. Fig. 241. This practice may vary according to the length of the risers. It is a common occurrence on this system to have one riser circulating faster and more thoroughly than the others. This is often due to the fact that the connections to the distribution main are improperly made, as for example, where three risers are connected to a distributing main, one being taken from a main with a tee set vertically, another tee being

set at a 45-degree angle, and a third tee set horizontally. The vertical tee affects the circulation of water at the 45-degree connection
and almost completely stops the circulation of water at the horizontal connection. Heated water tends to expand, and a riser which
provides easier direction for this expansion, in terms of liquid flow,
will interfere with the circulation of the other risers.

It is advantageous to a mechanic to be able to build his connections, using the various units in joining riser and main so as to
increase the equal circulation. This practice requires experience together with the utmost care and responsibility. For example, should
an installation consist of three risers of varying heights, the longest
riser can be connected to the main horizontally, Fig. 242. This practice retards circulation. Circulation in the short riser can be favored
by a vertical connection, Fig. 243. The other riser may be connected
with a 45-degree fitting, Fig. 244.

The flow riser is then passed vertically as near the fixture location as possible. Care should be taken to support it thoroughly,
and to allow for expansion by using swing joints to prevent breaking of the pipes. The flow riser must be equipped with a valve and
a drip at its base.

The circulating return is connected to a tee that is installed in
the riser below the highest fixture to overcome air lock, Fig. 240. The
return riser is usually one size smaller in diameter than the fixture
riser. It is connected to a return main suspended from the basement
ceiling. The return riser must be equipped with a drip and control
valve at its base.

The circulating main is suspended from the basement ceiling
and installed with a grade to a Y fitting installed in the return
connection between the storage tank and the heating unit. A valve
must be placed at this connection.

All valves used on the hot-water system should be of the gate
type. This is to assure a full water-way and to overcome trapped
water lines—a fault which occurs in the use of disc or globe valves.

OVERHEAD FEED AND GRAVITY RETURN SYSTEM

The overhead system of hot-water distribution is the most efficient method of delivering hot water to fixtures. It generally is used
in buildings of extreme height. Like the upfeed and gravity system,

the overhead feed method of water distribution is dependent on natural laws governing expansion and gravity. The mechanic must exercise extreme care in the construction of the system so that these laws may operate to their full benefit.

The overhead system of water distribution is more efficient than the upfeed and gravity return system. Its design is more favorable to the natural course of hot-water travel. It permits circulation even though there may be a defect in its mechanical construction. The more accurately a mechanic can apply the fundamental science involved in the circulation of hot water, the more efficiently will the overhead system operate.

The operating principle of the overhead system is based on the fact that in the closed system of piping water rises when heated. After it has reached the high point of the system, natural forces of gravity return it to the storage unit.

Circulating Return Main. The circulating return main of an overhead system of hot-water distribution is a line suspended from the basement ceiling. It is pitched and connected to a **Y** located on the return piping between the heater and the storage tank. All vertical drops or inverted risers from the overhead main are connected to the circulating return main. The circulating return main keeps the hot water in constant movement, which assures the occupants of the building a ready supply of hot water.

Construction of an Overhead System. The storage unit of an overhead feed system of water supply must be located at the lowest point of the distribution piping.

The overhead feed riser is connected to a tapping on the top of the storage tank. The tapping should be located close to the flow connection of the heater. The riser must be extended as direct and as free from offsets as is possible to the work space or the ceiling above the top floor of the building. This riser must have no fixture connections taken from it.

A distributing main is connected to the top of the riser, and is suspended from the ceiling or building framework by means of metal hangers. This main must be pitched away from the riser so that the water will flow to the last drop riser. The main should be located conveniently so as to make the horizontal runs of the riser as short and as equal in length as possible.

The horizontal riser branch is connected into the main by means of an inverted 45-degree fitting, and is pitched to the drop or vertical riser proper, as shown in Fig. 244. The horizontal riser branch must be equipped with a valve installed as close to the main as may be practical.

The riser is passed downward through the various floors of the building, and to it are connected the fixture or toiletroom water supply branches. The riser of the upfeed and gravity return system and that of the overhead system differ in one respect—on an overhead system the largest diameter of piping is at the top of the riser, the

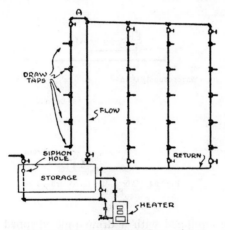

Fig. 245. Circulating Return Connection to
Storage Unit

size diminishing as it passes through the lower floors. On an upfeed system the largest diameter of pipe is at the bottom of the riser, the size diminishing as it passes through the upper floors of the building.

The vertical risers are connected to the circulating return main suspended from the basement ceiling. It must be equipped with a gate valve, Fig. 245, and a drip at its base. The return main is pitched downward to the storage unit, and is increased in size as risers are connected to it.

Relief Vent. The overhead system, because of its construction, is likely to become airbound. This condition prevents circulation of the hot water. An accumulation of air occurs at the highest point of the distribution piping. It must be eliminated in some practical way. There are two methods by which this is accomplished. The most practical method is to connect an uncirculated riser to the highest point of the overhead distributing main. The connection into the main must be made in a vertical manner, directly from the top of the pipe, as illustrated in Fig. 245 at *A*. This makes it possible to relieve the air lock from time to time, using the fixture that the riser serves.

Fig. 246 shows another method of air relief, using a patented air relief valve. The valve opens when accumulation of air occurs, and closes automatically when the air is eliminated. The relief

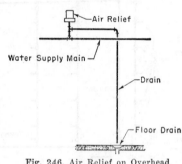

Fig. 246. Air Relief on Overhead System

valve must be equipped with a drain pipe, dripped into an open fixture.

PUMP CIRCUIT SYSTEM

The pump circuit system of water distribution is the circulation of hot water to the plumbing fixtures by means of a mechanical device. It is rapidly replacing natural methods of water distribution. A pump installed on a system of hot-water distribution greatly increases its efficiency.

Pump circuit installations generally are confined to the larger types of buildings. They are also installed where, due to construction difficulties, it is impossible to produce a natural circulation of hot water.

The one objectionable feature in a pump circuit system is the possibility of breakdown of the pumping equipment. However, because of the reliability of mechanical devices at the present time this objection is reduced to a minimum.

Circulating Pumps. The centrifugal type of pump is the most practical circulating pump for use on a hot-water system. This type of pump is especially dependable in water circulation because it has few working parts to get out of order, and it can be repaired easily should a breakdown occur. Furthermore, it is compact in construction and requires little space for installation. The motive power of such a pump usually consists of an electric motor that can be made to operate on any electric circuit. Another important reason why the centrifugal pump should be used is the fact that the rotary motion of the impeller creates an even movement of water in the piping

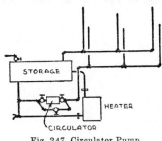

Fig. 247. Circulator Pump
Installation

system rather than a pulsating movement, such as that produced by a piston pump.

The circulating pump generally is used on installations that have inefficient circulation. This fault may be caused by building defects which compel the mechanic to trap runs of main piping. When scientific principles cannot be applied to produce circulation, a pump may be used to accomplish this end.

A pump may also be used on systems of the overhead feed and upfeed types. This practice will increase the efficiency of the system. The installation of a pump in such systems has an economic as well as a practical value, because the heated water is returned to the storage unit at a higher temperature. This naturally lessens the fuel expense and assures a faster circuit, giving the occupants a better supply of hot water.

Installation and Location of the Pump. The pump is installed on the circulating return main as close to the heating unit as possible, as indicated in Fig. 247. The circulating return is connected to the inlet side of the pump, and the outlet side of the pump is connected into the return to the heater. A gate valve must be installed on each side of the pump.

It is advisable to equip the pump with a by-pass, which is done by inserting tees of the same diameter as the circulating return ahead of the valves. The tees are connected and the line is equipped with a gate valve. Should the pump get out of order the control valves may be closed and the hot water will circulate around the pump into the return pipe of the heater. This practice serves as a temporary means of water circulation. When the by-pass is not in use, the valve with which it is equipped must be closed. The valves on either side of the pump must be open at all times when the pump is in operation.

CHAPTER XXIV

PRIVATE WATER CORRECTION

Water Softening Processes. Water taken from underground sources usually contains compounds of calcium (Ca) and magnesium (Mg) which render it inconvenient for domestic use. Forming its mineral content are calcium and magnesium sulphates, $CaSO_4$ and $MgSO_4$, called permanently hard water; or calcium and magnesium bicarbonates, $Ca(HCO_3)_2$ and $Mg(HCO_3)_2$, called temporarily hard water. Water containing these mineral elements reacts with the chemical elements contained in soap to form an objectionable curd, making it practically useless as a cleansing agent.

Because water containing mineral elements may be precipitated by heat greater than 135°F., and by centrifugal force, the precipitated salts form within pipes an objectionable scale that gradually reduces their diameter and consequently lessens flow capacity.

The process by which water is made soft was discovered in England accidentally. The water of a mountain stream, after flowing a short way, was found to be greatly softened. An examination of the gravel through which the water flowed showed that the change was due to the presence of a mineral called *permutt*, which later was given the trade name Zeolite. A large deposit of this mineral has since been found in southern California, and industry has supplemented the natural supply with a synthetic product which reacts in water in a manner identical to that of the mined mineral.

Chemistry of the Water Softening Process. Zeolite is known to the chemical engineer as a substance composed of the oxides of sodium (Na_2O), aluminum (Al_2O_3), and silicon (SiO_2). The mixture is expressed chemically as

$$Na_2O \cdot Al_2O_3 \cdot 2SiO_2$$

When water containing calcium or magnesium sulphates is passed through a bed of Zeolite, the calcium and magnesium are exchanged with the sodium content of the mineral, and the water thus treated, though it contains a compound of sodium, reacts favorably with soap. The chemical change which occurs is expressed as follows:

$$Na_2O \cdot Al_2O_3 \cdot 2SiO_2 \cdot CaSO_4 \rightarrow CaO \cdot Al_2O_3 \cdot 2SiO_2 \cdot Na_2SO_4$$

When all the sodium contained in the Zeolite has been exchanged for the mineral in the water it loses its softening qualities and must be reactivated. This is done by adding sodium chloride (NaCl), common table salt, to the water softener.

The chlorine of the regenerating compound forms a new salt with the removed calcium and magnesium, which is eliminated from the softener by thorough washing with water. The chemical excange is expressed as follows:

$$CaO \cdot Al_2O_3 \cdot 2SiO_2 \cdot 2NaCl \rightleftarrows Na_2O \cdot Al_2O_3 \cdot 2SiO_2 \cdot CaCl$$

Zeolite does not deteriorate rapidly unless the water contains elements which adhere to it and reduce its qualities of absorption. The cycle of reactivation may be repeated over and over again.

TYPES OF WATER SOFTENERS

There are three types of water softeners in use at present, namely, the manually reactivated type, in which the operator adds the regenerating compound; the semiautomatic, in which the sodium chloride brine is siphoned through Zeolite mineral manually; and the automatic which regenerates itself at periodic intervals. Each of these devices may be procured in either upflow or downflow design. The downflow type of water softener may be used to advantage in treating water which contains iron or sulphur.

Manually Reactivated Water Softener. The manually reactivated water softener, Fig. 248, consists of a heavy-gauge galvanized steel tank and an assembly of self-cleaning valves so arranged that water can be by-passed to accomplish its regeneration. The hard water inlet is located in the bottom of the tank, where a distributor fitting is provided to insure that the Zeolite is used to its full area and depth. A layer of coarse gravel, which acts as a fine screen serving to eliminate suspended materials, is spread over the dis-

tributor fitting and the tank is then filled with Zeolite to within a
few inches of the top. The conditioned water is drawn from above
the bed of Zeolite through a screened outlet to prevent small particles of mineral matter from being carried into the water distribution system. The diaphragm-actuated valves are of brass construction and have rubber seats.

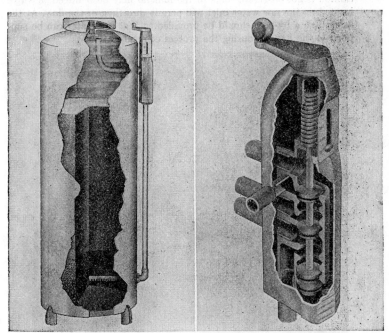

Fig. 248. Single-Tank Manually Reactivated Water Softener and
Diaphragm-Actuated Valves
Courtesy of Rheem Manufacturing Co., South Gate, Calif.

Another model has a single-crank, multiport valve that performs every function of the four or five single-duty valves of
earlier appliances of this type. The tanks of the water softeners
here described are lined with a plastic, microcrystalline sealing
material which cannot become abraded by movement of the mineral
grains contained in the water.

To regenerate a manually reactivated water softener it is only
necessary to follow the simple directions given on the supply valve.

Directions:

1. Turn crank to Position 1. Wash; leave 5 minutes.
2. Remove clamping screw cap. Pour in the required amount of sodium

chloride (pure granulated table salt) and allow it to remain in tank 8 to 10 hours for large softeners.

3. Turn crank to Position 3. Allow the salt water to run into the drain for a period of half an hour or until all taste of salt has been washed out of the water.

4. Turn valve to Position 1. The softener is now ready for use.

Caution: Do not let the raw water wash the salt water out too fast. If this caution is not observed, some of the mineral may be removed. A valve located on the raw water inlet will permit throttling of the flow. At the same time a by-pass should be provided so that raw water can be supplied to the fixtures during the process of regeneration.

Fig. 249. Semiautomatic Water Softener with Brine Tank
Courtesy of Stover Water Softener Co., St. Charles, Ill.

Semiautomatic Water Softener. Semiautomatic and manually reactivated water softeners are identical in construction except that the semiautomatic type has a brine tank and siphoning device, as shown in Fig. 249. A specified amount of sodium chloride is put into the brine tank and the tank is then filled with water to the required level. The semiautomatic water softener may be regenerated as here indicated.

Directions:

1. Move shift lever to Position 2, as you would the gear shift on your car. This allows the raw water inlet to close and the brine from the brine tank to flow through the mineral, thus removing all hard chemicals.
2. When the brine in the tank has been drawn off to the lower level, move the gear shift to Position 3. This action closes the flow from the brine tank and allows the raw water to flow through and remove all taste of salt from the water.
3. When all taste of salt has been removed (which usually takes from 10 to 15 minutes), shift the gear to Position 4. This action closes the drain and permits the water to flow into the brine tank.
4. When the water reaches the upper level of the tank, shift the lever to Position 1. This action closes both tank and drain and allows the softened water to flow into the line to be used as needed.

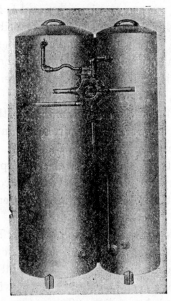

Fig. 250. Rear View of Automatic Water Softener,
Showing Automatic Valve
Courtesy of Rheem Manufacturing Co., South Gate, Calif.

Automatic Water Softener. The automatic water softener, Fig. 250, affords great convenience in that it requires scarcely any attention. The water drawn from it passes through a metering device which automatically operates the regenerating mechanism. The sole duty connected with operation of the appliance is that of keeping the brine tank filled with salt. Automatic water softeners are used extensively today owing to the fact that their original cost is little above that of the manually operated or semiautomatic appliances.

Connecting the Softener to the Distribution System. The water softener may be installed in the distribution system at any convenient location. The hard water inlet is provided with a gate valve. A third pipe is then run to the various fixtures which require soft water in the same manner as is piping of the ordinary distribution system. It is not advisable to supply water closets, lawn sprinklers, or similar devices with soft water.

DOMESTIC WATER FILTRATION

There are many buildings such as hospitals, schools, clubs, etc., which, for sanitary reasons, require water that is free from bacterial, organic, or suspended impurities. The responsibility of producing this relatively pure water is placed on the plumber, since it is he who must connect the various mechanical devices in the water distribution system which provide this protection. The pressure filter is one of these appliances.

The pressure filter is used in connection with circulated drinking water supplies, hospital equipment, swimming pools, and water used for the preparation of foods in industry. Even though the original source of the water used for these purposes is a safe one, there are many ways in which it may become contaminated in the piping arrangement of a large distribution system before it is consumed.

Types and Construction of Pressure Filters. There are two types of pressure filters, vertical and horizontal. Both consist of steel pressure tanks filled to about two-thirds their interior cubic content with various sizes of a good quality of quartz—which material tends to eliminate suspended materials. The unfiltered water usually is admitted through an opening in the top of the tank, and is permitted to percolate through the quartz material. The filtered water is outletted from the bottom of the stone bed and is then allowed to enter the distribution system. After a quantity of water passes through the filter, the quartz bed becomes fouled and the unit must be washed. This is done by reversing the flow of the water. A coagulant is added to hasten this process.

Coagulation Unit. The coagulation unit consists of a cast-iron vessel that is filled with the proper kind of coagulant. Alum generally is the compound used. This compound tends to form a flocculent mass of the materials that have accumulated on the top of the stone bed, and makes removal of these objectionable materials a simple process.

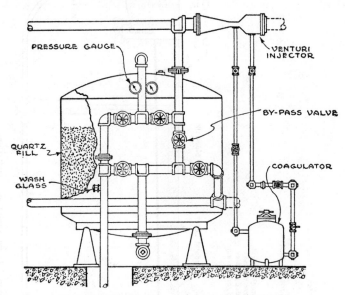

Fig. 251. Vertical Filter and Venturi Coagulant Feeder

The coagulant is injected into the filter by means of a venturi tube, or, on smaller installations, through a restricted orifice. The inequality in pressures on the inlet and outlet sides of the tube courses the water through the coagulation pot containing the compound and then into the filtration unit. These types of filters are very efficient and, because of their compact nature, are in constant demand. Fig. 251 illustrates a vertical filter and venturi coagulant feeder.

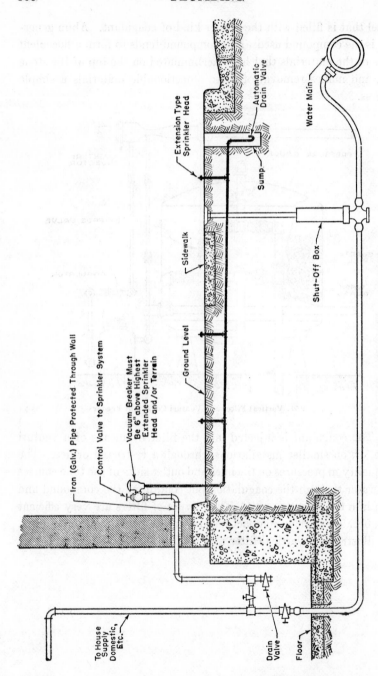

An Approved Lawn Sprinkler Installation
Courtesy of Department of Public Works, Bureau of Engineering, City of Chicago

Extension Type Sprinkler Head

Automatic Drain Valve

Water Main

Sump

Sidewalk

Shut-Off Box

Ground Level

Vacuum Breaker Must Be 6" above Highest Extended Sprinkler Head and/or Terrain

Iron (Galv.) Pipe Protected Through Wall

Control Valve to Sprinkler System

To House Supply Domestic, Etc.

Drain Valve

Floor

CHAPTER XXV

CROSS=CONNECTIONS

The possibility of contaminating drinking water supply by cross-connecting it with the waste disposal system until recently was considered improbable by authorities interested in sanitation problems of a general character. The plumber, however, has long realized the possibility of contamination from this source, and for a period of thirty years has advocated the adoption of codes and other devices for regulation of practice.

The average layman believes that water of standard purity at the source of supply is safe to consume, and that there is little danger of contracting an infectious disease by drinking such water. Water of this nature has been used for domestic purposes for many years with no apparent harmful effects, hence one can see why such a belief would prevail. Authorities, on the other hand, are aware that the possibility of contamination is present and know that the combined efforts of the medical and engineering professions as well as the services of the plumber are needed to safeguard health.

As was previously stated the plumber, with his background of practical experience and his knowledge of physical science, has been aware of the dangers of cross-connections for many years and has striven for recognition of the relation of scientific trade practice to public health.

Until recently, though, it seemed that the plumber's efforts to direct attention to drainage systems and devices with respect to their sanitary quality were futile. True, epidemics of typhoid and cases of dysentery and diarrhea occurred, but in the majority of cases the individual recovered and the source of the disease was ascribed to consumption of contaminated food or to contagion.

The situation has changed completely however. It has been proved beyond question that infectious diseases can be contracted from conditions originating in seemingly safe plumbing installations, and all the agencies associated with the sanitary needs of the consumer are aware of this and are doing their utmost to remove **exist-ing** evils.

In the last few years epidemics of amoebic dysentery have taken their toll among citizens of congested communities. In one of the larger cities it was established that faulty plumbing fixtures and improperly constructed water supply systems were responsible for an epidemic of amoebic dysentery which resulted in many deaths. Moreover, it is known that the possibility of epidemic of the disease is constantly present in cities owing to the fact that many appliances and practices which as recently as five years ago were considered adequate are today acknowledged to be potential agencies of contamination. These improper appliances and connections actually permit organic materials contained in the waste disposal system to enter the domestic water supply and pollute it. Whether these installations can be corrected is questionable, at least their correction will be a slow process. However, at this time, the consumer has assurance that all new installations are being carefully planned and are safe to use. Old installations will have to be changed in time, and it will take years to accomplish their correction and eliminate all possibility of pollution. To this end medical authorities are cooperating with sanitary engineers and plumbers. Inspections are being made of hospital, industrial, and residential installations, and dangerous forms of cross-connection are ordered to be removed. The engineering profession is using its laboratories to conduct tests, and, from this information, data of value to the plumber is being compiled. National and state associations of plumbers are spending large sums of money in research. With a cooperative program of this kind the menace of contamination will, in time, disappear and another victory for modern science will have been won.

SCIENTIFIC ASPECTS OF CROSS-CONNECTIONS

The reader may wonder how water can be contaminated through private supply systems and what can be done to avoid this possibility.

There are a number of ways in which pollution can occur. The most common form of water contamination results from back-siphonage of organic materials into the water supply piping. The scientific principle involved is that of the common siphon. It has

been demonstrated that water can be drawn from a vessel or source
of supply by an arrangement of piping so constructed that the length
of the outlet, or discharge leg, is greater than that of the inlet leg,
in which pressure of less than one atmosphere can be produced. To
illustrate the principle of an ordinary siphon more thoroughly the
accompanying illustration is offered.

Fig. 252 is a sketch which indicates how back-siphonage could
occur in the water piping of a plumbing system. The tank represents
any plumbing fixture or appliance with its content of contaminated
water. The water supply is delivered into the tank by means of a

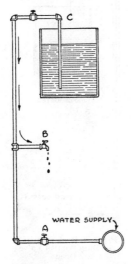

Fig. 252.   Back-siphonage
—Main Control Valve
Closed

faucet or other device, the discharge end of which extends into the
liquid content of the fixture. The pipe arrangement consists of a
riser that connects to the city water main, extends vertically to a
valve above the tank, and then turns downward into the tank. A
fixture supply connection is taken from the riser at a point below the
tank, for the purpose of serving a fixture installed at that level.

Suppose the entire pipe system were filled with water under
pressure and the water service control valve, indicated at A, were
closed to make repairs on the water faucet or valve, B. It is custom-

ary when making repairs to drain the content of the riser to prevent flooding of the building when the fixture or any part of it must be removed. Should the valve, at C, be defective, or opened to admit atmospheric pressure to the installation in order to drain off the water in the system above B, the content of the tank would be siphoned into the water supply system. This occurs because of an unequalization of atmospheric pressure. Withdrawing the water from the pipe at the low fixture level produces a vacuum or a pressure in the pipe of less than one atmosphere. The pressure of the

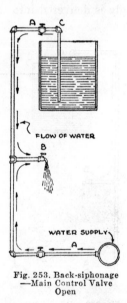

Fig. 253. Back-siphonage
—Main Control Valve
Open

atmosphere bearing down on the surface of the water in the tank is greater than the minus pressure in the pipe, and the contaminated water is forced into the pipe by this pressure.

It is not necessary to set up as favorable a condition as that described above to produce back-siphonage. Back-siphonage can occur even though the water supply is not shut off.

The water retained in a distribution system not in motion has a static or constant pressure practically equal to that present in the main. Drawing water at B in Fig. 253 would tend to draw water from points C and A. Practical installation and laboratory tests have proved that this arrangement is one in which back-siphonage

commonly occurs. It usually is the result of undersized water pipe
or heavy draw on the riser and can be eliminated only by careful
planning of the distribution system.

Another type of cross connection involves the principle of gravity
and occurs in plumbing installations where the water supply devices
are submerged. This objectionable condition is illustrated in Fig.
254.

The tank in Fig. 254 represents any plumbing fixture or appliance with its water supply entering the vessel below its highest
water level. The piping arrangements are identical with those of
Figs. 252 and 253, and may be any form of water distribution system. The water supply pipe, being below the contaminated tank
content, serves as a waste for the vessel should the proper condi-

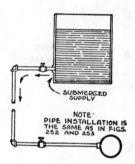

Fig. 254. Back-siphonage
—Submerged Supply

tion prevail. All that is necessary to permit the contaminated water
to enter the water supply system is to shut the valve at the base and
drain the line at any point below the level of the fixture. This condition may also occur with the water distribution system under pressure and may be the result of undersized or improperly installed
water pipe. Once bacteria enter the drinking water supply, serious
illness is likely to result.

CROSS-CONNECTIONS IN PLUMBING FIXTURES

Plumbing fixtures of every kind may constitute a cross-
connection unless the water supply devices are properly installed
and adequately protected against the possibility of siphonage. The
water closet is probably the greatest source of trouble.

Water Closets. A water closet cross-connection is shown in Fig. 255. It consists of a siphon washdown closet provided with a flush valve as the water supply medium. In order to become a cross-connection the fixture must have its outlet obstructed as shown in the illustration. The contaminated water has reached an overflow level and has completely submerged the flushing rim. When a condition of this kind occurs a perfect siphon has been formed and, in order to draw the content of the bowl into the supply line, water

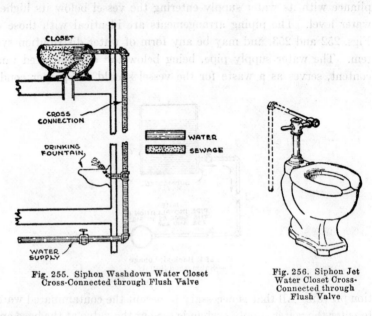

Fig. 255. Siphon Washdown Water Closet Cross-Connected through Flush Valve

Fig. 256. Siphon Jet Water Closet Cross-Connected through Flush Valve

must be drawn from the lower drawcock. The drawcock may represent another closet, wash basin, or drinking fountain. The content of the bowl could also be siphoned if any part of the riser was drained to make repair, as is illustrated in Fig. 252.

Fig. 256 shows a siphon jet water closet similar in design to that shown in Fig. 255. The liquid content of this fixture can be siphoned back into the drinking water supply line without the occurrence of stoppage. The jet extended into the dip of the trap forms a perfect siphon, and is extremely objectionable unless some means of protection against siphonage is employed.

Fig. 257 represents a siphon breaker which must be used on

various types of flush valve closets. It is a device installed between the flush valve and the closet bowl that is constructed in such a manner that an atmospheric condition is provided at this point at all times.

There are many kinds of siphon breakers of a practical nature. All of them operate on the same principle but vary in design. Some of these devices are provided with means to overcome spilling of the water onto the floor during the flushing interval. The siphon

Fig. 257. Siphon Breaker

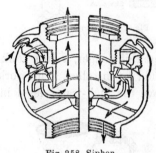

Fig. 258. Siphon Breaker with Air Ports

breaker shown in Fig. 258 has air ports which expose the interior of the flush pipe to atmospheric pressure once the flow of water into the bowl ceases.

Fig. 259 represents a cross-connection of a closet bowl that is provided with a flush valve of the submerged type. This type of valve should be prohibited under any circumstances, because the content of the closet bowl can be drained into the water supply line by gravity.

Fig. 260 represents the common flush tank used as a rule on residence installations. This device does not offer a serious form of cross-connection because the water in the tank is relatively pure.

However, the possibility that drinking water may become contaminated exists. The ordinary float valve is the device through which the water contained in the tank is drawn into the water supply. To

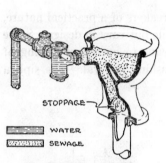

Fig. 259. Submerged Type
Flush Valve

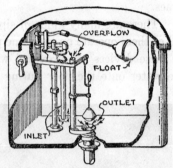

Fig. 260. Common Water Closet
Flush Tank

avert this danger, a flush tank containing an antisiphon float valve should be installed. The working parts of this valve are situated above the high-water level in the tank, and a siphon breaker is

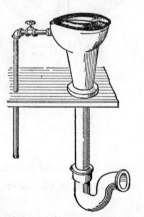

Fig. 261. Common Hopper Closet with
Hand-Operated Flush Valve

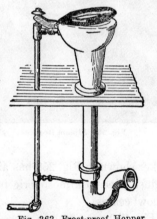

Fig. 262. Frost-proof Hopper
Closet with Automatic Flush Valve

built in. The level of water maintained in the tank is controlled by the *overflow* tube. Retention of an adequate amount of water in the bowl during the period of nonuse of the fixture is assured by the *afterfill* tube. (See Fig. 286, Chapter XXVII.)

Figs. 261 and 262 show two closet fixtures of the hopper type. Outmoded fixtures such as these are still found in some localities.

The common hopper closet has its valve placed below the overflow rim and is connected to a submerged inlet. Thus, in the event of the house drain becoming obstructed, this closet constitutes a serious cross-connection.

The frost-proof closet is another form of cross-connection. The fixture which makes it a frost-proof installation is the drain tube, which is cross-connected with the fixture waste to eliminate the water subjected to freezing temperature. To overcome this form of cross-connection, the tube must be dripped into a floor drain or similar device. The water supply into the closet should be run above the closet bowl and be provided with a siphon breaker at this point. This type of fixture at its best is undesirable, and should be used only when no other can be adapted.

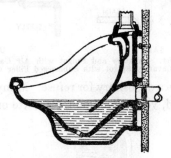

Fig. 263. Siphon Jet Urinal

Urinal. The siphon jet urinal that is equipped with a valve as the flushing device, Fig. 263, offers a possibility of cross-connection. Where it is used, a siphon breaker similar to that recommended for a water-closet flush valve should be installed. When foot-operated valves are used with it, these also must be protected.

The stall urinal, which is flushed by means of a tank, may also become a cross-connection where a float valve is used as the water delivery agent. When this condition exists, a siphon breaker should be installed, or the discharge end of the float valve must be at least 1 inch above the highest possible water level in the tank.

Lavatories. Lavatories may become a cross-connection when the discharge end of the spigot is below the flood-level rim of the fixture. This may occur if the basin overflow is plugged and the water has risen to the height of the rim of the fixture—a condition that must be guarded against. The end of the spigot should have an air gap between it and the flood-level rim of at least twice the diameter of the spigot orifice. Lavatories designed in such a way that the supply device is an integral part of the fixture are not considered safe. Fixtures of this type were common a few years ago, Fig. 264.

Drinking Fountain. The drinking fountain or bubbler that has a submerged orifice, Fig. 265, is a common form of cross-connection.

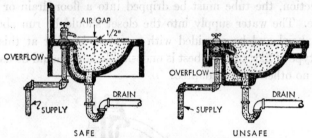

Fig. 264. Lavatory and Spigot with Air Gap (left),
and Lavatory Spigot with Submerged Filler (right).

The fixture may be a depository for refuse. Many people are careless in this respect and, very often, gum, cigarette or cigar butts, saliva

Fig. 265. Fountain
with Submerged
Drinking Orifice

and other objectionable matter are deposited in it. The waste may become stopped or back up into the basin of the fixture. Under these conditions the contaminated waste matter may be drained into the water supply.

The only safe drinking fountain is one which has the drinking orifices associated with the bowl of the fixture located above its overflow rim.

Bathtubs. The bathtub with the submerged bell supply, Fig. 266, was formerly considered an attractive and sanitary fixture. Up to a few years ago it was installed commonly and constituted one of the worst forms of cross-connection. Many homes and commercial buildings still have them, therefore they will continue to be a source of danger for many years. This type of fixture is, at present,

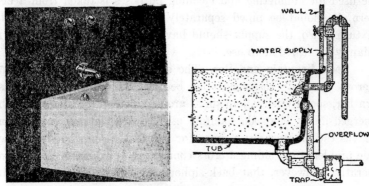

Fig. 266. Safe (*left*) and Unsafe (*right*) Bell Supply on Bathtub

seldom installed. It has been replaced by the over-rim filler type, which is much more satisfactory. The sitz bath and foot tub belong in the same category as the bathtub, because the supply devices on these fixtures are of similar kind.

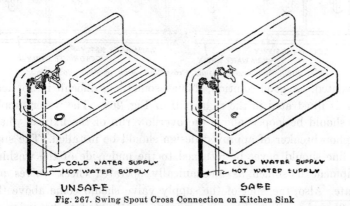

Fig. 267. Swing Spout Cross Connection on Kitchen Sink

Sinks. The attractive swing spout faucet used on sinks **may** be a source of water contamination when one side of the faucet is served by a polluted supply such as a cistern, Fig. 267. This is often the case where water is used that, because of its high mineral content, is unfit for washing purposes unless it has been treated previously. Cistern water was used extensively a few years ago, and it was supplied to the water supply pipe by means of a pump. Thus the danger of by-passing the cistern water into the cold-water supply through the swing-spout faucet was present. When water to be used for laundering and cleaning purposes is taken from a cistern, it should be piped separately and directly from cistern to fixture. Also, the supply should have an individual faucet that is plainly identified as to use.

Soda Fountain and Bar. The soda fountain provides a dangerous type of cross-connection because its compartments—glass washer, spoon washer, and sink—are supplied with water by a submerged inlet. The soda fountain usually is a compact fixture that leaves the plumber little space for run of the piping. This makes the problem of avoiding a cross-connection a difficult one. It is imperative, however, that back-siphonage of contaminated water be

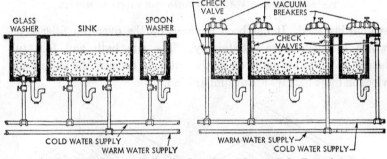

Fig. 268. Submerged Supply Cross-Connection of Soda Fountain

prevented. A good solution, though sometimes difficult in application, is illustrated in the right half of Fig. 268. The water supply line should be looped above the overflow rim of the fixture, where a siphon breaker of approved design should be installed. The supply line should never be returned to the underside of the washing equipment inasmuch as mechanically operated check valves are unsafe. Also, the inlet of the supply valve should come above the overflow rim of the fixture a distance calculated to be twice as great as the size of the inlet opening.

Hospital Fixtures. Many appliances installed in hospitals have, in recent years, been discovered to be dangerous from the standpoint of possible cross-connection. Sterilizers, aspirators, bidets, and bed pan washers have been shown to be the means by which harmful bacteria may be conveyed to the drinking water supply.

Manufacturers of hospital equipment are today cooperating with health agencies to eliminate this danger. Competent men are employed to design equipment that is absolutely safe. Older institutions are making changes, or are replacing their equipment to help prevent the contamination of water, and, if this program continues, which it undoubtedly will, the hospitals probably will be the first institutions to eliminate cross-connections between waste and water supply.

To show the reader just where to look for this fault in hospital

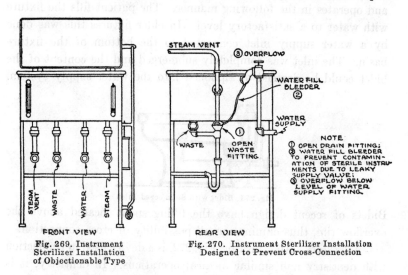

STEAM VENT

③ OVERFLOW

WATER FILL BLEEDER ②

WATER SUPPLY

WASTE   ① OPEN WASTE FITTING

NOTE
① OPEN DRAIN FITTING;
② WATER FILL BLEEDER - TO PREVENT CONTAMIN- ATION OF STERILE INSTRU- MENTS DUE TO LEAKY SUPPLY VALVE;
③ OVERFLOW BELOW LEVEL OF WATER SUPPLY FITTING.

STEAM VENT   WASTE   WATER   STEAM

FRONT VIEW
Fig. 269. Instrument Sterilizer Installation of Objectionable Type

REAR VIEW
Fig. 270. Instrument Sterilizer Installation Designed to Prevent Cross-Connection

equipment a number of hospital appliances are described and illustrated in the following paragraphs.

Sterilizer. The instrument sterilizer shown in Fig. 269 is one of the most common and dangerous types of cross-connection. The purpose of this device is to sterilize the instruments used in surgery before and after each operation. The process involved in sterilizing is to fill the sterilizer to a predetermined level with water delivered into it by a submerged supply. Steam, under pressure, tends to heat the water to a temperature of 212°F. in the open type sterilizer, and to a higher degree in the pressure installation. There is a definite possibility of drawing the content of the sterilizer into the drinking water under these conditions. Modern sterilizers are provided with vacuum breakers, and their waste is dripped into an open funnel connection in the waste line, which prevents backing up of the waste into the sterilizer proper. Fig. 270 illustrates a sterilizer of the nonpressure type together with its connections to the plumbing system.

Bed Pan Washer. The bed pan washer is an appliance used to sterilize the bed pan after it has been used by the patient. Its foot-action flush valve should be protected with a vacuum breaker.

Bidet. The bidet is a form of bath used in hospitals to wash the lower extremities of the body. The fixture is shown in Fig. 271 and operates in the following manner. The patient fills the fixture with water to a satisfactory level. In older fixtures this was done by a water supply inlet connected to the bottom of the fixture basin. The inlet was completely submerged and the content of the bidet could be drained or siphoned into the water supply system.

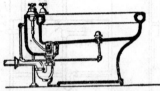

Fig. 271. Bidet with Submerged Inlet

Bidets of recent design have the filling spout located above the overflow rim, thus eliminating all possibility of cross-connection.

Aspirators. An aspirator, Fig. 272, is a device used in connection with dentistry and similar medical operations. In dentistry, it is used to convey saliva from the mouth and drain it to a waste terminal. The vacuum is created by a flow of water passing an orifice that is associated with the tube leading to the patient's mouth. Direct connection with the water supply and the waste makes this installation an objectionable one.

Fig. 273 represents the proper manner of connecting this unit to the plumbing system. Arresters are used to prevent saliva from being drawn back into the water supply. The waste must be dripped as the illustration indicates.

Commercial Appliances. There are many commercial appliances which may become cross-connections unless the individual who installs them uses good judgment in devising the piping layout. Under this head such devices as condensers, filters, water softeners, pumps, refuse cookers in slaughter houses, laundry equipment and so on may be catalogued. To go into specific detail with respect to each of these units would require many pages, and space does not permit such an extensive analysis. There are, however, a few precautions appropriate to all installations and these are to be rigidly observed.

The water supply to these units should be a pure and wholesome one and should be administered to the device from a source which does not come in direct contact with its content. It may be fed into the unit above its overflow rim, or pumped from an open surge tank that is provided with an over-rim filler.

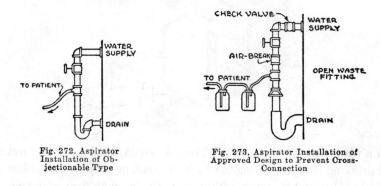

Fig. 272. Aspirator
Installation of Ob-
jectionable Type

Fig. 273. Aspirator Installation of
Approved Design to Prevent Cross-
Connection

The waste connection must always be of the dripped variety, indirectly connected to the waste disposal system. As an added protection the installation must be equipped with approved siphon breakers and check valves to give added protection.

Steam Cross=Connection. Another form of cross-connection commonly found in industry is the direct heating of the water in the distribution system by high pressure steam for toilet purposes. Serious scalding has resulted from careless regulation of this system of hot water supply by jack-of-all-trades maintenance men. The solution to the problem is to employ only licensed plumbers for maintenance work in office and factory buildings.

Fig. 274 illustrates how a cross-connection of this type is formed. The water supply is installed in the usual manner and serves a wash sink in an industrial toiletroom. It also serves water closets, bubblers, and lavatories.

The steam is taken from a high-pressure main and passed through a valve which reduces its pressure to just a few pounds. When the employees wish a supply of heated water, the controlling valve of the steam pipe is cracked to admit a sufficient amount of steam to produce the desired temperature of water. Should the steam valve be left open and the reducing valve fail, the high temperature of the steam would heat the content of the cold water line

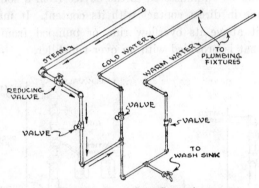

Fig. 274. Steam Cross-Connection

to a temperature above 212°F. Thus, a person using a water closet or drinking fountain during the interval would be subjected to a blast of exceedingly hot vapor rather than the normal flow of water. Painful scalding would result.

To pass over the possible harm resulting from cross-connection without indicating the need for correction and planning safe installations for the future is willfully to ignore the interests of public health and safety.

Precautionary measures of the kind discussed in this chapter apply also to lawn sprinkler systems and to cross-connected fire line installations.

CHAPTER XXVI

FIRE LINE INSTALLATION

The installation of fire lines or fire standpipes in tall buildings is one of vital importance, and construction of this fire control mechanism is often under the jurisdiction of the plumber. The problem of extinguishing a fire in a tall building is a complex one. The fire, especially in tall buildings, may begin in some isolated room which cannot be reached with water delivered by the department's equipment. Under these conditions, the installation of a fire standpipe within the building usually offers a solution. Any person of intelligence and initiative can manipulate the standpipe equipment and quickly bring the fire under control. The standpipe offers a convenient supply of water for firemen, making it unnecessary for the fire department to transport and use much expensive fire fighting apparatus. This device also avoids the costly repair that is necessitated by delay in obtaining water supply.

Fire insurance underwriters realize the advantages of standpipe installations and when a proper installation is provided the insurance rate becomes materially less. This economy reduces the cost and maintenance of a fire system over a period of years; thus the installation becomes an asset to the building.

Larger cities having local fire ordinances generally demand this form of fire protection, and most buildings used for commercial or industrial purposes provide it.

TYPES OF FIRE LINE INSTALLATION

The construction of a fire line installation is not difficult and does not require extended knowledge of engineering. There are a few simple rules established by fire protection agencies which must be observed, but beyond this, any mechanic with a background of pipe fitting can produce a good installation. As was previously stated the construction of a fire standpipe may be under the jurisdiction of the plumber. This is because of his knowledge of the water distribution system and his awareness of the serious consequences likely to result from cross-connection with a contaminated source of water.

The three types of systems generally approved are:

1. Wet standpipe.
2. Wet standpipe with siamese connection.
3. Automatic system.

Dry standpipe installations were common a few years ago, but they have become obsolete and today are seldom installed except on buildings erected for temporary or emergency use.

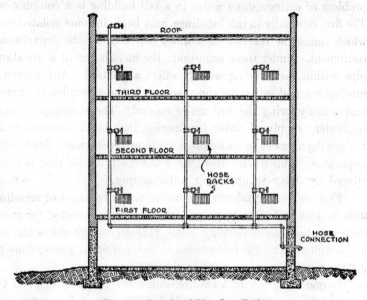

Fig. 275. Dry Stand-Pipe Installation

The dry standpipe offers a means for the local fire department to attach pumping equipment and quickly deliver a flow of water at the upper floors of a building. However, no water is available until the fire department arrives, and usually the time element is important in controlling the spread of fire.

The dry pipe installation, Fig. 275, consists of an arrangement of pipe having a connection on the outside wall of the building, to which the fire department may attach its equipment. From the connection a main is suspended from the basement ceiling to locations of the fire standpipes. The standpipes are run vertically through the upper floors of the building and are generally exposed in corridors, stairways, or places where they may be seen readily. Hose racks or stations are provided on each floor.

The number of standpipes is determined by the size of the building. Each rack is provided with from 50 to 100 feet of hose and serves an area which can be reached with a stream of water through this amount of equipment.

The diameter of the standpipe varies from 2 inch for buildings two to three stories in height to 6 inch for buildings more than six stories tall. For definite information as to the required diameter of dry standpipe installations, local fire control agencies should be consulted, because standards vary.

The standpipe is installed in much the same manner as the riser of the water distribution system. The same principles of support, change of direction, and general workmanship are applied.

Wet Standpipe. The wet standpipe installation is directly associated with the water distribution system and is under the same pressure. Water is immediately available, therefore the installation is a very practical one.

The wet standpipe does not have facilities by which fire department pumping equipment may be connected to it. This is a disadvantage, although not a serious one. This type of installation is primarily intended for smaller buildings which do not offer the problems in fighting fire that are common to tall buildings. Equipment can be so disposed as to cope with a blaze in a building of moderate height, and little difficulty is encountered in directing a stream of water where it is needed.

Fig. 276 shows the connection of a standpipe to the water distribution system. A branch taken from the domestic water distribution main is provided with a gate valve of the manual control type. The valve should be equipped with a tag stating that it serves a fire standpipe and should be allowed to remain open at all times. A drip connection at the base of the riser is essential. The fire standpipe is then extended vertically through the floors of the building. Cabinets situated on each floor are equipped with a hose and nozzle connected to a wheel-handle gate valve. The hose should be folded on a swinging rack of the automatic type, which allows the person

using it to grasp the nozzle firmly, turn on the water, and then make
his way to the room in which the fire is burning.

The hose used on this kind of an installation generally is of standard 2½-inch unlined linen. About 50 to 75 feet are essential, and a
nozzle of standard design (1 or 1⅛-inch diameter) is required. The
number of hose cabinets is determined by the size of the building,
and the units must be located so that all parts of the building are
protected.

The hose cabinet may be installed flush with the plastered surface of the wall and should not be more than 6 feet from the floor.

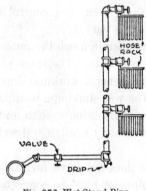

Fig. 276. Wet Stand-Pipe
Installation

This height permits persons of average size to reach the cabinet and
operate its equipment efficiently. The hose cabinet should be
labeled "Danger, for Fire Purposes Only." The hose itself should
be tested periodically to see if it has deteriorated, as often happens
owing to infrequent use of the equipment.

Wet Standpipe with Siamese Connection. The wet standpipe
with siamese connection to which the fire department can attach its
pumping equipment is considered one of the most practical types of
fire protection installations. This system is commonly used and
can be adapted to buildings of all types. The fire standpipes are
under constant pressure, being served by the domestic water supply
system, and, should additional water be needed, the fire department
pumper may be attached to the outside siamese connection to provide increased pressure.

There are some disadvantages in this type of installation. The connection with the domestic water supply system is objectionable, because a cross-connection may occur if the fire department sees fit to take its supply of water from a river, sewer, or drainage ditch. Under these circumstances, the water from the department's pumping unit could be injected into the distribution system of the building and be conveyed into the water mains throughout a large area, possibly contaminating them. The installation should be carefully

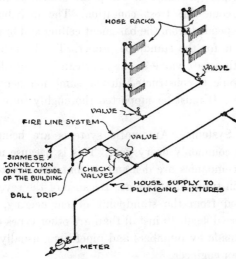

Fig. 277. Wet Stand-Pipe Installation with Siamese Connection

supervised by sanitary authorities and be installed by competent mechanics to avoid the possibility of water pollution.

Fig. 277 represents a wet system of standpipes with outside siamese connection, and illustrates the manner in which it can be connected to the water distribution system.

The connection extending between the domestic water system and the fire distribution main, indicated in the illustration, must be provided with two check valves and a manually controlled gate valve. The back flow devices prevent contaminated water, delivered into the fire standpipe system through the siamese connection, from being forced into the water distribution piping because of the tremendous pressure created by the pumping apparatus. The gate valve is an added safeguard and may be closed by the fireman in the routine of operations.

The portion of the fire distribution system that extends to the siamese connection must be a dry installation, because it is exposed to freezing temperatures. A check valve is provided to serve this need. This arrangement allows the standpipe installation to be full of water under normal city pressure at all times, and prevents the water from escaping through the outside siamese connection. A ball drip is provided at the point indicated on the illustration to overcome the frost condition. The distribution main may then be suspended from the basement ceiling and branches may be taken from it for the standpipe proper. The branch must also be provided with a gate valve so the riser can be controlled.

The standpipe is constructed in the same manner as are the wet installations. It must be supported thoroughly to avoid breakdown.

Automatic Systems. Automatic systems are being installed more and more commonly every year. This is because many cities and insurance companies are demanding that these systems be installed in certain types of buildings. Because of the several factors to be considered from the standpoint of engineering, automatic systems are more difficult to install than are other types of systems. Installation is made by plumbers and pipefitters, usually under the supervision of an engineer.

In installing an automatic system the first requirement is to determine the size of the main, which calculation is based on maximum demand load, with all outlets open. A main supply valve having pressure gauges and automatic control valves is installed before lateral lines are connected. These laterals branch out from a feeder pipe large enough to feed the entire number. The laterals run the length of the ceiling and are spaced 10 to 20 feet apart, the interval depending upon the fire potential of the area in question.

The pipeline, which reduces in size as it nears the end of the laterals is large enough in diameter to supply the entire sprinkling system. The heads of the sprinklers are similar to those of a lawn sprinkler system. A fusible strip of metal that melts at a predetermined temperature starts the system in operation. Thus a steady stream of water is immediately directed upon the area affected.

In the case of large office buildings and manufacturing plants, a separate control system is installed for each floor.

GENERAL DETAILS

Valves, Hose, etc. All valves of the fire system must be tagged properly to identify them from other controlling devices. Fire system piping usually is painted a bright red.

The hose cabinets generally are provided with automatic racks and 2½-inch unlined linen hose. A 1⅛-inch or 1-inch nozzle is standard. The hose must be of sufficient length to reach a point within 20 feet of any point of the building. The cabinets must be located so the entire building may be served.

The valves and equipment used on fire standpipe systems must be of approved design. They must fit the department's equipment in order to be effective.

Size of the Standpipe. Correct size is determined by the height of the building and the number of outlets provided. Specifically, the sizes are as follows: For buildings up to six stories in height (approximately 75 feet), use 2½- to 4-inch pipe. For buildings more than six stories in height (over 75 feet), use 6-inch pipe.

Privately Operated Systems. Some buildings, especially those where modern fire-fighting equipment is not available, have storage reservoirs and private pumping equipment to combat fire. Installations may be of the pneumatic air pressure variety or overhead supply and embody the principles of domestic water supply systems.

CHAPTER XXVII

PLUMBING FIXTURES

Plumbing fixtures are receptacles for wastes which are ultimately discharged into the drainage system. In the past 30 years the plumbing fixture has changed from a crude, unattractive device to one decidedly scientific in its operation, sanitary in every respect, and so designed and constructed that it adds much to the room in which it is installed. It can be obtained in practically any color to harmonize with the color scheme favored by the building occupants.

Modern plumbing fixtures are constructed of cast iron coated with a nonabsorbent enamel, vitreous china, or pressed steel.

The iron used in the construction of enameled fixtures is of good quality. It is heated to a molten state and then poured into molds. When the metal has cooled, the rough edges are chipped off and the entire fixture is sandblasted to make it smooth. The rough casting is then placed in a large oven and heated to a bright red. The enamel, which has a high silicon content, is powdered onto the red-hot casting with automatic shakers that are operated manually. Each fixture is given two rough coats and one finish coat and is then allowed to cool slowly. The process is an interest- ing one to observe. After the fixture has cooled, it is thoroughly in- spected for defects.

The vitreous china plumbing fixture is constructed of clay to which water has been added in sufficient quantity to give the mix- ture the consistency of soft cement. The mixture is then poured into plaster of Paris molds and allowed to dry. After it has dried but is still plastic, workmen remove the fixtures from the molds and work each piece until it is smooth. The fixture is then baked in a kiln and given a coat of enamel, after which it is again fired under high temperature. The finished product is an impervious and sani- tary plumbing fixture.

Plumbing fixtures must be handled carefully by the installer. They do not stand abuse, and roughness on the part of the mechanic usually results in breakage. This caution may be heeded by the consumer as well. The materials used in plumbing fixtures are on a par with those in fine chinaware and should be treated with care. Using plumbing fixtures for garbage depositories or as objects to support body weight is a costly habit. It is unwise to use scratchy, gritty soaps for cleaning purposes, and chemical compounds for the removal of accumulated scale are, as a rule, also objectionable.

In the following paragraphs a general description of the mechanics, design, flushing devices and installation procedures of plumbing fixtures is offered.

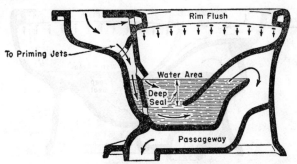

Fig. 278. Siphon Washdown Water Closet
Courtesy of Kohler Company, Kohler, Wis.

WATER CLOSETS

The water closet is the most commonly used fixture and, from the standpoint of sanitation, it is one of the most efficient. Up to 25 years ago, the crude, unscientifically constructed water closet was considered a luxury, and very few buildings were equipped with them. Sanitation standards, however, increased steadily, and, at present, very few buildings, even among those located in rural communities, are without water closets.

The water closet is a plumbing fixture used to convey organic body wastes to the plumbing system. It is made from imported clay of a quality equal to that used in the manufacture of fine china. The clay, which is obtained in powdered form, is mixed with water, and when it is of proper consistency it is poured into plaster of Paris molds. A water closet is cast in about thirteen pieces, all of which are molded together by skilled mechanics to form the closet bowl. As a final operation, the closet is treated with liquid glaze, placed in dry kilns, and fired to a temperature of 2500°F. This process renders it entirely impervious to moisture.

Types of Water Closets. There are two types of water closets used in standard plumbing installations; namely, the siphon washdown closet and the siphon jet washdown closet.

The siphon washdown closet, Fig. 278, is considered highly sanitary and generally is used for residence service. The siphon jet closet, Fig. 279, is similar to the washdown closet in appearance, but its design embodies an integral jet which produces a more rapid siphon action. The jet closet is not as quiet in operation as the

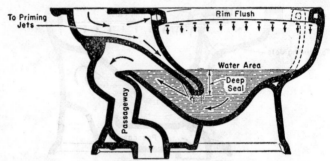

Fig. 279. Siphon Jet Water Closet
Courtesy of Kohler Company, Kohler, Wis.

washdown closet and generally is used for industrial installations.

Operating Principle of a Closet Bowl. *Siphon Washdown Closet.* The quality of a closet bowl is established by the efficiency with which it eliminates organic wastes. Its action, which is that of a siphon, is produced by a decrease in atmospheric pressure on the outlet side of the trap. When a closet is in a neutral position it contains a seal of water, and the atmospheric pressure is the same on both sides of the trap.

The construction of a water closet trap appears to be the same as that of an ordinary trap, with the inlet and outlet passageways of same diameter. This, however, is not the case. The diameter of the outlet passageway is smaller than that of the inlet passageway. It is built with short offsets and turns so that the flow of water through it is retarded, thus causing a head of water to be built up in the fixture. When the closet is flushed, the water passing through the outlet passageway eliminates the air it contains, thus producing a partial vacuum. Atmospheric pressure and the head of water retained on the inlet side of the trap force the collected organic solids from the fixture. Fig. 280 illustrates how this siphonic action occurs in a siphon washdown closet bowl.

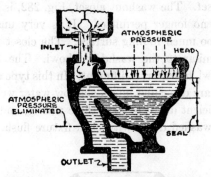

Fig. 280. Siphonic Principle of Reverse Trap Washdown Water Closet

Siphon Jet Washdown Closet. It is the objective of all water closet designers to produce siphonic action as rapidly as possible. For this reason some closets are constructed with an integral jet connected to the dip of the trap. This type of fixture is called the

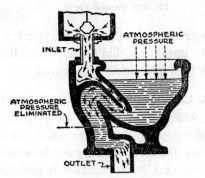

Fig. 281. Siphonic Principle of a Siphon Jet Water Closet

siphon jet closet. Its action is identical with that of a washdown closet, but it provides faster elimination of waste, because a stream of water is delivered immediately from the closet spud to the outlet passageway of the fixture, making it unnecessary to have a head of water in the bowl. Both types of closets are standard. Fig. 281

illustrates how the action of the jet induces flushing activity in the siphon jet closet bowl.

UNSANITARY CLOSETS

Washout Closet. The washout closet, Fig. 282, is obsolete and its installation is no longer permitted. It is very unsanitary because of having too much fouling surface. The closet from outside appearances is similar to the washdown bowl. The interior construction of the bowl, however, is different. In this type of fixture the organic materials are deposited into a shallow water well. The well does not have sufficient depth to float these wastes, and the sides of the bowl are not washed down with each fixture flush. Fouling of the fixture results.

Fig. 282. Washout Closet

Pan and Plunger Closets. Pan and plunger type closets were the first closets to be installed within the building. Both fixtures are obsolete and have been practically eliminated.

INSTALLATION OF THE CLOSET BOWL

The plumber, when called upon to set a closet bowl, follows definite procedure developed through experience. The connection of the closet bowl to the plumbing system must be water-tight and be finished to the extent that its appearance is satisfactory.

In setting a closet bowl, the floor of the bathroom is swept clean to avoid chipping the glazed surface of the fixture itself. Next, the floor flange is inspected and then the closet bolts are inserted. The bowl may temporarily be set on the closet flange to study its appearance—often the bowl is slightly warped from the kiln temperature and this defect is discovered by close inspection after it is set on the floor. Following this, a ring of grafting wax or bowl-setting wax, or Feder's oilless bowl-setting putty is placed in the recess at the bottom of the bowl. The next step is to wipe the bolt holes clean and put the bolts in place, after which the bowl is set on the closet ring and a bolt flange and nut are placed on each bolt.

The bowl is now drawn lightly to the floor by tightening the closet bolts. All excess of filler material is squeezed out in doing this. The material remaining is then wiped smooth to obtain a clean finish. For a smooth and flawless joint, white cement should be used to fill in around the base of the bowl.

Because the oil in putty is liable to discolor the floor around the bowl, this material should not be used on wood floors. Also, it should be kept in mind that in time the oil in the putty will dry out, leaving a "lifeless" joint.

Wax does not dry out, therefore it is recommended over other substances used for bowl setting. A bowl set with wax can be taken up after years of use and the wax used again in resetting the fixture. This filler can be relied upon to provide a lasting joint.

Sponge-rubber gaskets are sometimes used to secure the closet to the floor, but this procedure is not recommended.

When the flush elbow tank is used, the flush elbow is installed in the closet bowl and the tank, with flush valve, is leveled and connected to the fixture. The tank is supported by means of wood screws inserted into backing already provided in the partition. The water supply to the flushing device is also completed, the toilet seat attached, and the fixture is ready for use. The plumber may place china caps over the exposed nuts of the closet bowl to lend a trim appearance.

FLUSH DEVICES FOR WATER CLOSET

A flush valve is a device used for eliminating fecal and liquid wastes from plumbing fixtures. It is a very efficient unit and may be applied to water closets, urinals, and hospital fixtures. It is commonly installed in place of the flush tank for commercial and industrial installations because of its rapid flushing action.

A flush valve can be operated with satisfaction at 10-second intervals, which assures a sanitary condition. The efficiency of the valve is due to its small number of internal working parts. The principle involved in its operation is a simple one. Equalization of pressure on both sides of a relief valve provided in the body of the valve proper is the basis of its operation.

There are many types of valves in use today. Plunger, diaphragm, piston, and oil dashpot valves are those most commonly installed. Although they vary in outside appearance, they are similar in their working principle. All flush valves must be equipped with an antisiphon device.

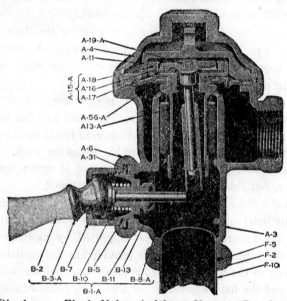

Diaphragm Flush Valve (without Vacuum Breaker)

Valve		Handle		Outlet	
A-19-A	Relief Valve	B-1-A	Handle Assembled	F-5	Slip Joint Rubber
A-4	Outside Cover	B-2	Grip	F-2	Coupling Assembled
A-11	Inside Cover	B-3-A	Shank	F-10	Outlet
A-15-A	Disc Complete	B-5	Bushing		
A-56-A	Segment Diaphragm	B-8-A	Plunger		
A-17	Relief Valve Seat	B-10	Spring		
		B-11	Packing		
		B-13	Packing Nut		

Fig. 283. Flush Valve
Courtesy of Sloan Valve Company, Chicago

Construction of a Flush Valve. The diaphragm type of valve, Fig. 283, furnishes the basis of design of many other flush valves. To avoid repetition it is well to discuss its construction in preference to that of other types.

The flush valve is connected directly to the water supply piping and its interior consists of two chambers built into a brass chrome plated body. The two chambers are separated by a relief valve mounted on a rubber diaphragm. The upper chamber is connected with the city water supply by means of a small by-pass, and the lower, or flushing, chamber is connected directly to the water supply. The relief valve is equipped with an extension piece which contacts the valve handle. Pressure on the handle causes the valve to function.

All flush valves should be provided with siphon breakers of approved design to avoid back-siphonage.

Operation of a Flush Valve. The valve is in a neutral position when the pressure of the city water supply is equal on both sides of the relief valve. Pressure applied to the handle of the flush valve releases the water under pressure on the inlet side of the

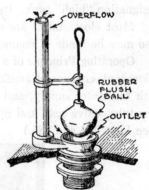

Fig. 284. Elevated Antisiphon Float Valve   **Fig. 285. Rubber Ball Flush Valve**
Courtesy of Scovill Manufacturing Co., Waterville, Conn.

relief valve. The inequality of pressure which now prevails lifts the diaphragm from its original position and admits water from the distribution system to the closet bowl. When the valve handle is released, the relief valve seats itself, and water, under city pressure, is admitted to the inlet side of the valve through a by-pass orifice of small diameter. This operation requires a 6- to 10-second interval, after which the rubber diaphragm is again forced down on its seat, shutting off the flow of water into the closet bowl. With each flush of the fixture the valve performs the operation.

THE FLUSH TANK

Flush tanks are used in connection with water closets and urinals, and are especially satisfactory in residences. They have the advantage of quiet operation and require less pressure and a smaller volume of water than do flush valves. Flush tanks may be served with smaller supply piping than flush valves use, which lowers the cost of installation.

There are three types of low flush tanks in common use, namely, the wall-hung type, with flush elbow, the close-coupled closet combination which rests on the bowl, and the one-piece combination, in which the closet and the bowl are a single unit. Low flush tanks operate by gravity flow.

The high tank, although still used for home installations in some few localities, and occasionally for industrial use, has long been regarded as obsolete so far as new installations are concerned. The fixture is installed about five feet above the bowl, hence the designation "high" tank. Its operation is based on siphonic action.

Most closet tanks are constructed of vitreous china, but they also may be made of enameled iron.

Operating Principle of a Low Tank. The working mechanism of a low tank consists of an elevated antisiphon float valve connected with the water supply, and a siphon valve, consisting of an overflow tube, valve seat, and rubber flush ball attached to a trip lever. (See Figs. 284 and 285.)

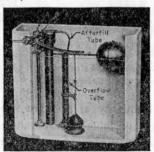

Fig. 286. Low Tank Water Closet
Courtesy of Kohler Company, Kohler, Wis.

The closet tank, Fig. 286, receives its supply of water by means of a device called a float valve. It contains a brass seat, onto which is faced a soft rubber compression packing. The seat usually is in an upright position. The rubber washer is fastened to the float valve plunger, which moves up and down over a brass seat. The plunger is connected to a brass rod equipped with a float valve.

Pressure on the flush handle starts the tank in operation. The float valve drops in the tank with the receding water. This lifts the plunger from the seat and allows water to enter the closet tank. The water level in the tank is controlled by the setting of the float valve. When the water attains a given height, the plunger is forced onto the float valve seat. This shuts off the water supply.

To flush the closet bowl, a handle on the outside of the tank is tripped. This operation raises the rubber tank ball from its brass seat. The water stored in the tank is rushed by gravity into the water closet with such velocity that siphonic action occurs in the bowl. The tank ball floats on top of the receding water. It is drawn onto the seat by the action of the water passing into the closet bowl. When the ball has seated itself, the flush is stopped and the tank again becomes water-tight and refills with water.

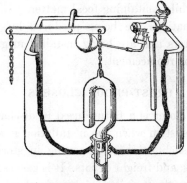

Fig. 287. High Tank Water Closet Mechanism Assembled

The flushing valve is constructed with an overflow pipe, which is by-passed into the closet bowl below the brass seat of the valve. The overflow takes care of any leakage of the ball cock, allowing the surplus water to enter the bowl rather than overflow the tank. The afterfill tube is also dripped into the overflow pipe.

It is essential that closet bowls be protected with an afterfill tube. Closets are flushed by siphonic action, which draws practically all the water seal from the trap. This would allow sewer air to enter the building were it not for the re-sealing of the closet trap through the afterfill tube, which is part of the float valve.

Operating Principle of a High Tank. The high tank mechanism consists of a copper float valve, which is practically identical with that of the low tank mechanism, and a flush valve, consisting of a gooseneck siphon equipped with a featheredged washer.

The high closet tank, Fig. 287, is similar to the low tank in operating principle, except that the flushing valve is in the form of a siphon. The high tank produces a forceful flush and sometimes is favored for industrial installations.

Combination Closet. A water closet of a distinct type known as a *combination* is commonly installed today. In this type of fixture, the bowl has an extension at the rear upon which the tank rests. Advantages of the combination closet are quiet operation and elimination of the flush bowl elbow and 2-inch slip-joint nuts. The tank is bolted to the floor by means of a cushion gasket that is situated between the bowl and the tank.

A disadvantage of fixtures of this type is that they are disposed to back-siphonage if, through failure of the water supply, the bowl becomes clogged while containing fecal matter.

Closet combinations are being produced by many plumbing fixture manufacturers and are being installed in great numbers in new homes, and for replacement.

FROST-PROOF CLOSETS

A frost-proof closet is a fixture used to discharge fecal wastes into the drainage system when the installation is subject to freezing temperatures such as those occurring in storage warehouses, refrigeration plants, and freight depots. It is generally of the hopper variety and is sometimes called a dry closet. Any form of dry closet is extremely unsanitary because it offers too much fouling surface. It should not be used except where conditions permit no other type of installation.

Construction of a Frost-proof Closet. The frost-proof water closet, Fig. 288 is composed of a number of parts:

1. Hopper   4. Tank
2. Trap
3. Seat-operated flush valve

The United States Bureau of Standards recommends the installation of frost-proof closets only in compartments that have no direct access to a building used for human habitation or occupancy. The soil pipe between the hopper and the trap should be of lead or cast iron enameled on the inside. The waste tube on the valve should not be connected to the soil pipe or sewer.

In most cases where a frost-proof closet must be installed, it is agreed that any water pollution which might occur is the responsibility of the installer, rather than the manufacturer. One type of frost-proof closet developed to prevent cross-connection is shown in Fig. 288.

The hopper closet is set on the floor flange with putty in the usual manner. It is constructed of cast iron or vitrified china, is cone-shaped, and has a flush rim connected to the flush valve.

The flushing device consists of a three-way valve operated by seat action. The water control valve is of the compression variety with an extended stem. The water valve is always installed below

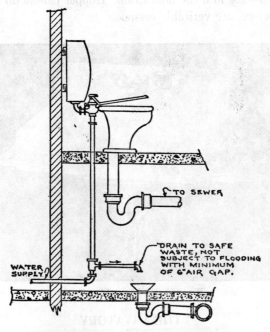

Fig. 288. Construction of an Approved
Type Frost-proof Water Closet

the frost line, and is connected to the public water supply. A pipe of 1¼-inch diameter extends vertically to the flush rim of the closet bowl and then to a vertical galvanized steel tank of about 8-gallon capacity. The tank serves as a storage and compression unit. The bleeder or drip pipe, which protects the valve against freezing when it is not in use, is connected to the compression stop on the fixture side and dripped into the floor drain.

Working Principle of a Frost=proof Closet. When the water closet is in a neutral position the seat is elevated about 3 inches above the closet bowl. Pressure on the seat opens the compression valve which in turn by-passes water under pressure into the compression tank. Water is prevented from entering the closet bowl by a three-way valve arrangement. Release of the pressure on the closet seat closes the compression stop and opens the waterway from the tank to the closet. Water drains from the flush rim and washes the closet bowl. The pipe that is exposed to frost is drained through the bleeder tube into the floor drain. Hopper closets do not siphon, and, therefore, are veritable cesspools.

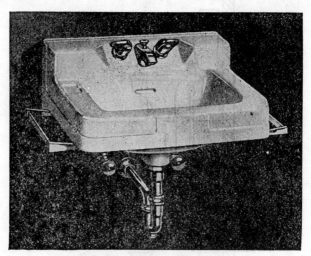

Fig. 289. Wall-hung Wash Basin

THE LAVATORY

The lavatory or wash basin is a fixture of common installation. Lavatories are manufactured in many styles and colors and are equipped with hot and cold water supply and a waste connected directly with the drainage system.

The plumber can add much to the attractiveness of a bathroom by centering the lavatory properly in relation to the other fixtures in the room. He can increase the efficiency of the fixture as well by setting it at a proper height to serve all members of the family. As a general rule 30 inches from the finished floor is adequate, as this height permits the individual to stand in a somewhat bent position so the water tends to drain from the arms into the basin.

TYPES OF LAVATORIES OR WASH BASINS

Wall=hung Basin. The wall-hung basin, Fig. 289, is constructed of enameled iron or vitreous china. The enamel is applied to the red-hot casting in powdered form. The fixture is then allowed to cool slowly to prevent rapid contraction of the metal. Vitreous china fixtures are constructed of clay poured into plaster of Paris molds and finished by skilled workmen.

The wall-hung lavatory illustrated by Fig. 289 is used extensively today. Its integral soap receptacle, overflow, and shelf-back contribute to the convenience of the user and to sanitation as well. The towel bars lend decorative enhancement.

Fig. 290. Pedestal Wash Basin

Dental Lavatory. This fixture is a small, wall-hung basin designed to serve the needs of dental hygiene. It is not ordinarily installed in homes of moderate cost.

Pullman Lavatory. The Pullman or boudoir lavatory has a flat rim and is installed in the same manner as the flat-rim sink. A tile or composition commode-type counter flanks the basin itself.

Pedestal Basin. The pedestal basin, Fig. 290, is constructed of enameled cast iron or vitreous china as previously described. It is distinguished from the wall-hung basin by its pedestal, which rests on the floor and serves as a support for the fixture. The pedestal basin is an attractive fixture for residential installations.

SUPPLY DEVICES

Individual Compression Faucets. The compression faucet is the most commonly installed water supply device. It is constructed of a brass body and has a threaded tailpiece and curved spout. The body contains a machined seat and, in the more expensive types, the seat is removable.

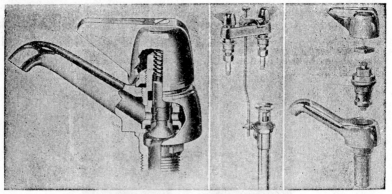

Figs. 291 (*left*), 292 (*center*), and 293 (*right*)
Courtesy of Crane Company, Chicago

A brass stem equipped with a hard rubber washer is fitted into the body of the faucet and is raised and lowered on the seat by a screw thread. The faucet is sealed against water leak between the stem and body by a stuffing box. The exterior of the valve usually is chrome plated.

The newest type of lavatory faucet is that shown in Fig. 291. This faucet closes *with* the water pressure instead of *against* it, as does the compression faucet. Internal working parts are included in an interchangeable cartridge that can be quickly replaced. The stem is of one-piece construction and is lubricated. The packing below the stem threads prevents lime from becoming deposited on the threads and keeps the lubricant from being washed away.

Combination Compression Faucet. The combination faucet shown in Fig. 292 generally is used in connection with pedestal wash basins. It consists of two compression valves joined together with a common spout. The operating principle and construction of the valves are the same as those of ordinary compression faucets.

Self=Closing Faucet. Self-closing faucets, Fig. 293, are used on wash basins in commercial or public buildings. Their use offers economy in that the operation of shutting off the water is automatic. The faucet is of the compression variety and is similar to the ordinary faucet. It is built with a spring assembly and sliding stem instead of a screw stem. Pressure on the faucet handle puts tension on the spring, lifting the washer from the seat. Releasing the handle closes the waterway. Faucets of this variety tend to cause water hammer when they are improperly installed.

Fig. 294. Fuller Faucet   Fig. 295. Lift Lavatory Waste Plug
Courtesy of Crane Company, Chicago

Fuller Faucet. The Fuller faucet, Fig. 294, derives its name from its inventor. The faucet consists of a brass body fitted with a machined seat. An eccentric or cam that is fitted with a handle raises and lowers a Fuller ball, which is attached to a brass stem from the seat. This allows water to pass through the faucet spout. The faucet is screwed into a brass tailpiece. It is made water-tight with a stuffing box. This type of faucet is now infrequently installed.

WASTE CONNECTIONS

Manual Waste. The manual waste (chain and stopper), Fig. 295, is used on low-cost wash basin installations. It consists of a rubber stopper that fits into the waste outlet. Removal of the stopper from the outlet allows the water to drain from the fixture.

Mechanical Pop=Up or Lift Waste. The lift waste, Fig. 296, consists of a brass waste outlet into which a sliding metal stopper is fitted. The stem passes through the overflow and is equipped with a lever handle. Turning the handle over opens and closes the stopper.

Model Waste. The model waste, Fig. 297, is used on the older two-piece wash basins. It consists of a brass pipe equipped with a fitting, into which a seat is machined. The seat is below the water outlet of the basin. A brass tube equipped with a rubber washer is fitted into the seat. Raising or lowering the tube controls the water in the fixture. The interior of the tube serves as an overflow.

Trap. A trap which is connected to the waste pipe on which the basin is installed should be of the **P** variety, Fig. 298. It must be of 1½ inches or more in diameter and of approved design.

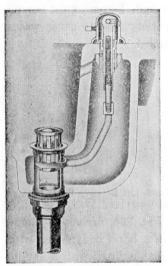

Fig. 296. Bathtub Lift Waste
Courtesy of Crane Company, Chicago

Fig. 297. Model Waste

Fig. 298. **P**-Trap

INSTALLATION OF A WASH BASIN

After the walls of the bathroom have been decorated, the lavatory may be set by the plumber. First the proper height is established and then the basin is centered with the wall mirror customarily associated with it. The hanger from which the basin is suspended may then be fastened to the wall by means of screws which are driven into wood backing that is secured to the rough studding. In tile walls, expansion shields and bolts may be used. The hanger must be leveled carefully so the lavatory will hang correctly.

The faucets and trap tailpieces must now be mounted on the

fixture, using putty under the flanges of the faucets to seal the opening through which they are attached to the fixture.

The trap offset is also screwed to the drainage system at this time. The lavatory may now be placed on the hanger and should be forced down into the slots of the hanger to make it secure. To complete the job, the water supply and trap may be connected to openings roughed in the wall for this purpose. After careful inspection for defects in the fixture, the lavatory is ready for use.

BATHTUBS

The bathtub is a necessity of the modern home. Its contribution to comfort and health is of inestimable value. Bathtubs are constructed of enameled cast iron or steel in colors, and of vitreous china. The enameled iron fixture is the one most commonly used because of its many advantages over the vitreous china tub.

The bathtub, though designed on sanitary principles, does not always live up to the high standard its appearance might suggest. This is especially true when it is used in public baths by many individuals, some of whom may be carriers of infectious diseases. For public use it should be replaced by the more satisfactory shower bath. For residence use, the bathtub is very satisfactory and usually is favored over other types of bathing facilities.

When recommending a bathtub, a few essential qualities should be considered. The tub should have a smooth hard surface and possess a flat bottom. Many accidents occur each year because of the round bottom contained in bathtubs of obsolete design. The outlet should be large to assure rapid discharge of the waste and should be provided with a device for retaining adequate water level. The height of the fixture from the floor is also an important element of design.

The plumber can add much to the appearance of the bathroom by careful installation of the bathtub. He also may increase the sanitary quality of the fixture by using good judgment in the design of its waste.

TYPES OF BATHTUBS

Built=In Bathtub. Various styles of bathtubs are available to domestic needs, and there are special designs to suit the purposes of private and institutional use. The square, recess-type bathtub, **Fig. 299,** has a ledge seat that affords convenience in bathing small children, and also serves to make the fixture useful as a foot bath. An overhead installation is controlled by a combination over-rim bath and shower valve.

Fig. 299. Built-in Bathtub with Ledge Seat

The built-in bathtub is a favorite of home planners. It is constructed of enameled cast iron or vitreous china. Fixtures of this type are available in sizes proportioned to the room. One having an apron on two sides can be installed in a corner space; another is recessed, with all sides concealed except that facing into the room.

Fig. 300. Steel Enamel Bathtub with Over-rim Tub Filler, Shower, and Trip Waste
Courtesy of Briggs Manufacturing Company, Detroit, Mich.

In a color chosen to harmonize with the decor of the room, the built-in bath forms a unit composing fixture, wall, and floor. Besides saving space, this fixture is safer and generally more convenient than the old-style, leg-based bathtub.

The steel enamel, safety bottom bathtub with over-rim tub filler, shower, and trip waste plug is shown in Fig. 300. This fixture has a corrugated bottom that serves to avert the danger of slipping in the bath. Water is diverted from filler to shower head by lifting

Fig. 301. Foot Tub

the small button lever on the tub filler. When the water is turned off, the button lever drops to the filler position.

Foot Tub. Another form of bath is the foot tub, shown in Fig. 301. This type of fixture is seldom installed in homes. It is well

Fig. 302. Sitz Bath

adapted for hospital purposes, however, and its installation is one of fairly common demand.

Foot tubs are constructed of vitreous china or enameled iron, in both built-in and leg styles. The bath has a depth of about 20 inches. An over-rim filler and trip waste plug remove the danger of contamination.

Sitz Bath. The sitz bath, Fig. 302, was at one time a commonly installed fixture, but today it is rarely found outside hospitals and similar institutions. It is constructed of enameled cast iron or vitreous china, and is used chiefly in the care of invalids and the infirm.

Foot tubs and sitz baths sometimes come equipped with a standing waste and submerged inlet. Since installation of these devices is prohibited in most cities, they should never be used. New and approved types of baths of this kind are now available.

Bidet. The bidet, Fig. 303, is another form of bath which is installed chiefly in hospitals. It is constructed of vitreous china or

Fig. 303.   Bidet
Courtesy of American Radiator & Standard Sanitary Corp.

enameled iron. At one time bidets were commonly installed in residences.

Hospital Baths. There are many forms of special bathtubs used in modern hospitals. These are all special fixtures and would require too much space to describe adequately. The principles involved in connecting these fixtures to the plumbing system are the same as those which apply for bathtubs of ordinary use.

WATER SUPPLY DEVICES AND WASTE CONNECTIONS

Combination Overrim Tub Filler. The combination overrim tub filler, Fig. 304, is used in connection with built-in bathtubs. It consists of two compression valves joined with a mixer spout. The construction of the valves is identical with that of the compression

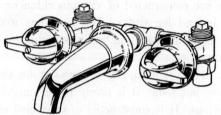

Fig. 304. Combination Over-Rim Tub Filler

faucet. The water supply device should be installed above the rim of the bathtub to avoid possible back-siphonage. Spouts or bells within the tub should not be used, as such devices encourage back-siphonage.

Combination Waste and Overflow. The combination waste and overflow, Fig. 305, generally is used in connection with leg-based bathtubs. It should be constructed of brass of at least 17-gauge

Fig. 305. Combination Waste and Overflow

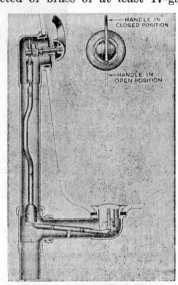

Fig. 306. Concealed Bath Waste and Overflow

quality, and usually is provided with a rubber stopper. The fitting is connected to the bathtub and putty is used to provide a water-tight joint between fixture and fitting.

Lift Waste. The concealed bath waste and overflow, Fig. 306,

is used with built-in bathtubs. The connecting mechanism between
the lever at the end of the tub and the stopper in the outlet at the
bottom is enclosed in a brass tube, the bottom of which connects
with the waste pipe. Turning the handle to the right raises the stopper, thus allowing the water to drain rapidly from the fixture. Turning the handle to the left seals the waste and allows the tub to fill
with water. All parts of this device are easily accessible through
the overflow or waste outlet.

Drum Trap. This type of trap, Fig. 307, is used in many localities where venting is not required by local ordinance. This device
permits the waste to flow through it rapidly. For the purpose, however, it is recommended that either the adjustable Durham deep-

Fig. 307. Drum Trap

seal **P**-trap or the soil **P**-trap be used, as these can be vented adequately. Venting is the means by which atmospheric pressure is
maintained and trap siphonage prevented.

INSTALLATION OF A BATHTUB

The bathtub of the built-in type is usually set in place before
the finish coat of plaster or tile is applied to the bathroom. The
tub should be inspected carefully for defect, as it is difficult to remove and replace the fixture once it has been walled in. A 2x4
plate must be spiked to the rough studs of the bathroom wall to
serve as a rest for the back rim of the tub and prevent it from breaking away from the plaster or tile. The tub can then be set in place
with the back rim fitted on the leveled support, its apron resting
on the rough floor of the bathroom. The tub should also be plumb
with floor and wall surfaces of the room.

The waste and water supply may now be connected to the fixture through the medium of a pipe cabinet installed on the water
supply and waste end of the tub. During the course of building

construction, the tub must be protected against damage by workmen. One effective method for protecting the tub is by covering it with corrugated board.

SHOWER BATH

The shower bath is an installation that is favored by many people. There is no question that a shower bath does offer advantage over the bathtub from a sanitary viewpoint. This makes it a more popular and advisable installation in buildings of public character, such as hotels, schools, gymnasiums, etc., or where more than one person uses the same fixture. Shower baths are also considered a desirable unit of the bath in private residences and apartment buildings. In addition to its appeal from the standpoint of hygiene are the benefits of the invigorating needle spray now provided.

Installation of a shower bath on the upper floors of a building was formerly considered very impractical. Modern methods of tile and marble work have eliminated earlier difficulties and it is now practical to place the shower on any floor of a building.

The plumber can add to the efficiency of a shower installation in a number of ways. He must use care in locating the fixture spray head and mixer valve of the installation. It should be possible to reach the mixer with the right hand from outside the shower enclosure. The spray head must be set so the full volume of the water does not strike the individual in the face, making it difficult for him to see.

TYPES OF SHOWER BATHS

. Shower baths cannot be classified independently. They may consist of one shoulder spray or a series of sprays installed in an enclosed space ranging from 4 feet square for small residential showers to rooms of large dimension in industrial buildings, school gymnasiums, clubs, etc.

There are two distinct types of residential showers. One consists of a metal enclosure and the other is constructed of glazed tile.

Metal Shower Bath. The metal shower bath enclosure, Fig. 308, is constructed of bonderized, galvanized steel to eliminate rusting. The precast stonetex receptor provides a solid, permanently watertight base. All edges are smooth. This shower enclosure can be fitted with a glass and metal door, as shown in the illustration. The size of the shower is 32"x32"x76".

Tile or Marble Shower. The tile or marble shower bath enclosure has been used for residential purposes for a long time. This type of shower bath may be built in almost any form to fit the general pattern of the bathroom and is an attractive installation.

The tile shower bath may be a source of serious trouble. Tile or marble backed with cement mortar is not necessarily water-tight.

Fig. 308. Metal Shower Bath Enclosure
Courtesy of Fiat Metal Manufacturing Co., Franklin Park, Ill.

The walls may crack, because of various circumstances, and allow water to seep through them. To avoid this defect the plumber must install sheet lead safing in such a manner that the escaping water will be confined and conveyed to a safe terminal.

WATER SUPPLY DEVICES

Valve Combination. The individual hot and cold mixer valve, Fig. 309, is sometimes used on industrial type showers or less expensive residential installations. The combination consists of two compression stops cast in one body, which is fitted with a spray delivery pipe. The temperature of the water being outletted by the spray is regulated manually. This method is an objectionable one and many people have been scalded because of a lack of experience in the control of the water. The danger may be overcome by using an approved temperature control device on the hot water supply. A unit of this kind allows cold water to mix with the hot until the water is no hotter than may be used safely.

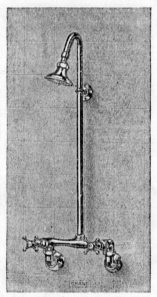

Fig. 309. Hot and Cold Water
Mixer Valve

Mixing Valve. The most desirable type of water supply device for a shower bath is the mixing valve shown in Fig. 310. This is a unit constructed of a brass mixing chamber into which is fitted a hot and cold sliding valve that is provided with jets which allow hot and cold water to pass through them when the handle of the mixer is operated. These valves have been carefully designed to provide safety.

Traps and Strainers. The shower bath may be provided with a strainer-fitting screwed into a 2-inch drum or an ordinary 2-inch **P**-trap. A receptacle of special design, Fig. 311, is commonly used for this installation. The trap must be of normal 2-inch seal so located that a minimum amount of resistance to rapid discharge of the shower waste is provided.

INSTALLATION OF SHEET LEAD SAFING

When preparing a shower enclosure for water proofing with sheet lead, the floor must be swept clean. After this has been done, it is advisable to lay a few thicknesses of building paper on the rough floor to serve as a cushion for the ductile sheet lead. Sheet

Fig. 310. Shower Mixer
Valve

Fig. 311. Shower Strainer
Fitting

lead of about 4-pound weight may be used. A box equal in size to the layout of the shower may be constructed. The corners of the pan may be folded or soldered to make it water-tight. If the manufactured type of shower receptacle is used for the drain fitting, the sheet lead may be dressed into the receiving pan of the trap and fastened with the clamp ring to make it water-tight. Fig. 311 illustrates this assembly. Should an ordinary **P**-trap be used, a sleeve made of sheet lead may be soldered to the strainer fitting and lead pan to make the box water-tight.

A lead box should be provided with a tell-tale or drain, which may discharge any accumulated water at a safe terminal. Because of the tendency of shower baths to leak at the corners, they too should be water-proofed with sheet lead. This is done by making a 90-degree angle of a piece of sheet lead 12 inches wide and 5 feet long and tacking it in place on the rough studs of the shower bath walls. The bottom end of the sheet lead corners must extend into the pan so that the water can drain into it. It is important that the sheet lead water-proofing be protected from damage during construction.

SINKS

A sink is a very common and important fixture. Its installation requires good judgment and careful planning on the part of the plumber. Correct positioning of the fixture is important, as improper adjustment with respect to height causes fatigue. Inasmuch as the mistress of the house uses the fixture most frequently, it is advisable that she be consulted as to the height best suited to her comfort and convenience.

A reasonable height at which to set the sink may be determined by allowing the person who will use it most frequently to stand before the fixture in normal posture. This practice eliminates many backaches and promotes efficiency.

The location of the sink fixture in the room is also important. It should be placed in a position which affords the best light. Artificial light must be provided, but at no time should the person using the sink be compelled to work in a shadow.

The design of the sink must also be considered. If selection is left to the plumber he should consult the person who will use the fixture as to her own preference and needs. It is extremely awkward for some persons to use the left hand, others the right; hence both right and left drain boards are made to accommodate individual requirements.

The plumber should be careful to choose a sink that is sanitary. An element that enters into this important phase of sink design is a well-pitched bottom, in which the strainer is located centrally.

The drain board should have sufficient pitch to permit water to run off into the bowl. The sink should be of ample size to avoid cramping. A back of adequate height serves to prevent soiling of wall or counter surfaces when fruits or vegetables are prepared, and when dishes and utensils are washed. The supply fixtures should be of good quality and standard design.

MODERN KITCHEN SINKS

Sinks are manufactured in designs suited to specific uses. The most commonly used sink is the kitchen sink, which is made in two patterns: the one-piece sink, with roll or apron rim, and the flat-rim sink.

One=Piece Enameled Iron Sink. The one-piece enameled-iron sink, Fig. 312, is both practical and sanitary. It is constructed of iron that is coated with a good quality of glazed enamel, and is

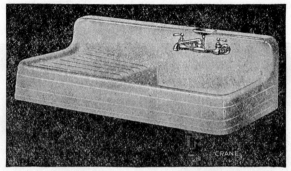

Fig. 312. One-Piece Enameled Iron Sink

obtainable in single and double drain board patterns. The basin varies in size from 16x20 inches to 24x36 inches. The over-all length

Fig. 313. Wilshire Ledge Sink with Double Drain Boards
Courtesy of Kohler Company, Kohler, Wis.

is from 42 to 72 inches. Because of its trim appearance, the apron rim style is favored more than the roll rim.

Flat=Rim or Two=Piece Sinks. The sink of best utility is the one-piece, flat-rim enameled-iron fixture, with or without self-contained drain boards. The fixture proper forms a unit with the cabinet or work base into which it is dropped. See Fig. 313. Sinks of this type are manufactured in sizes ranging from 12x12 inches to 24x48 inches in single patterns, and from 20x32 inches to 24x48 inches in the double style.

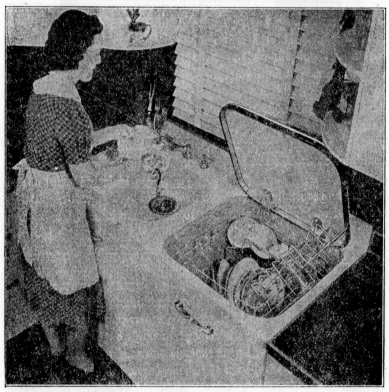

Fig. 314. Electric Sink
Courtesy of General Electric Company, Schenectady, N.Y.

The top or counter of the cabinet or work base may be tiled, or linoleum or composition rubber may be applied to the surface. An edge trim of stainless steel or chrome provides a neat finish.

Another material, called *Formica*, forms a durable and good-looking counter top. This material is installed hot and spread smoothly over the base to which it is applied. *Formica* provides a sanitary and waterproof work surface of high gloss and needed hardness.

Electric Sink. The electric enameled-iron sink is popular because of the convenience it affords the housewife. This sink is provided with a dishwasher and garbage cutter. See Figs. 314 and 315.

Fig. 315. Food Disposal Units

Dish-washing appliances should not be directly connected to the drain as, in case of stoppage of the line, sewage could back up into the dish-washing compartment, which is difficult to clean. An

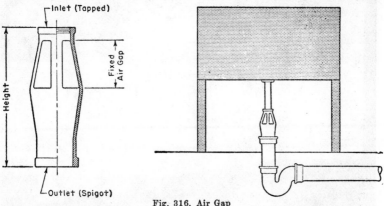

Fig. 316. Air Gap

air gap, Fig. 316, between washer and trap should be provided at the time the appliance is installed to prevent contamination of fixtures through direct waste connections. Then, should stoppage occur, the effluent will flow into the floor drain and not into the dish-washing compartment.

When a garbage disposal unit is used in conjunction with a

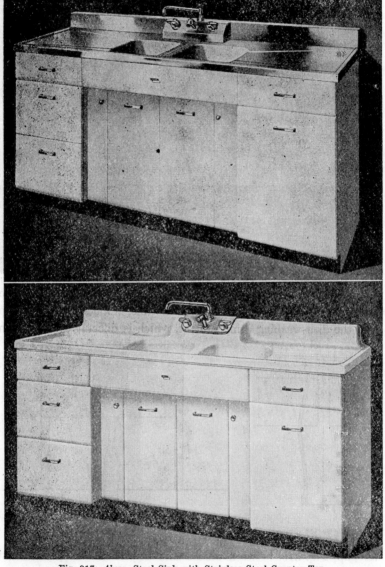

Fig. 317. *Above,* Steel Sink with Stainless Steel Counter Top.
Below, All-Steel Enameled Sink

septic tank having two or more compartments, it is recommended that the first compartment be larger than any adjoining, as this provision makes frequent cleaning unnecessary.

Steel Cabinet Kitchen Sinks. Stainless steel cabinet sinks and all-steel enameled cabinet sinks, Fig. 317, with double waste receptacles and newly designed supply fixtures, are found in many modern homes. Their acid-resisting double drain boards are a much appreciated convenience. These sinks tend to minimize the dishwashing chore and are exceptionally easy to keep clean.

Fig. 318. Slop Sink

Another sink of this type is the **monel metal** sink, which is constructed of steel of high nickel content. This sink is popular because of its beauty and its excellent resistance to acids, though it is comparatively expensive.

SERVICE SINKS

Slop Sinks. Sinks of many varieties are available to specific needs. One of these is the slop sink, Fig. 318, which is used chiefly for the disposal of scrub water. It is made of enameled cast iron and is constructed with a trap standard to the floor or to the wall. The basin is deep enough to accommodate a scrub pail. These sinks require a waste of large diameter for proper drainage.

Scullery Sink. The sink shown in Fig. 319 is a pot sink and is used in connection with restaurant and large kitchens for the paring of vegetables and the washing of pots and pans. These sinks usually are constructed of heavy gauge sheet metal.

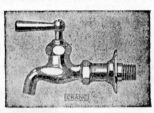

Fig. 320. Compression Faucet

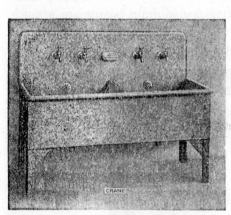

Fig. 319. Scullery Sink

Fig. 321. Combination Faucet

Supply Devices for Sinks. The water supply devices for kitchen sinks consist of individual compression and combination swing-spout faucets, Figs. 320 and 321. These are similar in operating principle to those used on a lavatory. For sinks such as pantry, hospital, and industrial wash-up sinks the water supply devices are of special design.

Installation of a Sink. Wall-hung kitchen sinks are suspended from the wall by use of two or more cast-iron or pressed steel hangers. For large sinks, leg supports are used.

Cabinets in which flat-rim sinks are mounted are secured to the wall and floor with screws of appropriate strength to insure stability.

Waste and Trap Connections. Three types of sink strainers are manufactured. These are the *plug type* strainer with rubber stopper; the *regular* strainer; and the *basket crumb cup strainer*. The latter is the most popular of the three. It requires a 4-inch opening and is set low enough to accommodate the crumb cup.

Strainers are fastened to the sink by a locknut and gasket. A ring of putty placed under the strainer helps to protect the enamel. A tailpiece with locknut connects the strainer to the slip joint of the trap. See Fig. 322. Double sinks utilize the same trap.

URINALS

The installation of a urinal is common in industrial, commercial, and public toiletrooms. The fixture is of sanitary design, but it seems that the purpose for which it is used makes it an installation not always up to the standard of other plumbing fixtures as far as sanitation is concerned.

It is important that the urinal be equipped with an adequate water flushing device, as most of the difficulty arises because of a lack of this facility. Urine is more objectionable as a waste than organic substances considered from the standpoint of possible communication of infectious disease. It has a tendency also to foul the fixture very rapidly. An odor that is extremely offensive results.

The flushing devices of a urinal should be automatic in design as negligence on the part of users is common. This eliminates a certain amount of the danger of fouling.

Not always is the fixture the factor of sole importance in making a sanitary installation. Very often a urinal is set into a floor of absorbent material, a condition to which there is strong objection. Such a floor becomes fouled quickly, because it is next to impossible to confine the urine waste to the fixture. The design of the fixture is characteristically at fault, and it is doubtful whether this condition can be altogether avoided. The plumber can help to solve the problem by recommending a nonabsorbent floor of adequate pitch.

Should it be necessary to make repairs on a urinal waste, it is advisable to provide protection for the hands. Rubber gloves of the surgical variety are recommended for this purpose. The mechanic may subject himself to infection if this precaution is not observed.

TYPES OF URINALS

Trough Urinal. The trough urinal of vitreous china is widely accepted and a suitable installation where conditions do not warrant the more expensive battery-type arrangement. Cast-iron enameled trough urinals, such as that shown in Fig. 323, are disapproved for

Fig. 322. Sink
Trap Connection

Fig. 324. Pedestal Urinal

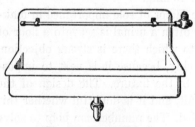

Fig. 323. Trough Urinal

the reason that wearing away of the enamel produces an unsanitary condition.

Wall-hung types of urinals are (1) the siphon jet fixture with integral trap; (2) the fixture with outside trap; and (3) the siphon jet fixture with side shields. All are proper for public buildings having tile or other kinds of impervious flooring in which a drain may be installed.

Pedestal Urinal. The pedestal urinal, Fig. 324, is permissible and may be adapted to certain restricted installations. It is not considered as satisfactory as other types of urinals because its height from the floor offers fouling possibilities. Its flushing device should be automatic or of the foot-operated variety. If hand-operated flush valves are used, they must be placed well above the urinal.

Fig. 325. Stall Urinal

Stall Urinals. The stall urinal, Fig. 325, is a fixture which is generally approved. It is constructed of vitreous china and is set into the floor in such a manner that wastes intended for the urinal may drain into it. The important sanitary feature of this fixture is the fact that its interior surface is completely washed with each discharge of the flushing device. Stall urinals may be set in batteries when three or more are required. The spaces between them should be sealed with impervious cement or plaster of Paris.

WATER SUPPLY DEVICES FOR STALL URINALS

The automatic flush tank is the most satisfactory flushing device for the stall urinal. This unit consists of an enameled-iron tank provided with a gooseneck siphon similar to that used in a high closet tank. The tank is also equipped with a pet cock which may be regulated to admit water so the tank will discharge at any required interval. There are other varieties of flush tanks which contain working mechanisms such as float valves, tilting buckets and similar devices. Any one of these is satisfactory provided it answers the purpose for which it is intended.

Traps. Urinals may be connected to the drainage system by a 1½-inch **P**-trap if the fixtures are of the small, wall-hung type, and by a 2-inch connection if they are of the integral trap type. For stall-type urinals, it is recommended that a 3-inch trap and drain be used because of the liming characteristic of these fixtures, and also because of the accessibility of the trap.

INSTALLATION OF A STALL URINAL

The stall urinal is usually set against the wall of the toiletroom in a depression cast into the rough concrete floor. The urinal waste outlet must be checked carefully so no difficulty is encountered when placing the urinal on it. The depression may then be filled with fine-grained sand to a point flush with the rough floor. After the strainer fitting has been mounted on the urinal, the fixture may be temporarily set in place so the strainer opening is directly over the waste pipe. Once this has been accomplished the excess sand can gradually be removed, which allows the urinal to lower slowly into its permanent position. After the fixture attains its proper position, the space from which the sand has been removed should be filled with a quick-setting mixture of concrete. Should the pipe stub between the trap and strainer be of galvanized steel, brass, or copper, the union may be made water-tight with a calk joint. If the stub is of lead pipe it may be soldered to the urinal strainer. The flushing pipe can now be connected to complete the job. Finally, the automatic flushing device must be regulated to wash the fixture at short intervals. When batteries of urinals are required, careful matching of the fixtures is important.

LAUNDRY TUBS

The laundry tub or tray, as it is sometimes called, has a definite place in every household. This fixture is an important one and yet it receives only secondary consideration. A well-equipped and attractive laundry is appreciated by every housewife, and, when opportunity permits, the plumber should always recommend the best quality of fixture.

Fig. 326. Laundry Tub

The plumber can increase the efficiency of the laundry by placing its fixtures in a position favorable to proper light, both artificial and natural.

Laundry tubs usually are constructed of a rich mixture of non-absorbent concrete, vitreous china, enameled iron, or stone of a satisfactory quality. They are of single, double, or multiple design, selection depending upon the individual requirement. A double compartment laundry tray is shown in Fig. 326.

Note the sturdy steel stand by which the tubs can be made level and adjusted to the right height. This unit often incorporates a raised ledge which keeps articles from being pushed off the shelf. For convenience the tub should be not more than 2 feet 8 inches in height.

A BATHROOM DESIGNED FOR CONVENIENCE AND EFFICIENCY

Water Supply Devices. The usual water supply device is a rough-plated combination compression swing-spout faucet similar to the kitchen sink swing faucet in construction and working principle. Individual compression faucets may also be used, but they should be attached to the fixture above its overflow rim so as to avoid a dangerous cross-connection.

Trap. The compartments of a multiple unit laundry tub may be connected with a continuous waste pipe which discharges its content into a 4 x 5 or 4 x 8 drum trap, or a Durham or slip-joint **P**-trap.

Installation of a Laundry Tub. The first step in setting a laundry tub is to assemble the tub stand and set it in its permanent location. Level the stand carefully so that the tub also will be level when it is in place. Convenient, rust-proof levelers are provided for this purpose.

The laundry tub may now be placed on the stand and its water supply and waste connected. The connection of the waste pipe to the fixture often is made with a flared tailpiece and putty joint. The joint is made water-tight with the aid of a sink collar.

DRINKING FOUNTAINS

Drinking fountains are a convenience which most building operators provide for the public. These fixtures may be located conveniently in corridors, rest rooms, workshops and other places. As a rule, the drinking fountain or bubbler, as it is sometimes called, is connected to the domestic cold-water distribution system, which provides its supply. In some cases elaborate and expensive drinking water equipment is installed. This equipment consists of a filter, cooling tank, sterilizer, and circulator pump, which keeps the water at a constant temperature, making it pleasant to drink.

Drinking fountain design has greatly improved in the last few years, because sanitary authorities came to realize that the older types of fixtures were a source by which disease might be communicated. The older forms of fountain consisted of a basin with a bubbling head which allowed the water to drain over it. Variations in pressure and other circumstances often reduced the flow so the person using the fountain would contact the bubbler head with his mouth. This condition is decidedly dangerous.

The new fountains are provided with a drinking orifice, above the overflow rim, that directs a slanting stream of water into the waste basin. The orifice is protected in such a manner that the person using the fountain is prevented from direct contact with it.

Wall Fountain. Fig. 327 illustrates a wall-hung drinking fountain with a slanting stream drinking orifice. This type of fixture should be set from 30 to 36 inches from the floor.

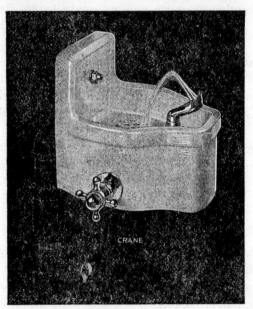

Fig. 327. Wall-hung Drinking Fountain   Fig. 328. Pedestal Drinking Fountain

Pedestal Fountain. The pedestal drinking fountain, Fig. 328, is constructed of vitreous china and is designed in many styles. The fixture is an elaborate one and its use is confined to public places.

Supply Devices. The water supply device of a drinking fountain consists of a self-closing compression stop and a stream regulating valve attached to the drinking orifice. A self-closing valve is considered practical because it eliminates excessive waste of water. The trap of a drinking fountain usually is of 1¼-inch diameter.

Installation Procedure. The installation procedure of wall-hung pedestal types of drinking fountain depends largely on the general design of the fixture. Drinking fountains are not standard and the method of installing them is distinctly individual.

CHAPTER XXVIII

ILLUSTRATIVE EXAMPLE

In this chapter two distinct types of plumbing are represented. One is based on the minimum requirements often found in sparsely populated and rural areas. The other deals with more rigid requirements prevalent in incorporated cities and certain sections of the United States. It is important that these differences be recognized, as in many instances a definite line is drawn with respect to installations and practices, as these relate to specific regions of the country.

Two sets of drawings are used to show the differences which exist. The first set (Figs. 336, 342, and 343) is illustrative of the first type of installation which meets minimum requirements. The second set (Figs. 346, 347, and 348) is illustrative of types of installations based on more rigid requirements.

It should be remembered that the style of dwelling in greatest demand today is the bungalow type home. In large areas of the country, dwellings of this type do not require storm drainage except for roof drains which, if desired, may be extended to the curb for conveyance of rain water to the street. Frequently, gutters and downspouts are not required. When they are specified, the downspout usually spills upon a cement slab which is graded in such manner that the rain water is drained away from the building foundation.

Owing to the fact that throughout the South and West heavy snows are virtually unknown, except in areas of high altitudes, piping systems designed to drain saturated soil are unnecessary. Modern homes have built into them appliances designed to lighten household tasks; for example, automatic dishwashers and food waste disposal units. It is the plumber's job to install such appliances, and to use his knowledge and skill to do so in a manner that will not endanger the health of occupants of the building.

In the working drawings or blueprints which architects prepare as a guide to the construction of houses, there are seldom, if ever, any details of the required plumbing system given other than those which show the fixture symbols in their proper places and the location of one or more 8-inch partitions through which 4-inch soil pipes are to be run. Designing of the piping systems, therefore, must be done by the plumbing contractors before they can make estimates, bid on such jobs, or actually install the systems.

In this chapter, by means of an illustrative example, the design of a typical plumbing system is illustrated and explained. The house selected for this purpose is one of medium price, the like of which can be found in any locality throughout the East and Middle West.

In the following, the blueprints for the typical house, the specifications for it, the roughing drawings, the illustrations used, and the design and installation of the system are explained in regular order.

Blueprints. The blueprints upon which the illustrative example is based begin on the following page and are designated and numbered as Plates 1 to 8.

All drawings indicating floor plans and elevations were originally done according to the $\frac{1}{4}'' = 1'0''$ scale. The detail drawings, which appear on some of the plan and elevation plates, were drawn to the $\frac{1}{2}'' = 1'0''$ or $1\frac{1}{2}'' = 1'0''$ scales.

The blueprints were originally drawn full size, but they have been reduced in size in order to fit the page size of this book. Therefore, although the blueprints are all in correct proportion, they cannot be scaled.

A study of the various floor plans will show that the architect has indicated all fixtures and their locations and the partitions where soil pipes are to be run. The symbols used to indicate soil stacks in partitions are circles. In addition to the foregoing information the architect has shown the plans for the house sewer. The plans indicated by the architect should be followed exactly unless obstacles arise, in which case the architect should be consulted before any changes are made.

The blueprints, as indicated in succeeding pages, should be studied so that a perfect visualization of the entire house and its

PLATE 1

SANITARY SEWER 8'-0" DEEP

STORM SEWER

CENTER LINE OF STREET

SUNSET ROAD

CURB LINE

15'-0"

5'-0"

6"

12'-6"

2'-0"

4'-0"

8'-0"

LINES

50'-0"

70'-0"

SIDEWALK

GRASS

GRAVEL DRIVE

3'-0"

FRONT ENTRANCE PLATFORM

CONCRETE

12'-?"

FOUNDATION 35'-10"

22'-2"

TERRACE

RESIDENCE

GARAGE

CONCRETE

150'-0"

LOT LINES

N
E
S
W

PLOT PLAN

SCALE: $\frac{1}{16}$" = 1'-0"

0 5 10 15 20 25

NOTE:

DIMENSIONS ON THESE PLANS ARE GIVEN
AS FOLLOWS:
EXTERIOR:— TO OUTSIDE SHEATHING LINE —
TO FACE OF MASONRY.
INTERIOR:— TO CENTER LINE OF STUD
PARTITIONS AND FACE OF PLASTER, WHERE
SO NOTED.

BLUEPRINTS

FOR USE WITH
CHAPTER XXVIII

Although these reproductions are smaller in
size than the original blueprints, the scales found
on each drawing can be used in determining measurements, since the scales themselves have been
reduced proportionately.

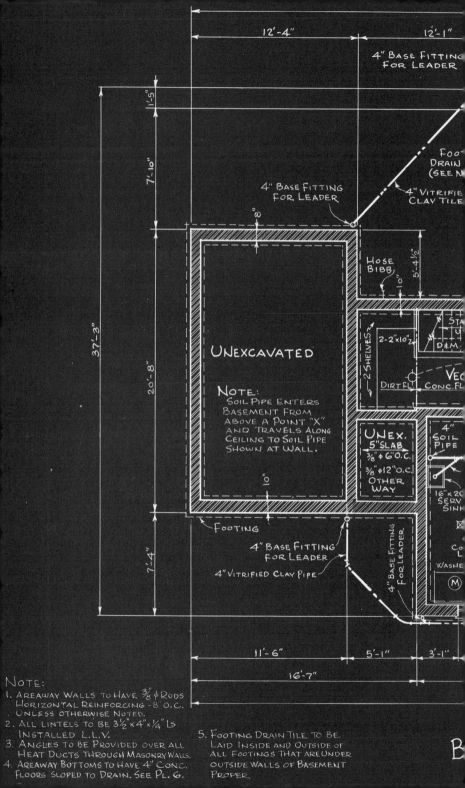

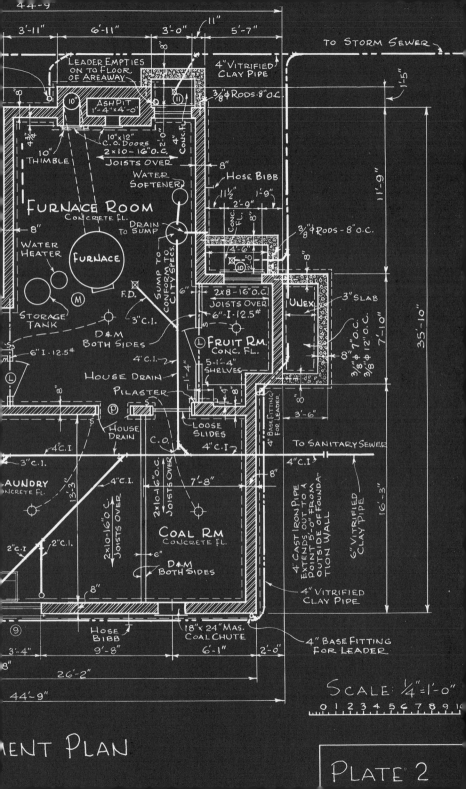

SCALE: 1/4"=1'-0"
0 1 2 3 4 5 6 7 8 9 10

ᴇɴᴛ PLAN

PLATE 2

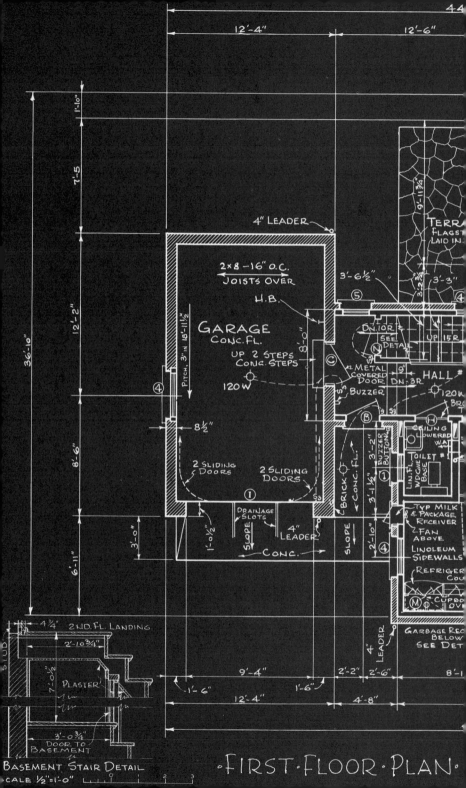

FIRST·FLOOR·PLAN

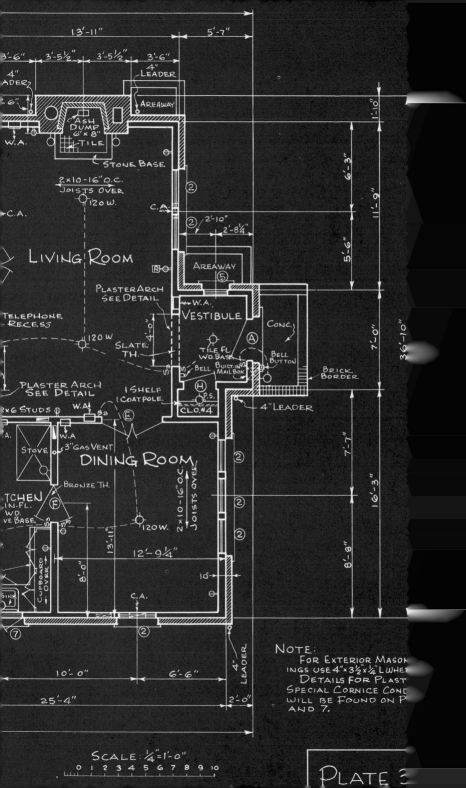

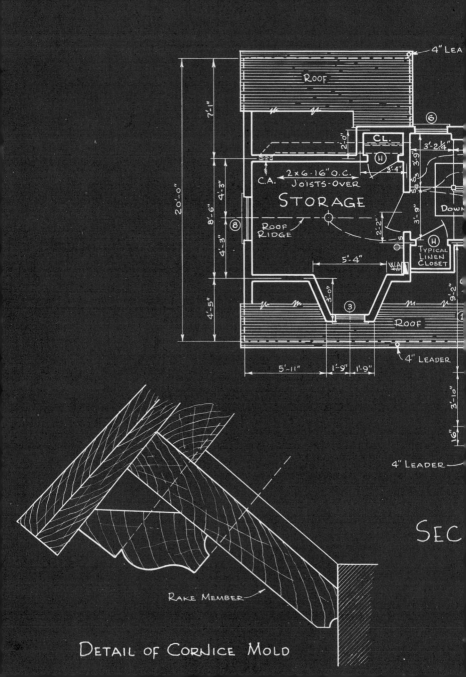

4" LEA

ROOF

6

CL.

3'-2¼"

2'-0"

H

7'-1"

3'-9"

C.A.

2×6-16"O.C.
JOISTS-OVER

3'-4"

STORAGE

4'-3"

20'-0"

8'-6"

8

ROOF
RIDGE

2'-2"

3'-9"

DOWN

4'-3"

H

TYPICAL
LINEN
CLOSET

5'-4"

9'-2"

W.A.

4'-5"

3'-0"

3

ROOF

4" LEADER

5'-11"

1'-9"

1'-9"

3'-10"

16"

4" LEADER

SEC

RAKE MEMBER

DETAIL OF CORNICE MOLD

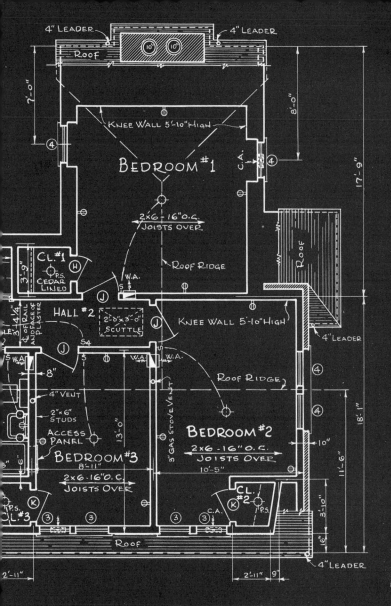

4" LEADER   4" LEADER

ROOF

7'-0"

8'-0"

KNEE WALL 5'-10" HIGH

BEDROOM #1

17'-9"

C.A.

2x6 - 16" O.C.
JOISTS OVER.

ROOF RIDGE

CL.#1
P.S.
CEDAR
LINED

3'-9"

ROOF

HALL #2

2'-0"x3'-0"
SCUTTLE

KNEE WALL 5'-10" HIGH

4" LEADER

℄ OF RAIL
AND FACE OF
PLASTER

3'-4¼"

W.A.

ROOF RIDGE

4" VENT

2"x6"
STUDS

ACCESS
PANEL

BEDROOM #3

13'-0"

BEDROOM #2

2x6-16" O.C.
JOISTS OVER

10'-5"

18'-1"

10"

11'-6"

2x6-16" O.C.
JOISTS OVER

8'-11"

3' GAS STOVE VENT

CL.
#2

C.A.

P.S.

3'-10"

1'-6"

CL. #3

ROOF

2'-11"

2'-11" 9"

4" LEADER

·FLOOR·PLAN

SCALE ¼"=1'-0"

0 1 2 3 4 5 6 7 8 9 10 11 12 13 14 15

PLATE 4

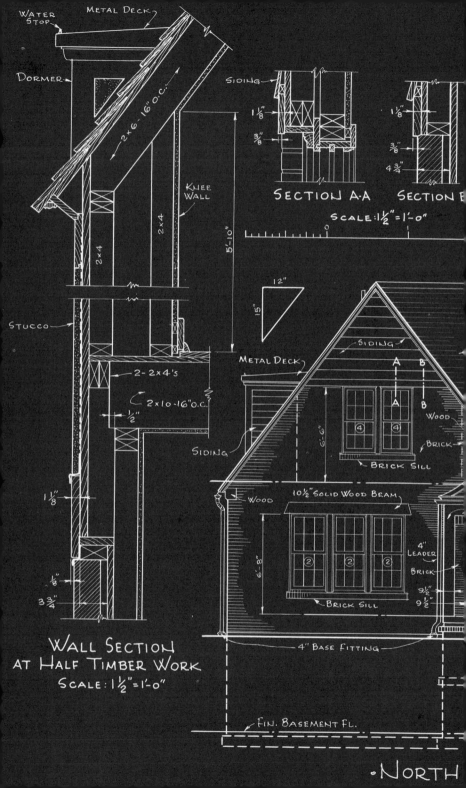

WATER STOP

METAL DECK

DORMER

2×6-16" O.C.

KNEE WALL

2×4

2×4

2×4

5'-10"

STUCCO

2-2×4's

2×10-16" O.C.

½"

1⅛"

⅛"

3¾"

SIDING

1⅛"

⅜"

SECTION A·A

SCALE: 1½"=1'-0"

1⅛"

⅜"

4¾"

SECTION B

12"

15"

METAL DECK

SIDING

SIDING

A

B

A

B

WOOD

BRICK

④

④

6'-6"

BRICK SILL

WOOD

10½" SOLID WOOD BEAM

6'-8"

②

②

②

4" LEADER

BRICK

9½"

9½"

BRICK SILL

4" BASE FITTING

WALL SECTION
AT HALF TIMBER WORK
SCALE: 1½"=1'-0"

FIN. BASEMENT FL.

·NORTH

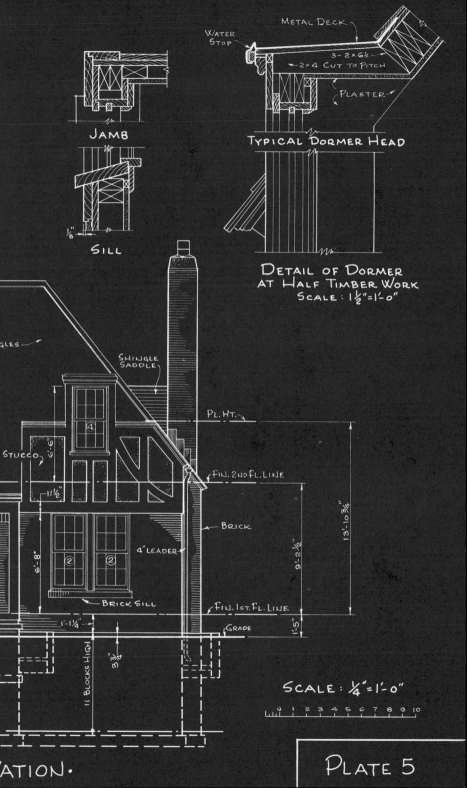

JAMB

SILL

1/8"

WATER STOP

METAL DECK

3 - 2 × 6'S

2 × 4 CUT TO PITCH

PLASTER

TYPICAL DORMER HEAD

DETAIL OF DORMER
AT HALF TIMBER WORK
SCALE: 1½"=1'-0"

GLES

SHINGLE
SADDLE

PL. HT.

STUCCO

6'-6"

11½"

FIN. 2ND FL. LINE

BRICK

4" LEADER

13'-10¾"

9'-2½"

④

②   ②

6'-8"

BRICK SILL

FIN. 1ST. FL. LINE

1'-1¼"

3¾"

GRADE

1'-5"

11 BLOCKS HIGH

SCALE: ¼"=1'-0"

0 1 2 3 4 5 6 7 8 9 10

ATION.

PLATE 5

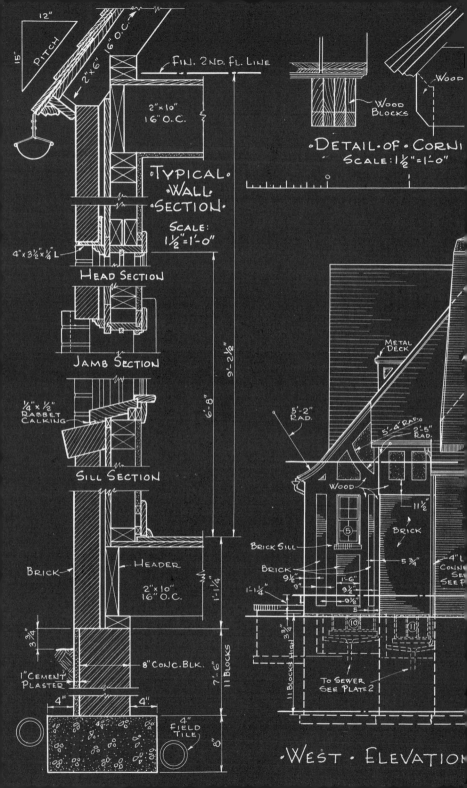

12"

15" PITCH

2"x6" 16" O.C.

FIN. 2ND. FL. LINE

2"x10"
16" O.C.

·TYPICAL·
·WALL·
·SECTION·

SCALE:
1½"=1'-0"

4"x3½"x¼"L

HEAD SECTION

JAMB SECTION

¼"x½"
RABBET
CALKING

SILL SECTION

9'-2½"

6'-8"

BRICK

HEADER

2"x10"
16" O.C.

1'-1¼"

3¾"
3"

1" CEMENT
PLASTER

8"CONC.BLK.

4"   4"

11 Blocks

7'-6"

4"
FIELD
TILE

8"

·DETAIL·OF·CORNI
SCALE:1½"=1'-0"

WOOD

WOOD
BLOCKS

WOOD

METAL
DECK

5'-2"
RAD.

5'-4" RAD.

2'-5"
RAD.

WOOD

5

11½

BRICK

BRICK SILL

BRICK

5¾"

4" L
COLN
SE
SEE P

9½"

9"

1'-6"

9½"

9½"

5

10

11

1'-1¼"

11 Blocks High 3¾"

11 Blocks

TO SEWER
SEE PLATE 2

·WEST · ELEVATION

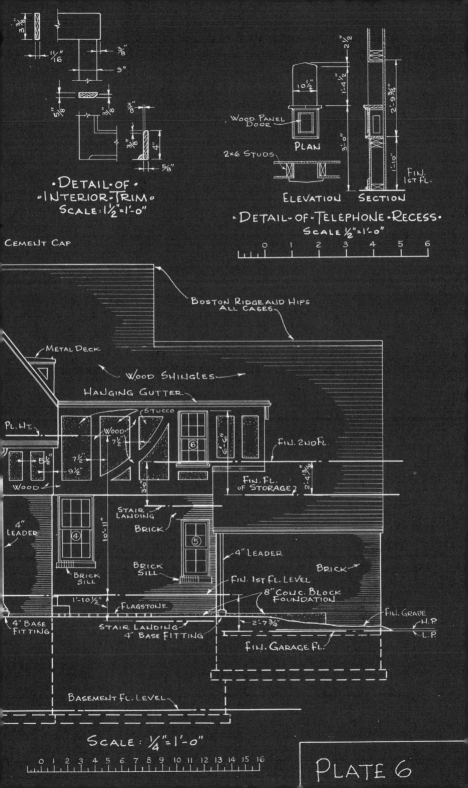

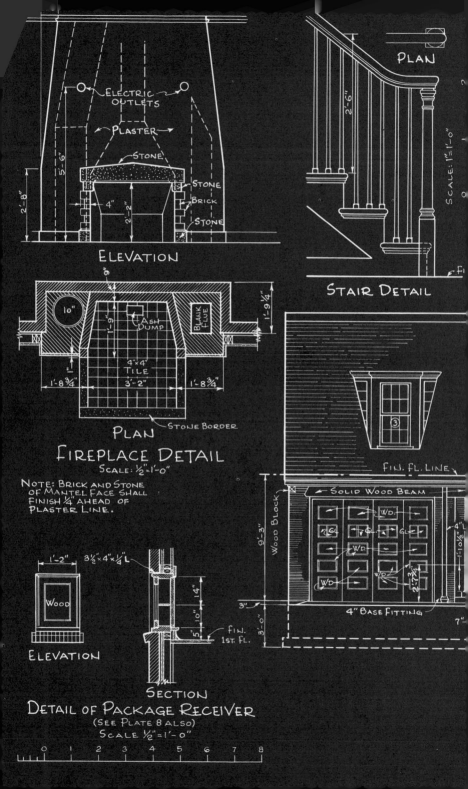

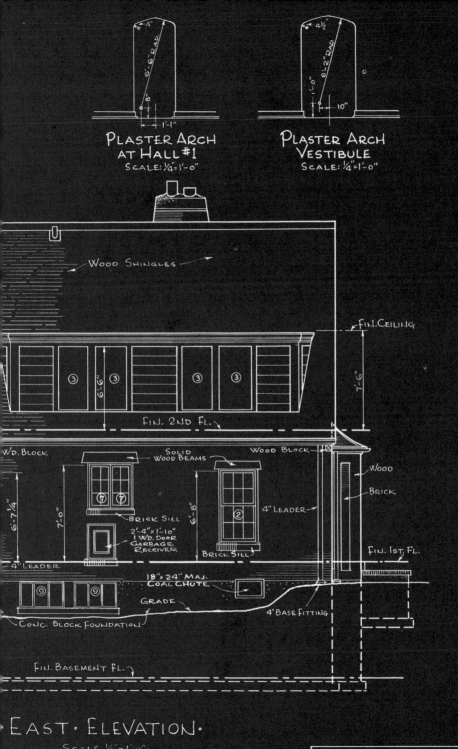

PLASTER ARCH
AT HALL #1
SCALE: ¼"=1'-0"

PLASTER ARCH
VESTIBULE
SCALE: ¼"=1'-0"

WOOD SHINGLES

FIN. CEILING

7'-6"

③   ③   ③   ③

9'-6"

FIN. 2ND FL.

WD. BLOCK   SOLID WOOD BEAMS   WOOD BLOCK

WOOD

BRICK

7'-0"

6'-8"

6'-7¼"

⑦ ⑦

②

4" LEADER

BRICK SILL

2'-4" X 1'-10"
1 WD. DOOR
GARBAGE
RECEIVER

BRICK SILL

FIN. IST. FL.

4" LEADER

⑨   ⑨

18" X 24" MAJ.
COAL CHUTE

GRADE

4" BASE FITTING

CONC. BLOCK FOUNDATION

FIN. BASEMENT FL.

EAST · ELEVATION ·
SCALE ¼"=1'-0"
0 1 2 3 4 5 6 7 8 9 10 11 12 13 14 15

PLATE 7

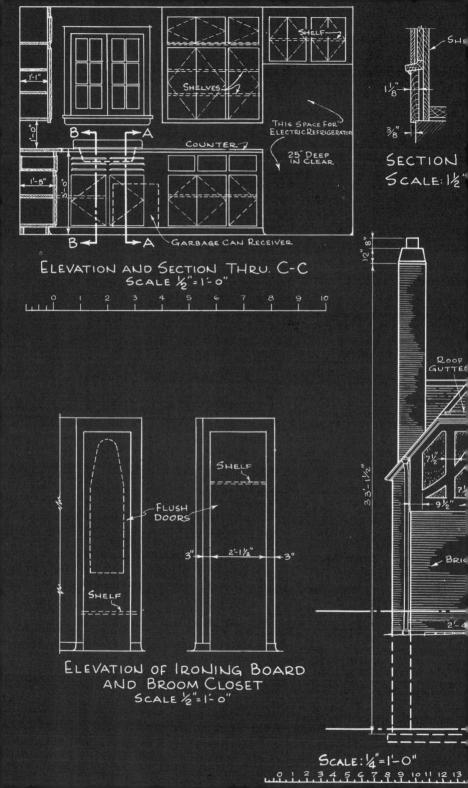

SHE̶

$1\frac{1}{8}$

$\frac{3}{8}''$

SECTION
SCALE: $1\frac{1}{2}$"

1'-1"

1'-0"

B ◄───── A

COUNTER

1'-5"

3'-0"

B ◄───── A

SHELF

SHELVES

THIS SPACE FOR
ELECTRIC REFRIGERATOR

25" DEEP
IN CLEAR

GARBAGE CAN RECEIVER

ELEVATION AND SECTION THRU. C-C
SCALE $\frac{1}{2}$"=1'-0"

0 1 2 3 4 5 6 7 8 9 10

FLUSH
DOORS

SHELF

SHELF

SHELF

3"   2'-1$\frac{1}{2}$"   3"

ELEVATION OF IRONING BOARD
AND BROOM CLOSET
SCALE $\frac{1}{2}$"=1'-0"

12"8"

ROOF
GUTTER

33'-1$\frac{1}{2}$"

$7\frac{1}{2}$"

$7\frac{1}{2}$

9$\frac{1}{2}$"

BRI

2'-4

SCALE: $\frac{1}{4}$"=1'-0"

0 1 2 3 4 5 6 7 8 9 10 11 12 13

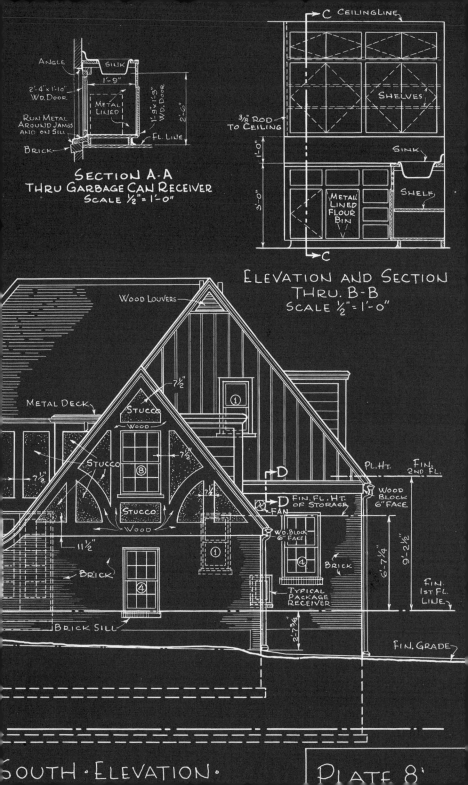

ANGLE
SINK
2'-4"x1'-10"
WD. DOOR
METAL
LINED
RUN METAL
AROUND JAMBS
AND ON SILL
BRICK

1'-9"
1'-9"x1'-3"
WD. DOOR
2'-6"
FL. LINE

SECTION A-A
THRU GARBAGE CAN RECEIVER
SCALE ½"= 1'-0"

C CEILING LINE

SHELVES

3/8" ROD
TO CEILING

1'-0"

SINK

3'-0"

METAL
LINED
FLOUR
BIN

SHELF

C

ELEVATION AND SECTION
THRU. B-B
SCALE ½"= 1'-0"

WOOD LOUVERS

METAL DECK

7½

STUCCO
WOOD

STUCCO

7½

7½

7½

STUCCO
WOOD

11½"

BRICK

BRICK SILL

①
② D
  D
FAN

FIN. FL. HT.
OF STORAGE

WD. BLOCK
6" FACE

TYPICAL
PACKAGE
RECEIVER

PL. HT.

FIN.
2ND. FL.

WOOD
BLOCK
6" FACE

9'-2½"

6'-7¼"

BRICK

④

2'-7¾"

FIN.
1ST FL.
LINE

FIN. GRADE

SOUTH · ELEVATION ·

PLATE 8

Window Schedule

Location Number	Quality of Sash Needed	Sash Size	Remarks
①	2 PR.	1'-10" x 2'-0"	Double Hung, Obscure Glass
②	6 PR.	2'-7" x 2'-9"	Double Hung
③	5 PR.	2'-3" x 2'-4"	" "
④	7 PR.	2'-4" x 2'-5"	" "
⑤	2 PR.	1'-10" x 2'-0"	" "
⑥	1 PR.	2'-3" x 2'-0"	" "
⑦	1 PR.	1'-6" x 3'-0"	Casement
⑧	1 PR.	3'-0" x 2'-5"	Double Hung
⑨	2	3'-2" x 1'-9"	Steel Basement Sash
⑩	1	2'-9" x 1'-3"	" " "
⑪	1	2'-9" x 1'-11"	" " "

Door Schedule

Symbol	Size	Remarks
Ⓐ	3'-4" x 6'-8" x 2¼"	Front Entrance · See Elevation
Ⓑ	2'-8" x 6'-8" x 1¾"	Hall #1 - Glazed
Ⓒ	2'-8" x 6'-8" x 1⅜"	5 Panel - Metal Covered Door
Ⓓ	2'-4" x 6'-8" x 1¾"	L.R. - French · See Elevation
Ⓔ	2'-4" x 6'-8" x 1¾"	L.R. to D.R. - French · See Elevation
Ⓕ	2'-8" x 6'-8" x 1¾"	D.A.D. Type - Interior - 2 Pan.
Ⓖ	2'-8" x 6'-8" x 1⅜"	Typ. Int. 2 Pan.
Ⓗ	2'-4" x 6'-8" x 1⅜"	" " "
Ⓘ	4 - 2'-3½" x 7'-6" x 1¾"	Garage Sliding · See Elevation
Ⓙ	2'-6" x 6'-8" x 1⅜"	Typical Interior
Ⓚ	2'-4" x 6'-6" x 1⅜"	" "
Ⓛ	2'-6" x 6'-8" x 1⅜"	5 Pan. - Basement
Ⓝ	2'-8" x 6'-6" x 1⅜"	Typ. Int. 2 Pan.
Ⓟ	2'-8" x 6'-8" x 1⅜"	5 Pan. Basement

Special Abbreviations used on these Plates

Typ.	Typical
Int.	Interior
F.D.	Floor Drain
D&M.	Dressed and Matched
B.T.	Bronze Threshold
S.T.	Slate Threshold
F.S.	Full Size
L.L.V.	Long Leg Vertical

structural details can be obtained, all of which must be considered as the design of the plumbing system is being carried on.

Specifications. The function of written specifications is to supplement the blueprints. In other words, the specifications give instructions and explanations which are not or cannot be shown on the blueprints. Therefore they should be studied as carefully as the blueprints and every item noted with extreme care. For example, the blueprints show the locations of water closets, lavatories, etc., but only the specifications describe these items as to their quality, manufacturer, etc. In like manner the blueprints show tile, soil pipes, etc., but the specifications must be studied to learn the required quality, kinds, etc.

The specifications for use with the blueprints used here are found in succeeding pages.

Some specifications are also shown on the blueprints in the form of notes. These notes are as important as any of the symbols, plans, or items in the written specifications.

PLUMBING SPECIFICATIONS

GENERAL CONDITIONS. The plumbing contractor will be subject to the "General Conditions" of the contract as approved by the American Institute of Architects as far as their content can be applied to this branch of the work. A copy of these may be obtained from the architect in charge of building construction.

EXECUTION OF CONTRACT. The contractor shall execute his branch of the work completely, even though some requirements are not contained in the specification or indicated on the plan.

He must accomplish the work as rapidly as possible and cooperate with other tradesmen so the progress of construction will not be delayed.

EXTRA LABOR. Cost of extra labor or material will not be paid by the owner unless it has been authorized in writing by the owner before the material has been installed or labor performed.

BIDS. The owner reserves the right to reject any or all proposals based on the specifications or drawings; or to accept any proposals in part or whole.

SCOPE OF THE WORK. The contractor shall complete a plumbing and drainage system as shown on the drawing and specified in the specification in accordance with the building and plumbing codes of the city and state.

He shall connect all fixtures with water and terminate them into the soil and waste system with the final drainage terminal at the city main.

Roof leaders are also to be connected to a separate installation of drainage indicated on the plan and terminated at the storm sewer in the street.

The contractor shall furnish and perform all skilled and unskilled labor, as well as all material, piping, fittings, valves, fixtures and appliances necessary and required in doing and completing the entire drainage installation. All excess materials, refuse, etc., are to be removed from the premises by him upon completion of his contract.

All material, piping, fittings, valves, fixtures and appliances hereinafter specified and shown on the plan and those which are essential and have not been specified shall be new and of the highest grade and quality, free from defect, such as breaks, flaws, or other imperfections.

All materials installed are to be concealed in building partitions wherever possible.

The joists and studs or wooden framework through which the pipe must pass are to be drilled in such a way that the building will not be structurally weakened.

The contractor is also to furnish the gas piping from the meter to the locations shown on the plan. Its installation is to conform with the rules of the local utility which is to furnish the gas.

PERMITS. The contractor shall give the proper authority all the notice required by law relative to the work in his charge. He shall obtain all official licenses and permits for use of street, alleys, taking up and relaying pavements, etc., and shall pay all fees for licenses, permits for sewer and water and other privileges necessary to perform and complete his work and contract.

The contractor in entering into contract with the owner agrees to bind himself to protect and keep the owner harmless from all liability and damage arising from his violation of local and state ordinances, codes, and laws by workmen, agents, sub-contractors and others employed by the contractor on or about the building, building site and premises.

SANITARY HOUSE SEWER. From the connection in the street run a 6" vitrified clay drain to a point indicated on the plan within 5 feet of the north building wall. The joints to be made of a cement mortar consisting of a mixture of 50%-50% cement and sand.   Excavate for all pipe.   The pipe is to be laid on solid earth and the trench refilled, watered and tamped.   The excavation in the street to be filled with a good quality of washed gravel.

INSIDE SANITARY HOUSE DRAIN. The house drain is to be constructed of four-inch cast iron pipe of extra heavy quality with fittings of the same quality.

A cleanout with brass screw cover is to be provided immediately inside the north foundation wall and at such intervals that the entire drain is accessible.   All soil and waste stacks are to be provided with a cleanout at their base.   The joints are to be made with oakum and molten lead using at least 4 lbs. of lead for each joint.   The pipe is to be laid on solid earth and the trench refilled with earth, flooded and tamped.   The drain is to be left exposed and tested with water under 5 lbs. of pressure.

STORM DRAIN. The storm drain is to be constructed of four-inch vitrified clay pipe and fittings and extended from the connection in the street to the base of all conductor outlets shown on the plan.   The joints are to be made of cement mortar consisting of a 50%-50% cement and sand mix.

The base fitting of each conductor is to be made of cast iron pipe and $\frac{1}{4}$ bend of extra heavy quality and extended to a point six inches above the finished grade.

The pipe is to be laid on solid earth in excavations made by the plumber and then refilled with surplus earth.

SOIL – WASTE – VENT. All soil stacks are to be constructed of 4" cast iron extra heavy pipe and fittings well supported and extended through the roof where they are to be flashed with 4-lb. sheet lead.   The joints are to be made with oakum and molten lead and calked to prevent leaking. At least 4 lbs. of lead per joint must be used.

All waste pipes are to be constructed of galvanized steel pipe of acceptable quality and recessed cast iron drainage fittings. Both materials must comply with state and local codes, and be sized to ensure a self-cleansing flow and meet plumbing ordinance requirements.

The pipe must be well supported to prevent sagging and all joists and studs through which it must pass are to be drilled to allow for settling of the building. All exposed threads must be painted with acid-resistant paint.

The vent pipes are to be constructed of galvanized steel pipe with plain cast iron fittings well tarred.   The general conditions of its installation are the same as for waste pipe.

The soil, waste, and vent pipes must be provided with cleanouts to make them accessible and when completed be subjected to a water test and be inspected by local or state authority.

HOUSE SERVICE. The house service is to be extended from the city main with 3/4" lead pipe of extra strong quality and laid on a shelf in the trench of the sanitary sewer. It is to be equipped with a corporation stop, a curb stop with stop box extended flush with the grade and a meter stop. The joints to be wiped with a mixture of 60% lead and 40% tin solder. A 3/4" meter is to be provided at a location on the north wall of the laundry. (This type of house service is not used in western states.)

WATER DISTRIBUTION SYSTEM. The water distribution system is to be constructed of $\frac{3}{4}$" and $\frac{1}{2}$" galvanized wrought iron pipe and screw thread malleable fittings of accepted quality.   It is to consist of cold water hard, cold water soft, and hot water soft.

The following fixtures are to receive:
Cold Water Hard
1. All water closets
2. Kitchen sink
3. Lavatory second floor
4. Sill cocks

Cold Water Soft and Hot Water Soft
1. Bath tub
2. Kitchen sink
3. Lavatory first floor
4. Laundry tub
5. Slop sink in laundry

All basement mains and rising lines to bathroom groups are to be of $\frac{3}{4}$" size. Runs of pipe to single fixtures may be constructed of $\frac{1}{2}$" pipe.

The plumber is to make all connections to the water softener, furnace coil and gas heater and provide valves of the stop and waste variety at the following locations:
1. House side of meter
2. All first and second floor risers
3. Laundry tubs
4. Cold supply to storage tank
5. Water softener

The plumber is to connect three sill cocks as indicated on the plan with a $\frac{3}{4}$" supply of cold hard water and provide each with a stop and waste valve just inside the foundation wall.

All piping must be substantially supported to overcome sagging and be so arranged that it may be easily drained.

The pipe must be tested to withstand a working pressure of not less than 100 pounds.

GAS PIPE. The plumber is to install gas pipe of $\frac{3}{4}$" size and black steel quality. The fittings are to be of black malleable screw thread type. The gas is to be connected to the auxiliary heater in the furnace room and a $\frac{3}{4}$" opening in the kitchen indicated on the plan for the gas stove. All stops are to be of the ground core variety. The pipe is to be strapped to the underside of the first floor joists in a workmanlike manner.

APPLIANCES. The plumber is to provide and connect the following appliances to the drainage and water distribution system:

1 – Storage tank of 40–gallon capacity of extra heavy quality connected to the heating furnace by means of $\frac{3}{4}$" black steel pipe coil.

1 – Triple coil, asbestos–lined gas–fired tank heater of standard design and acceptable to owner.

1 – 1200–gallon semi–automatic water softener of standard design acceptable to the owner and connected to the water distribution system in a workmanlike manner.

2 – Floor drains consisting of deep seal **P** traps and 8" strainers at locations shown on plan.

1 – Combination floor drain and drain tile receiver provided with a brass flap trap.

DRAIN TILE. The plumber is to lay a 3" open jointed clay drain tile line around the outside and inside of foundation and connect both with bleeder pipes laid in the foundation at about ten foot intervals.   The tile is to be laid on a bed and covered with six inches of coarse gravel and be connected to the drain tile receiver provided for this purpose in the boiler room.

FIXTURES. The plumber is to furnish and install in a workmanlike manner the following list of fixtures:

2 – Model A T/N one piece water closets, provided with a white seat to be approved by owner, closets to be provided with chrome plated supplies and angle valves.

1 – Crane C 19–348 porcelain enameled cast iron flat rim sink with center outlet and C 33880 waste and C–32767 chrome–plated three water swing spout faucet and $1\frac{1}{2}$" **P** trap.

1 – Crane C 754–V 12 Norwich vitreous china lavatory 24" x 21" with Rainier supply and indirect lift pop–up waste trimming, chrome plated supplies and $1\frac{1}{4}$" **P** Trap.

1 – Crane C 2251–V 12 Carolina porcelain enameled cast iron lavatory V 12 trimming.   Supply, pop–up and trimmings to be the same as above.

1–C 3330–C 11 Tarnia enameled cast iron reversible recess bath tub L H.   Outlet and 2 R 11 trimmings.

Tub to be provided with Crane shower mixer C 4792 and spray and Crane C 99 curtain rod assembly with snaps and hold–backs and colored water–repellent curtain.

1 – Crane C 21360 enameled cast iron service sink with C 21420–A **P** trap standard.   Faucet to be chrome–plated swing spout design.

1 – Cast cement, two compartment 24" x 24" x 16" laundry tub C 21294 with swing spout faucet and 4" x 5" drum trap.

GUARANTEE. The plumbing contractor does hereby guarantee the entire work, mechanical apparatus, plumbing, gas and
drainage system installed by him against mechanical defect,
for a period of one year from the completion of the work.
He also agrees to replace all defective materials, and make
good any damage done to the building in the performance of
this contract without cost to the owner.

Roughing Drawings. The drawings shown in Figs. 329 through
334 are what mechanics call roughing drawings or roughing-in-
measurements.

By studying these drawings it can be seen that they show all
of the necessary dimensions for wastes, vents, and supply pipes
going to and from the various fixtures. The purpose of the drawings
is to show the mechanics where to install the various pipes in exactly
the right places, positions, spacings, etc., to fit the fixtures when
they are installed during the finishing operations, which are carried
on after the wall coverings and floors have been placed.

Roughing drawings can be secured by plumbing contractors
when they place their fixture orders with manufacturers.

*It should be kept in mind that roughing drawings apply only to
the specific job for which the manufacturer issued them.* Such drawings should be obtained with each order placed with the manufacturers because various of the important dimensions change frequently as improvements and other changes are made in the fixtures
by their manufacturers.

Isometric Drawings. In this chapter, as in some of the previous
chapters, the principles of isometric projection are employed in the
presentation of many of the illustrations to help the reader visualize
the conditions more clearly.

In isometric drawings all pipes or structural parts which are
to be installed in a horizontal, or nearly so, position are drawn at
30°, whereas all vertical pipes, etc., are drawn actually vertical.
In other words, all slanting lines in an isometric drawing really
represent horizontal lines and all vertical lines represent actual
vertical lines.

METHOD OF INSTALLATION

House Sewer. The house sewer is usually the first phase of
the plumbing layout that must be installed. It is essential to do
this first in order that the excavation for the building may be drained

Model A T/N

ONE PIECE WATER CLOSET

(PATENTED AND PATENTS PENDING)

ROUGHING IN MEASUREMENTS

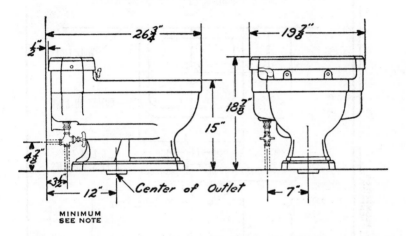

MINIMUM
SEE NOTE

NOTE: 12″ from wall to center of outlet is minimum measurement. We recommend 14″ roughing which gives more space behind fixture for cleaning purposes. The T/N, however, can be roughed at any measurement 12″ or greater from wall to center of outlet. In roughing further than 12″ from the wall consideration should be given measurement from wall to center of supply pipe.

W. A. CASE & SON MFG. CO.

Vitreous China Ware Measurements subject to a variation of plus or minus one-quarter inch. Measurements shown are for this job only.

Fig. 329

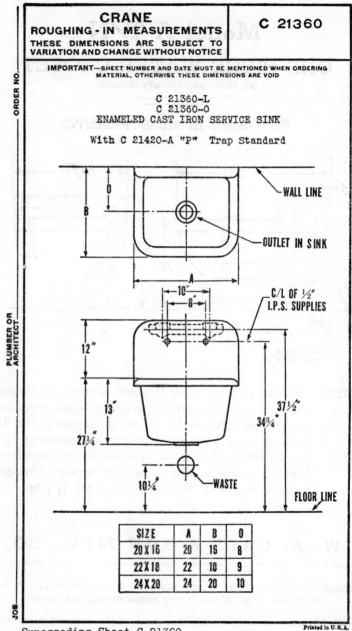

CRANE

ROUGHING - IN MEASUREMENTS
THESE DIMENSIONS ARE SUBJECT TO
VARIATION AND CHANGE WITHOUT NOTICE

C 21360

IMPORTANT—SHEET NUMBER AND DATE MUST BE MENTIONED WHEN ORDERING
MATERIAL, OTHERWISE THESE DIMENSIONS ARE VOID

C 21360-L
C 21360-0
ENAMELED CAST IRON SERVICE SINK

With C 21420-A "P" Trap Standard

SIZE	A	B	0
20 X 16	20	16	8
22 X 18	22	18	9
24 X 20	24	20	10

Superseding Sheet C 21360

Printed in U.S.A.

Fig. 330

CRANE
ROUGHING - IN MEASUREMENTS
THESE DIMENSIONS ARE SUBJECT TO
VARIATION AND CHANGE WITHOUT NOTICE

C 3330-C 11

IMPORTANT—SHEET NUMBER AND DATE MUST BE MENTIONED WHEN ORDERING.
MATERIAL, OTHERWISE THESE DIMENSIONS ARE VOID

C11-C12-R11-R12 TRIMMING
TARNIA ENAMELED CAST IRON REVERSABLE RECESS BATH
R.H. Outlet, Overrim Supply Trimming and Accesso
or Connected Waste and Overflow

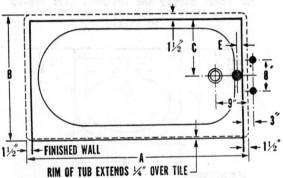

RIM OF TUB EXTENDS ¼" OVER TILE

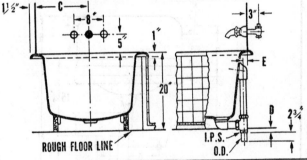

ROUGH FLOOR LINE

NOTE:
For L.H.
Installation
Reverse Di-
mensions
Shown.

SIZE FEET				WASTE			
				1½"		2"	
	A	B	C	D	E	D	E
5 FT.	60¾	32	14½	⅞	1¼	1¾	1
5½ FT.	67	32	14½	⅞	1¼	1¾	1

"Tarnia" Baths Are Furnished with a Wood Skid
Support, Which Should Be Used in Installation;
in Addition Provide Cement or Brick Supports at
Each Leg for Support in Case Wood Skid Shrinks.

Printed in U.S.A.

Superseding C 3336-R 30

Fig. 331

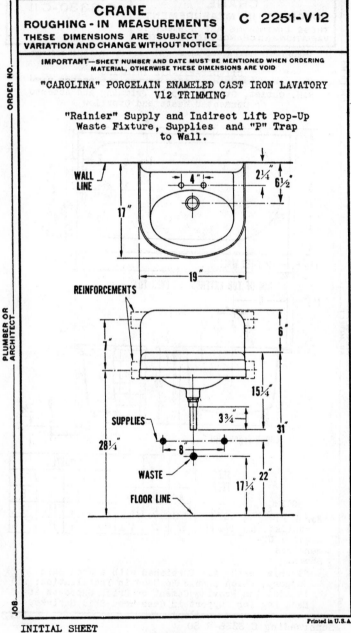

ORDER NO.

PLUMBER OR ARCHITECT

JOB

CRANE
ROUGHING - IN MEASUREMENTS
THESE DIMENSIONS ARE SUBJECT TO VARIATION AND CHANGE WITHOUT NOTICE

C 754-V12

IMPORTANT—SHEET NUMBER AND DATE MUST BE MENTIONED WHEN ORDERING MATERIAL, OTHERWISE THESE DIMENSIONS ARE VOID

"NORWICH" VITREOUS CHINA LAVATORY

"Rainier" Supply and Indirect Lift Pop-Up Waste
Trimming, Supplies and "P" Trap

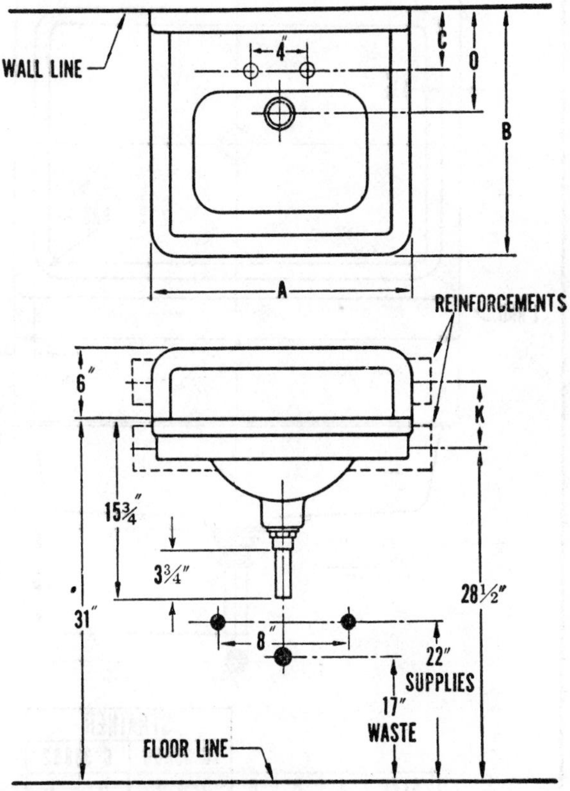

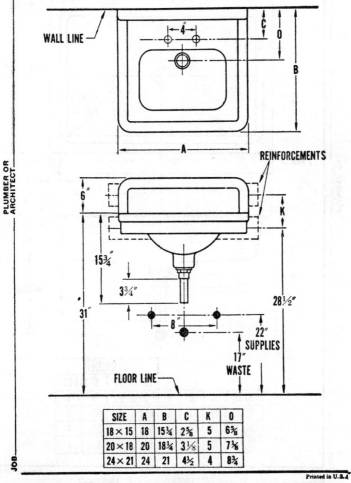

SIZE	A	B	C	K	O
18 × 15	18	15¼	2⅝	5	6⅝
20 × 18	20	18¼	3⅛	5	7⅛
24 × 21	24	21	4½	4	8¾

Printed in U.S.A.

Superseding C 757

Fig. 333

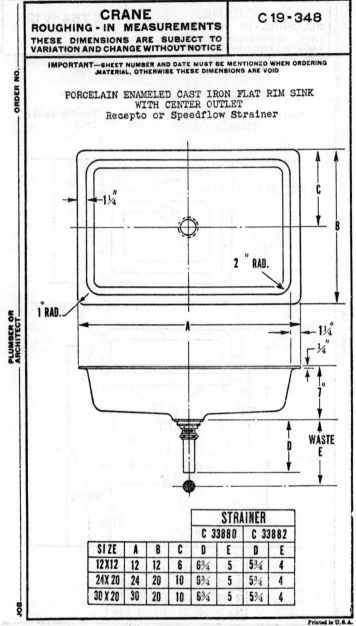

CRANE

ROUGHING - IN MEASUREMENTS

THESE DIMENSIONS ARE SUBJECT TO
VARIATION AND CHANGE WITHOUT NOTICE

C 19 - 348

IMPORTANT—SHEET NUMBER AND DATE MUST BE MENTIONED WHEN ORDERING
MATERIAL, OTHERWISE THESE DIMENSIONS ARE VOID

PORCELAIN ENAMELED CAST IRON FLAT RIM SINK
WITH CENTER OUTLET
Recepto or Speedflow Strainer

SIZE	A	B	C	STRAINER			
				C 33880		C 33882	
				D	E	D	E
12X12	12	12	6	6¾	5	5¾	4
24X20	24	20	10	6¾	5	5¾	4
30X20	30	20	10	6¾	5	5¾	4

Printed in U.S.A.

Superseding Sheet C 19348 and C 19348-X

Fig. 334

in event of inclement weather and also to serve as the starting point for the house drain. The last reason is only of secondary importance.

Before any actual work can be accomplished, the mechanic must study the basement and plot plan of the building, Plates 1 and 2, because he must formulate in his mind a general layout of the house drain as well as the building location relative to the piece of land on which it is to be constructed. It is also important that he familiarize himself with the Typical Wall Section details (see Plate 6), and the elevation drawings.

Building of any one portion of the plumbing system is dependent on all other parts of it and cannot be accomplished independently. The plumber must analyze the general requirements the architect has shown and incorporate these in his installations as well as the requirements of state and local codes. The specification must also be read carefully and all details of it carefully noted and adhered to.

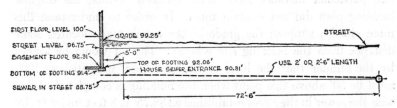

Fig. 335. Cross Section Showing Levels and Grades

To facilitate a better understanding of the many points which must be clearly understood before a plumber can proceed, a cross section drawing of the house sewer is shown. See Fig. 335. This drawing shows the sewer in the street, the house sewer, and the numerical grades necessary to complete the installation. Reference to this drawing is frequently made in the following paragraphs to explain construction details more clearly.

Planning the House Sewer. Before any consideration can be given to the actual laying of pipe for the house sewer, it becomes necessary to establish a number of factors which were explained in Chapter V, Page 56, such as the depth of the house drain outlet, the depth of the public sewer in the street, and the grade at which the house sewer must be laid. A careful analysis of the general plans of the building is essential as a beginning step to determine these details.

The blueprints, in most instances, indicate grades by means of whole numbers or whole numbers and decimals which have a definite relationship to a grade referred to by building engineers as a bench mark. The bench mark, which is usually a small copper plate mounted on a concrete pier, indicates the height of the plate above sea level or a previously established grade. All land within a given area of this marker can be surveyed and may be found to be above or below the original mark. City engineers and architects signify the depth of public sewers, street grades, water main grades, sidewalk levels and other public utilities, as well as building foundations, floor levels, landscape contours and many other phases of building construction, as either above or below the fixed bench mark. For example, by referring to Fig. 335 the grade of the first floor has been established as 100', the street as **96.75'**, the foundation as 92.06' and the sewer in the street as 88.75'. The figures in this particular instance have been assumed because the original building plan did not contain them. In order to understand this more clearly, suppose the grade of the bench mark located some distance from the building site was 50'. Then it would be known by the builder that the level of the first floor, indicated as 100', must be 50' above this point when the building is completed. Likewise the sewer in the street established as 88.75' is 8 feet under 96.75' or 11.25' below the level of the first floor. These grades very often confuse an uninformed reader, because they are given in decimals and our standards of measurement are in inches and feet. This is surveyors' practice and the contractor must transpose them.

Depth of the House Drain Outlet. In Chapter V, page 56, the second paragraph under "House Sewer Installation" explains in a general way how to determine the depth of the house drain outlet. By referring to Fig. 336 and scaling the branch to the laundry tub connection, which is the longest branch on the plan, from a point 5' outside the foundation wall, a measurement of 32' is obtained. The drain which is to be installed at $\frac{1}{4}''$ pitch per foot would then have $32' \times \frac{1}{4}''$ or 8" of pitch. To this must be added 16" which is the amount required to make up the drain tile receptor located in the furnace room of the basement. The drain terminal at a point five feet outside the foundation walls would be two feet under the level of the floor indicated in decimals as 90.31'.

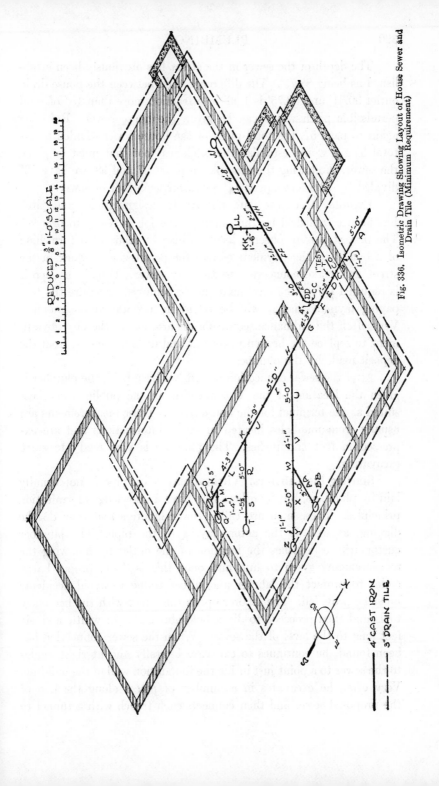

Fig. 336. Isometric Drawing Showing Layout of House Sewer and Drain Tile (Minimum Requirement)

REDUCED $\frac{3}{8}$" = 1'-0" SCALE

4" CAST IRON

3" DRAIN TILE

The depth of the sewer in the street has previously been established as being 88.75'. The difference then between the house drain outlet 90.31' and 88.75' is 1.56' or a fraction more than 18" of total permissible pitch which can be given the house sewer. Referring again to page 57, the total pitch of the house sewer divided by its total length would establish the pitch per foot that must be given the sewer. Following the example on pages 57 and 58, we have 18" divided by 72.6 feet equals $\frac{1}{4}$" pitch for each foot of sewer.

Permits. Once the data relative to design, grades and pitch has been established by the plumber, the job is ready to proceed. The first step is to observe local ordinances relative to furnishing of a bond and the payment of fees for permission to work in the street and make connection to the city sewer. Usually a deposit is required before the pavement can be removed to assure the taxpaying layman that it will be relaid in a workmanlike manner. Very often the ordinance requires a department of the city government to replace the broken pavement and deduct the cost from the deposit made by the plumber.

After the sewer connection permit has been paid, the plumber is given the location of the **Y** connection in the public sewer best suited as the terminal for the house sewer. These measurements are usually from manholes located in street intersections and are expressed in feet and inches. The plumber is now ready to start excavating.

Starting the Laborers. Vitrified clay sewer pipe is not usually laid by plumbers, but by others who have a knowledge of sanitation principles. These men are called sewer laborers and they do the digging, as well as the actual laying of the pipe. The plumber customarily establishes the location of the outlet by following the measurements given to him and once this is done, proper barricades to protect the laborer and direct traffic to avoid accidents must be provided. The laborer opens the street with the necessary tools and then proceeds to dig a hole about 2'6" in width and six feet long to the level of the sewer. When the sewer connection has been found, he continues to excavate laterally and at right angles to the sewer to a point just inside the foundation wall of the building. Very often he excavates in a number of places along the line of the proposed sewer and then connects each trench with a tunnel to save labor cost. The trench is dug to the depth of the previously established grade to assure a substantial footing for the piping which is to constitute the sewer. The plumber must watch and supervise this activity very closely in accordance with the method explained in Chapter V.

During the time the laborers are at work, the material for the sewer can be ordered.

Laying the Pipe. After the excavation is completed, the laborers can lay the pipe. They begin at the public sewer by removing the clay disc from the connection and in its place inserting the 6-inch clay curve, making a cement joint, as explained in Chapter IV, page 43. The pipe is laid one length at a time on the floor of the trench, care being taken to make substantial joints that will withstand reasonable tests should tests be required. Each foot of the sewer must be pitched ¼″ and carefully checked at about 10-foot intervals.

House Service. The house service was discussed in a general way in Chapter XVIII. In order to avoid duplication, the reader should review the chapter for specific details not common to every installation, and to acquire understanding of the trade terms involved in installation of the service. The usual practice is to install the house service in the same trench as the house sewer. This practice avoids additional excavation expense.

The house service is usually the second phase of the plumbing system to be installed, because a supply of water is needed by the building contractor for general masonry work and also because the trench for the house sewer is open and can be conveniently utilized.

The plumber again must analyze carefully all the details of the specification and determine a logical place to locate the water meter. The building specification in this problem calls for a ¾″ lead service pipe to be connected to the city main and equipped with the necessary stops required by local ordinances. The plumber pays for all tapping and street permits and also furnishes a temporary connection for the use of the building contractor.

Careful study of the plans shows that the house sewer enters the building under the coal room and because the water service is installed in the same trench it, too, would enter under the **same** room. Since all water must be metered, it would be impractical to locate the meter in a room of this kind. The meter might be covered by coal a good part of the year and could not be read conveniently by city employees assigned to this duty. The water service therefore must be extended into the laundry against the north partition of the room, at which point the meter can be connected and made accessible.

Location of the Water Main. The water mains in large cities are found on the north side of streets running east and west, and on the east side of streets running north and south, and from 6′ 0″ to 10′ 0″ from the curb. The main usually is seven or more feet below the street grade. Referring to the plot plan on Plate 1, it can be seen that the property runs slightly northwest. The water main

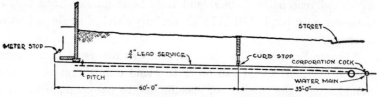

Fig. 337. Cross Section Showing Location of Main and Water Service

in such case can be found on the side of the street opposite the building and about nine feet from the north curb of the street. Fig. 337 indicates the location of the main and water service.

Permits. The payment of fees for necessary permits must again be made by the plumber before he can proceed with the installation of the water service. These consist of a fee for tapping the main, which is done by city authorities; payment for the brass stops to be used on the service; and a small fee for opening the street. The deposit for repair of the street usually is included in the house sewer deposit previously paid.

Installing the Pipe. The previously prepared water service can now be installed in the sewer trench on a shelf, as indicated in Fig. 338. It usually is started into the trench from the tap hole. Care must be used not to kink the service because carelessness in this respect reduces the flow of water through it, and may even cause the service to leak. After the service has been drawn into place and laid on the shelf prepared for it, the tailpiece can be made fast with a fiber washer to the corporation cock.

The curb box can now be put in place so that the curb stop is accessible at all times. Fig. 194 illustrates this procedure.

Backfilling. The trench can be filled with earth after the proper tests and inspection of the service and sewer have been made. To prevent breakage care must be exercised not to permit heavy rocks or large pieces of earth to fall directly on the plumbing work. In some localities the trenches in the street must be filled with gravel so the street can be repaired immediately without danger of settling and creating holes which may create traffic hazards. It also is necessary to flood the filled trench with water and then tamp the filled-in earth firmly.

Drain Tile Installation. The installation of drain tile shown in Fig. 339, around the basement foundation, is simple in layout. It is important, however, because its purpose is to keep the basement dry, and care must be exercised in the organization and laying of the drain. The installation of the drain tile discussed in Chapter VII, page 86, is typical of most buildings, and applies to the problem in this chapter as well.

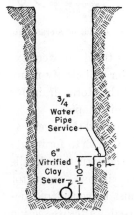

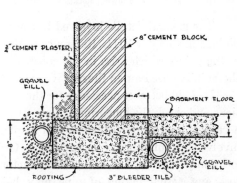

Fig. 338. Cross Section of Sewer Trench Showing Shelf for Water Service

Fig. 339. Cross Section of Foundation and Footing Showing Installation of Bleeder Pipe

Laying of the Drain Tile Pipe. Bleeder pipes laid in the foundation for the basement walls are the first to be laid, and in the manner shown in Fig. 339. This work is done, as a rule, by the general contractor, but it is supervised by the plumber. The purpose of this tile pipe is to allow the surface water accumulated by the outside tile drain to pass into the drain provided on the inside of the foundation wall.

The outside drain is laid in an open trench, dug around the building foundation in such manner as to allow a space of about ¼ inch between each pipe. The connections to the various bleeder openings are made by clipping the ends of the tile pipe or by cutting a hole into it to allow the water to flow from it. The pipe is covered with washed gravel, see Fig. 339, to prevent the entrance of mud and foreign materials. The inside drain is laid in the same manner. The inside and outside drains are both graded downward to the drain tile receptor.

Storm Sewer. This phase of the drainage system requires as much analysis and study as does the house sewer, the house drain, or any other part of the plumbing layout.

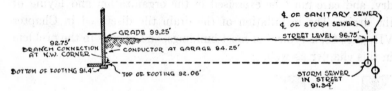

Fig. 340. Cross Section Showing Location of Storm Sewer and Layout of Storm Drain

The general details involved in constructing the drain were explained in Chapter VIII. Reference to the chapter will clarify many of the points which arise in the construction of the storm drain for the building used an an illustrative example in this chapter.

It again becomes necessary for the plumber to read the specifications and check the blueprints (Plate 2) carefully so as to formulate a general plan of procedure as well as plan the construction details involved in building the storm drain. Grades, windows, conductor locations, trenches, and many other factors must be considered before any actual laying of pipe may be started. A cross-section sketch, Fig. 340, shows the location of the sewer, the storm drain, the general construction, and the grades at which the drain may be laid from the sewer to a point just within the porch line, where it branches in two directions to serve the various conductor or roof leader outlets. Fig. 341 shows the drain from this junction to the locations of the conductor bends which are eventually to serve the rain water leaders.

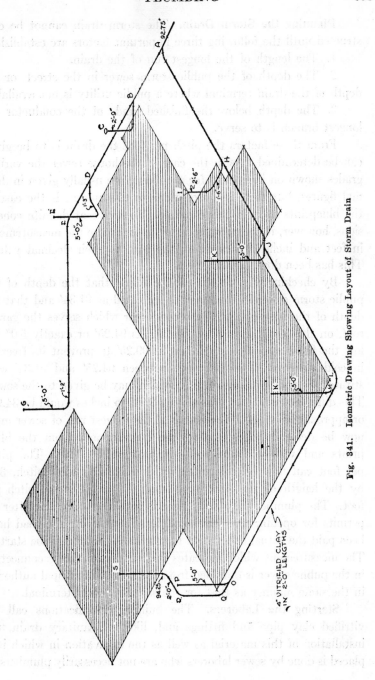

Fig. 341. Isometric Drawing Showing Layout of Storm Drain

Planning the Storm Drain. The storm drain cannot be con- structed until the following three important factors are established.

1. The length of the longest run of the drain.

2. The depth of the public storm sewer in the street, or the depth of the drain terminal where a public utility is not available.

3. The depth below the finished grade of the conductor the longest branch is to serve.

From these factors the pitch per foot the drain is to be given can be determined. As in the case of the house sewer the various grades shown on blueprints for a building are usually given in deci- mal figures, but there are exceptions to this rule as is the case of the blueprints used in this example. To arrive at definite conclu- sions, however, it becomes necessary to transpose the measurements in feet and inches indicated on the plan to their decimal values. This has been done in Fig. 340.

By checking Fig. 340 it can be found that the depth of the public storm sewer in the street is indicated as 91.34′ and that the depth of the base fitting of the conductor which serves the garage roof, on the east side of the building, is 94.25′ or exactly 5′0″ be- low the finished grade established as 99.25′ to prevent its freezing in cold weather. The difference between 94.25′ and 91.34′ con- stitutes the total amount of pitch which may be given to the sewer. This difference is 2.91′. Transposing 2.91′ to inches would be 34.92″ or approximately 35″. The length of the longest run of sewer must now be determined by scaling the measurement from the blue- prints and will be found to be approximately 140′. The pitch per foot can now be calculated by dividing the total pitch, 35″, by the length of the sewer 140′ and the result is ¼″ pitch per foot. The plumber may now proceed with the work and after all permits for opening the street and fees for relaying the road have been paid the actual excavation for the house drain may be started. The measurement which indicates the location of the **Y** connection in the public sewer is obtained from the proper municipal authority in the same manner as that for the sanitary sewer terminal.

Starting the Laborers. The building specifications call for vitrified clay pipe and fittings and, like the sanitary drain, the installation of this material as well as the excavation in which it is placed is done by sewer laborers who are not necessarily plumbers.

The excavation in the street is usually the first to be completed. This is done with the same precision as for the sanitary sewer terminal connection. Very often the same trench may be used for both the storm and sanitary sewers, which naturally is economical because many feet of costly excavation and the payment of dual road replacement fees can be eliminated. It may be well for the plumber, where the building plans do not indicate this procedure, to question the architect and make suggestions to him about combining both drains in one trench. There may be, however, a specific reason for separating the drains, and in such case there can be no alternative but to construct them as indicated.

The trench for the storm drain must be dug at right angles to the public storm sewer and to such depth as will assure a substantial foundation. It must be graded at ¼″ per foot so the drain, when completed, will have the previously determined pitch. The trench may be extended to the first branch shown in Fig. 341 and then along the next building wall to a point opposite the south corner of the furnace room. At this interval the trench changes direction at 45 degrees to the main run and finally terminates at the location of the conductor bend on the west side of the garage.

The trench which accommodates the drain on the north and east side of the building may now be started, taking advantage of the excavation made by others to complete the foundation of the building. This procedure lessens the amount of digging required to complete the drain and saves the plumbing contractor a substantial amount of labor. The trench is finally completed at the location of the conductor bend at the east end of the garage, as indicated in Fig. 341.

Laying the Pipe. After the trench has been completed, the laborer can lay the pipe. He must begin by removing the clay disc in the **Y** connection in the street and by cementing a 4″ clay curve in the side opening of the fitting. The pipe is laid one length at a time from the connection in the street to the first **Y** branch *A* located near the N.W. corner of the building (Fig. 341). From this junction he can continue to lay the pipe along the west wall of the building to the **Y,** at point *B*, which is to serve as a terminal for a basement window area drain, *C*, which may be completed by

calking a 4″ ¼ bend into a piece of 4″ cast iron pipe 2′ long.
The entire assembly is now cemented into the side opening of the
Y, at point *B*, and then brought up to the floor of the window areaway with a piece of 4″ cast iron pipe approximately 2′ 9″ long
which is provided, at floor level, with a 4-inch bar strainer.

The laborer can then lay the pipe to the **Y**, at point *D*, and
cement a short piece of 4″ clay pipe and a 4″ curve to serve as a
terminal for the roof leader at *E*. The base fitting for the conductor
is composed of a 4″ ¼ bend calked into a piece of 4″ cast iron single
hub to bring the connection above the finished grade. The entire
assembly may now be cemented into the curve to complete the connection for the conductor. Two lengths of pipe and a curve, *F*,
may now be cemented into the top opening of the **Y**, at point *D*,
and then pipe may be laid to the base fitting which is to serve the
garage roof conductor *G*. The base fitting for the conductor *G* consists of a 4″ ¼ bend calked into a piece of pipe 1′ 2″ long and is
brought to a point above grade with a length of 4″ cast iron pipe.
This completes the run of storm drain on the west side of the building. All the openings must be sealed with cement mortar.

The laborer can now lay the sewer alongside the north building line by cementing a clay curve into the **Y**, at point *A*, and
laying pipe at a ¼″ grade to the **Y**, at point *H*, into which a curve
has been cemented to serve as a terminal for the areaway drain
at point *I*. The area drain is made up of a 4″ ¼ bend calked into a
piece of 4″ cast iron pipe 1′ 6″ long which is brought up to the
floor of the areaway with a piece of 4″ pipe approximately 2′ 6″ long,
provided with a 4″ bar strainer at floor level. Four-inch clay pipe
is now laid to the **Y** at point *J*, into which a 4″ ¼ bend is cemented,
and the fitting provided with 4″ cast iron pipe to serve as a base
fitting for the downspout at *K*. The side opening of the **Y** at point
J is now extended with pipe to the N.E. corner of the building
where two curves, *L*, are installed to accommodate the drain on the
east side of the building. The **Y** at *M*, is now cemented into the
curves and the conductor, *N*, may be completed by cementing a
¼ bend into the **Y** at point *M*, and then extending the ¼ bend to
the grade line with 4″ cast iron pipe.

The drain may now be extended to the **Y** at point *O*, into
which a 4″ ¼ bend, provided with a length of 4″ cast iron pipe

to bring the connection to the grade, has been cemented. This connection serves as a base fitting for the conductor at P.

The drain now changes direction with the aid of two curves at Q and R and two lengths of pipe to accommodate the base fitting of the conductor S, which is made up of a ¼ bend calked into a piece of 4" cast iron pipe 2' long, and extended to the finished grade with a length of 4" cast-iron pipe. The entire drain has now been completed and after all openings have been sealed the trenches may be backfilled with earth. In some localities inspection and test are required by ordinance.

House Drain. The installation of the house drain may be made any time during the construction of the building. Most plumbers do this part of the work after the roof has been completed although this practice is not an accepted or fast rule. There is some logic, however, in planning of this kind. Quite often it becomes impossible to work on the upper floors of a building because of severe weather conditions or unforeseen difficulties which may cause loss of time and waste of money or moving to another job. Under these circumstances the plumber can go to work on the house drain and in this way minimize such difficulties as exist.

The house drain was discussed in full detail in Chapter VI. The explanations presented on those pages are common to most installations. It is a simple process to carry the planning of runs of pipe, grade, change of direction, materials and layout set up over to the installation of the house drain detailed on the plan of the building used in this chapter.

For the convenience of the reader a scaled perspective drawing of the house drain, Fig. 336, is shown to facilitate accuracy and also to convey an idea of how the drain will appear once it has been completed. An experienced plumber does not need a sketch of this kind because from practice he has formed the habit of visualizing this part of his work. Another reason for making this drawing is to impress the reader with the fact that blueprints are not always accurate and when scaled may vary a number of inches. This inaccuracy usually is the result of shrinkage of the paper on which the blueprints are made. Because of the possible discrepancy between scaling of the plan and the dimensions which are indicated thereon, an established building construction rule is that the dimensions given on the plan must be used even though they do not actually coincide with the scaled size of the building.

Before any pipe can be laid the specifications must be studied and the various locations of the soil, waste, and fixture terminals must be definitely established. This has been done in part to make the installation of the house sewer, but rechecking is essential to be sure that no costly errors are made. The specifications for this job call for cast-iron pipe of extra-heavy quality, and the finished drain must meet the approval of state and local code administrators. The joints between pipe and fittings must be made with oakum and an adequate amount of molten lead to make them substantial and watertight.

The plan indicates the manner in which the drain is to be installed and unless the building designer agrees to any change suggested by the plumber, the general layout must be followed. In this case the location of the laundry tub waste terminal must be relocated to a point between the windows on the east wall of the laundry, to facilitate ventilation of the trap outlet. The drain tile sump must also be changed to a drain tile receptor in order to comply with plumbing regulations. This automatically alters the water-softener connection because it may be drained indirectly into the combination floor drain and tile receptor. Other than these minor changes the general layout of the sewer is satisfactory.

Starting the Laborers. After the details of pitch and design have been agreed upon, the basement floor grades can be obtained from the construction foreman, and the laborer can begin to excavate for the installation of the house drain. As a general rule he excavates to approximately the established depth of the house drain and then grades the trench upward to the points of the soil and waste terminals. Once the trenches have been completed, the plumber is ready to lay the pipe.

Laying the Pipe. The first thing that the plumber must do before any pipe is laid is to check the grade of the drain carefully, and also check over the material which he ordered to be sure that it is on hand. After this has been done he is ready to install the pipe.

A space must be cleared of loose earth so that some of the work can be done outside the trench. The drain is generally calked together in sections and then placed in the trench. This practice permits the making of vertical joints and also shortens the time it takes to complete the job.

Referring to Fig. 336, a length of single hub pipe, *A*, is now placed in the hub end of the house sewer which was previously installed, graded ¼″ for each of its 5′, and then substantially joined with cement mortar to the clay house sewer. One foot, one inch of 4″ pipe, *B*, cut from a double hub length is set vertically on the cleared space and into this is calked a 4″ cast-iron **Y**, at point *C*, which is to be used for the cleanout (required by ordinance) immediately inside the foundation wall. A 4″ cast-iron tee *D* is set into the **Y** fitting and also calked, both side openings facing in the same direction. The purpose of the tee is to make testing of the completed drain possible. One foot, two inches of pipe, *E*, is cut from a double hub length and a four-inch **Y**, at point *F*, is calked into its hub in a vertical position. This **Y** is to accommodate the branch which is to serve the drain tile receptor located in the boiler room. The completed piece *FE* is now calked into the piece *DCB* previously assembled so the **Y** fitting faces in the proper direction after it has been placed in the trench. The complete assembly can now be laid in the trench, and after it has been graded carefully can be calked into the hub of the pipe already installed. A 4″ **Y**, at point *H*, is now calked into a piece of 4″ double hub pipe 4′ 3″ long, *G*, to accommodate the laundry tub and sink waste branch. Both pieces are laid in the trench, graded, and calked into the **Y**, at *F*. A 4″ x 3′ **Y**, point *K*, is then calked into a piece of four-inch pipe 2′ 9″ long, *J*, cut from another length of double hub pipe, the joint being made in a vertical position and calked to a length of single hub pipe, *I*, on the top of the trench. This can be accomplished by placing the two pieces of pipe and the **Y** in a horizontal position on blocks so as to raise it above the ground to permit making a horizontal joint. It can now be placed in the trench and joined to the **Y** fitting, at point *H*, using a horizontal calk joint to complete the task.

The next step is to cut two pieces of pipe from a 4″ double hub, one 4′ 3″ in length, *L*, and the other, *N*, 5″ long. A 4″ **Y**, point *M*, is calked into the 4′ 3″ piece in a vertical position. The 5″ piece, *N*, into which a 4″ ¼ bend, at point *O*, has previously been calked, can now be joined. The entire assembly can be laid into the trench, and after it has been graded the joint at the **Y** fitting K can be completed.

A piece of pipe 1′ 4″ long, P, can now be cut from a double hub length to complete the branch which is to serve as a terminal for the slop sink located in the laundry. A 4″ ¼ bend, Q, can be calked into it in a vertical position and then it may be fitted into the side opening of the **Y** at point M, and the joint made to complete the branch.

The next step is to cut a piece of pipe, 1′ 5½″ long, S, from a length of 3″ double hub pipe and calk a 3″ **P**-trap, T, into it. Both pieces can be joined to the length of 3″ single hub, R, which has been previously placed in a vertical position. The entire assembly can be placed in the trench, graded carefully and then be joined to the side opening of the 4″ x 3″ **Y**, at point K, installed in the main run of drain. The branch for the laundry tub and kitchen sink waste can now be completed. A piece of pipe 4′ 1″ long, V, can be cut from a double hub and the **Y**, at point W, calked into the length, making a vertical joint on the fitting. This assembly can be calked into a 5′ 0″ length of pipe, U, on top of the trench and when the joint has been adequately calked, it may be laid in the trench and joined to the side opening of the **Y** left in the main run of drain for this purpose, H. The piece of pipe, Y, 11″ long, left over from the last double hub length can now be used and a ¼ bend, Z, to serve as a terminal for the laundry tub must be calked into it. Both of these pieces are calked into a 5′ 0″ length of single hub pipe X and then laid into the trench and joined to the **Y** branch W as indicated on the plan. The sink waste branch can be fitted up by calking a ¼ bend, BB, into a piece of 4″ pipe, AA, 3′ 0″ long, which has been cut from a piece of scrap left over from the previously cut double hubs. The waste branch is then joined to the side opening of the **Y** at point W.

A short piece of 4″ pipe, 4″ long, CC, cut from a double hub is fitted with a 4″ ⅛ bend, DD, and calked into the **Y**, at point F, to accommodate the drain for the boiler room. Two pieces of left-over double hub pipe are cut to make up a length of 6′ 0″, EE and FF, and the 4″ x 3″ **Y**, at point GG, can be calked into them on the top of the trench. The assembled unit is now laid in the trench and graded. The **P**-trap is then calked into the 4′8″ pipe (II)

and the finished piece calked into the 2′ 3″ piece of scrap, *HH*. After this has been done, and the pipe *JJ* has been screwed in, the combination is laid in the trench and completed. This leaves only the floor drain branch to be completed.

One foot, six inches, *KK*, of the 3″ scrap piece of double hub can be fitted with a 3″ floor drain trap, *LL*, and then calked into the **Y** at point *GG* to complete this branch.

To complete the house drain, the next step is to bring the fixture, stack, and waste terminals up to the grade of the finished floor. The 4″ x 2″ reducers can be calked into the ¼ bends *Q*, *Z*, and *BB* to accomplish this. The floor drain strainers are calked into the **P**-traps, *T* and *LL*, so that they are flush with the finished floor grade at these points. The 4″ cast-iron cleanout plug may be calked into a piece of scrap 4″ pipe and then fitted into the **Y** at point *C*, previously installed for this purpose. The drain tile receiver is the last fixture left to be completed. Referring to page 87, Chapter VII, the reader will obtain an idea of how this should be done.

Testing the Drain. The house drain is now ready for testing. This is done by sealing the tee left for this purpose with a test plug or a clay bag as explained in Chapter XIV, page 210. The drain is filled with water through one of the waste openings and after it has been inspected for leaks and correctness of design by proper authority, the water may be drained from it and the test tee sealed with a 4-inch cast-iron plug. The opening of the drain should now be sealed with cement mortar and the trenches backfilled to permit laying of the concrete floor.

The Soil Pipe. In small residences the soil pipe is usually installed after the rafters of the building have been placed. This practice allows the plumber ample time to place the pipe within the tentative partitions of the house with a minimum of labor.

It is important that the plan of the building be carefully considered before installation is begun. The location of partitions in which the pipe is to be installed must be accurately established. The relationship of the drainage system to the work of other craftsmen is decidedly important and must be given consideration. The specification which indicates the type and quality of material to be used must also be read to assure that no costly errors are made. To make this task easier for the reader, a scaled isometric drawing of the plumbing layout of the building used in this chapter has been made. See Fig. 342.

The installation of plumbing for this building is not a simple one. It requires planning and good judgment to make the standard type of fittings fit into the building partitions. Careful study of Plates 3 and 4 reveals that the partition of the first-floor toilet room, in which the soil pipe is located, is not directly under the partition of the second-floor bathroom. This layout requires some thought because the available fittings do not permit an offset to be made to overcome this difficulty and still provide ample room to rough in the waste opening for the water closet on the first floor without exposing the offset near the ceiling of the kitchen. Details of this kind must be discovered and corrected before any pipe is placed. The usual procedure is to take up the points in question with the proper building authority and suggest a remedy. In this case, the soil pipe can be installed without making offsets by extending the partition of the broom closet to conceal the pipe and still not affect the plan of the kitchen to any appreciable extent. Another point which must be given consideration is the building of the drainage system so it may be installed with minimum cutting of rough carpentry work. The removal of sections of joists or studs is not recommended unless the carpenter is consulted and gives his approval. Very often it becomes necessary for him to header out sections of the joist frame work to permit the installation of pipe without weakening the building to any great degree.

A bill of material can now be made so the proper fittings will be on hand when the plumber needs them. This must be done carefully because labor costs mount as the result of delay.

Installation of Soil Pipe. The first fitting to be installed as a part of the soil pipe stack (Fig. 342) consists of a 4″ test tee *A* which is calked in a vertical position into the house drain opening previously provided for this purpose. A 4″ ¼ bend *C*, is now calked into a piece of soil pipe, *B-1*, 2′ 2½″ long cut from a double hub. Both pieces are fitted into another length of pipe, *B-2*, 3′ 3″ long, which has been cut from a 4″ double hub. The entire assembly is set into the hub of the test tee, *A*, and the joint completed after the pipe has been plumbed.

The horizontal run of pipe on the basement ceiling is now made

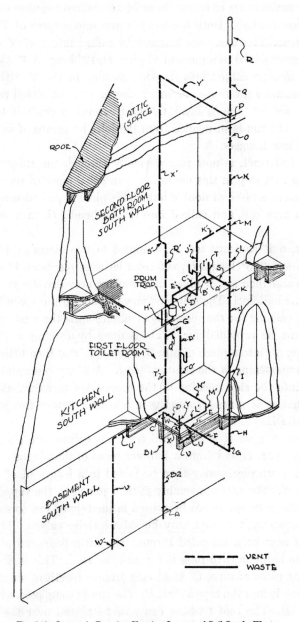

ATTIC
SPACE

ROOF

SECOND FLOOR
BATH ROOM
SOUTH WALL

DRUM
TRAP

FIRST FLOOR
TOILET ROOM

KITCHEN
SOUTH WALL

BASEMENT
SOUTH WALL

— — — — VENT
——————— WASTE

Fig. 342. Isometric Drawing Showing Layout of Soil Stack, Wastes.
and Vents

up. This run consists of two 4″ ⅛ bends, *G*, calked together to make a long sweep fitting. Both bends are fitted into a piece of 4″ pipe, *F*, 1′ 7″ long. This assembly can now be calked into a 4″ **Y** fitting, *E*, previously joined to a piece of 4″ pipe, *D*, 12″ long. A 4″ ⅛ bend, *U*, must also be calked into the side opening of the **Y** fitting, *E*, to accommodate the first-floor water closet. The completed run can now be fitted into the ¼ bend, *C*, being careful to grade it ¼ inch per foot. The run is suspended from the joist by means of substantial band iron hangers.

The soil stack is now run vertically through the ridge of the roof. The first step in this operation is to cut a piece of pipe, 2′ 3″ long, *H*, from a piece of double hub. This piece is passed through a four-inch pipe rest and calked into the ⅛ bends, *G*, in a vertical position.

Next, a 4″ sanitary tee, *K*, is joined to two pieces of 4″ pipe, 3′ 7″ (*J*) and 3′ 5″ long, (*I*), cut from leftover pieces of 4″ double hub pipe and then set in place in the space provided for it on the plan of the building. The soil pipe now extends to the second floor. The vent opening for the second-floor bathroom is the next opening which must be provided for. This is done by calking a 4″ x 2″ tapped tee, *M*, into a piece of pipe 3′ 2″ long, *L*, and then fitting this into the top opening of the sanitary tee *K*. A 2″ try-piece should be screwed into the side opening of the tapped tee to assure the mechanic that it is placed correctly in the wall before the joint is finally calked.

At this point a 4″ x 2″ tapped tee, *P*, is calked into a piece of pipe 10½″ long, *O*, cut from a leftover piece of double hub, 1′ 11″ in length. Both pieces may then be joined to a length of 4″ single hub pipe, *N*. The entire assembly is then joined to the tapped tee, *M*, and after a try-piece has been used in the tapped tee, to see that it faces correctly, the joint may be substantially calked. The soil stack has now been extended through the attic floor and all that remains to be done is to pass it through the roof. This is done by cutting the piece of pipe, *Q*, 4′ 4″ long from a length of single hub and calking it into the tapped tee, *P*. The run is completed with the increaser, *R*. The roof flashing can now be placed over the stack terminal so the roofer can properly install it.

A closet bend, S, with a 2-inch right-hand tapping is now fitted into the sanitary tee, K, and calked into place. The bend must be provided with a floor flange, T, to facilitate fastening the closet bowl to the plumbing system. The closet connection on the first floor is now completed with a 4-inch ¼ bend, V, a 4″ x 2″ tapped sanitary tee, W, and a piece of pipe 6″ long, X. The entire assembly is fitted into the ⅛ bend, U, previously provided to serve the first-floor water closet. A 4″ floor flange, Y, is now installed.

Installation of Waste and Vent Pipe. After the soil pipe has been completed, the waste pipe is the next phase of the plumbing system to be finished. The underfloor work of the second-floor bathroom is the logical place to begin the work. Again referring to Fig. 342, it is seen that a 2″ 60° drainage ell is screwed into a short nipple A′ and then made up to the side opening of the closet bend. Screwed into this fitting is another 2″ 60° drainage ell, B′, attached to a 3″ nipple. A 2″ x 1½″ long-turn 90° tee pattern **Y** is made up to a piece of pipe C′ 1′ 6″ long and is then screwed into the 60° drainage elbow. A 1½″ long-turn 90° ell to accommodate the basin waste is now screwed on to a 1½″ x 3″ galvanized pipe, D′, 8″ long, and then made up to the side opening of the long turn 90° drainage **Y**. A 45° drainage ell is now attached to a piece of 1½″ galvanized pipe provided with a bushing on one end. The ell is then screwed into the drainage **Y**.

A 4″ x 5″ drum trap is now screwed onto a piece of 1½″ pipe, F′, 7″ long and is made up into the 45° drainage ell. The bottom opening of the drum trap is provided with an ell into which a 1½″ plain cast-iron ell fitted to an 8″ piece of pipe, G′, is fitted. A short nipple is then brought up to the finished floor H′. A special 1½″ tee pattern **Y** is now screwed to a piece of 1½″ pipe, I′, 9¼″ long and fitted to the 1½″ long-turn ell previously installed.

A 1½″ plain cast-iron ell is screwed to a piece of 1½″ pipe, K′, 2′ ¾″ and then attached to the 4″ x 2″ tapped tee in the soil pipe stack to serve as a vent for the bathtub and wash basin. The vent for the wash basin can now be completed by screwing a piece of 1½″ pipe, J′, 19″ long in between the special tee pattern **Y** and 1½″ elbow on the horizontal run of the vent. The calk joint can now be made on the tee pattern **Y** to complete the job.

The underfloor work for the first-floor toilet room may now be installed. A piece of 2″ pipe, 20″ long, L′, is fitted with a 2″ 45° drainage ell and screwed into the 4″ x 2″ sanitary tee, M′. Into this is fitted a 2″ x 2″ galvanized nipple to which a 45° drainage ell has been attached. A 2″ long-turn ell is now screwed to a piece of 2″ pipe 19″ long, N′, to serve as a terminal for the basin waste. Next, a special tee pattern **Y** fitting is joined to a piece of 2″ pipe 19¼″ long, O′, and then screwed into the vertical opening of the long-turn elbow to serve as the basin trap terminal.

The work now shifts to the attic of the building where a piece of 2-inch pipe, Y′, cut 5′ 2″ long from the end to the center of a 2″ plain ell, is screwed into the 4″ x 2″ tapped tee left in the soil pipe stack. A 2″ x 1½″ x 2″ tee is next screwed to a 10′ 2″ piece of pipe, R′, into which a 5″ offset, S′, made with elbows has been built. To this is attached a 10′ 4″ pipe, X′, the assembly being fitted into the 2″ ell above the attic floor.

The vent is here extended to the center of the basin waste by attaching a 2-inch ell to a length of pipe measuring 8¼ inches (Q), which is then screwed into the side opening of the 2″ x 1½″ x 2″ tee. The vent for the basin is then completed with the installation of a 15½″ piece of 2″ pipe, P′, inserted into the special tee pattern **Y** fitting and then made fast with a calk joint.

The vent pipe for the basement sink is the final portion of the soil pipe stack to be completed. A piece of 1½″ pipe, T′, is fitted with a 1½″ plain ell, after it has been cut to a length of 4′ 3″ long and screwed into the bottom opening of the 2″ x 1½″ x 2″ tee. A piece of pipe, U′, 3′ 8½″ long is provided with a 1½″ plain ell and screwed into the piece of pipe T′. The vent for the basement sink is completed by screwing a short pattern tee to a piece of 1½″ pipe 6′ 1¾″ long, V′, these members being made up with the 1½″ ell provided for this purpose. The piece of pipe W′ is cut to a length of 10″ but cannot be installed until the final test has been made. The soil pipe installation for both toilet rooms has now been completed except that it must be tested with water and then inspected to make certain that it is without fault.

Sink and Laundry Tub Waste. The sink waste is generally installed immediately after the soil pipe has been completed. The

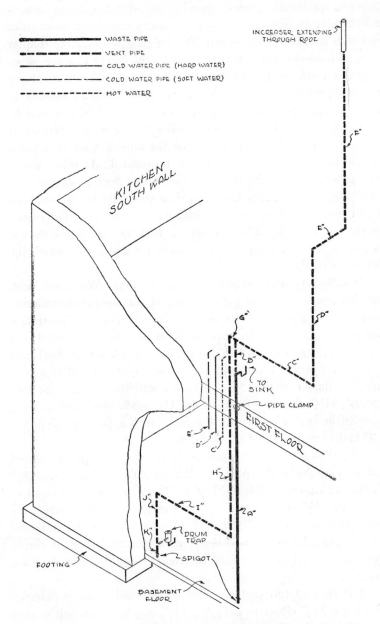

Fig. 343. Isometric Drawing of Sink and Laundry Tray Waste and Vent Pipes and Risers for Kitchen Sink

plans and specifications must again be studied to find a way to conceal the pipe in partitions with a minimum amount of damage to the rough building construction. The building used in this chapter offers an intricate layout which requires some study to solve, especially in the construction of the vent pipe system. The layout, however, is represented in Fig. 343, and is provided with a number of offsets so the pipe may be extended through the roof and still be concealed. The vent pipe rises vertically from the sink trap outlet and is then extended horizontally on the outside wall to a point directly under the north partition of bedroom 3. At this point it extends vertically to the center of the second floor joists where it again offsets horizontally and then rises vertically in the partition and terminates with an increaser fitting above the roof. The reason the stack is organized in this manner is that the bedroom dormer is set back a few feet from the east wall of the kitchen on which the sink is installed.

Installation of Pipe for Sink and Laundry Tray—Waste and Vent. The sink waste stack and laundry tub vent is a separate installation from the soil pipe stack on the building used as an illustrative example in this book. The first piece of pipe to be installed is a length of 1½" galvanized 9′ 6″ long, A″, to which a 1½″ short pattern tee has been screwed and set in place with the aid of a pipe clamp resting on the first floor. Into this installation a piece of 1½″ pipe, B″, 14½″ long and fitted with a 1½″ plain tee, is screwed and extended horizontally through the studs with a piece of pipe, C″, 3′ 9″ end to center of a 1½″ plain ell.

The vent pipe is then extended vertically through the second floor with a piece of 1½″ pipe, D″, 5′ 8″ from end to center of a 1½″ ell. Into this piece is fitted 17″ of 1½″ pipe, E″, measured end to center of a 1½″ plain elbow. The vent is now extended through the roof with a piece of 1½″ pipe 6′ 5½″ long, F″, to which a 1½″ x 24″ roof extension fitting has been fitted. A roof flashing is placed over the terminal to be used by the roofer to flash the waste stack properly.

The laundry tub vent can now be installed. This is done by screwing a 1½″ elbow to a 1½″ x 2½″ nipple, G″, which is then screwed into the 1½″ tee previously installed. The vent is then passed into the basement by screwing a 1½″ plain ell on a piece of pipe, H″, 8′ 2″ long. The vent is then run horizontally to the center of the laundry tub waste with a piece of pipe, I″, 3′ 9″ long, cut center to center of two 1½″ elbows. A 1½″ short pattern drainage tee is then screwed to a piece of 1½″ pipe, 2′ long, J″, to accommodate the laundry tub trap. This completes the job for test. However, the piece K″, which consists of a 1½″ x 3″ nipple and a spigot, is actually necessary to finish the job after the test has been completed.

Testing the Soil Pipe. Inspecting and testing of the roughed-in plumbing can now be done. The various kinds of tests and the manner in which they are made were previously discussed in Chapter XIV. The test-tee in the base of the stack can be plugged by use of an angle test plug (Fig. 186). The tester must be equipped with a valve and hose end so a rubber hose can be attached to admit water to the stack. The closet openings on the first and second floor must be sealed. It is necessary to block or wire the tester to the rough floor because the pressure of the water in the stack very often blows it out and ruins or delays the test routine.

The smaller waste pipe openings may be sealed by screwing a nipple and cap into them. The waste terminals into the sewer are closed in the same manner.

The stack can now be filled with water and inspection of it for leaks must be started from the roof terminal downward. Leaking joints on the cast-iron pipe may be calked gently in an effort to stop them. Defective material, however, must be removed and replaced.

The same procedure is followed on the sink waste stack and after the system has been made watertight, the plumbing inspector must be notified so his approval of construction details may be obtained. Once this has been accomplished the water can be drained from the stack and waste pipe. The water pipe which is to serve the bathroom fixtures may then be installed.

Water Pipe for Bathroom Groups. The specification for the building used in this illustration of an actual plumbing job calls for a three-pipe system of water distribution to be constructed of galvanized wrought iron pipe and galvanized malleable fittings. This work must be placed in the partitions immediately after the soil and waste pipes have been assembled to permit other craftsmen to go on with the building construction. There is no established practice in the setting up of the distribution piping, nor is there any standard of design. The installation of one mechanic will vary from that of

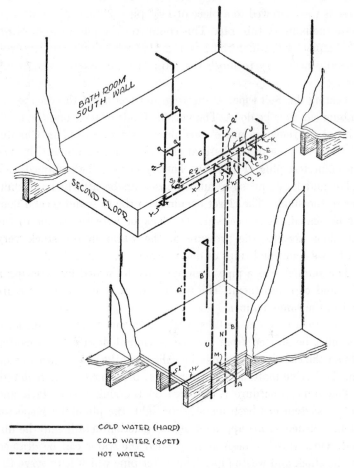

Fig. 344. Isometric Drawing of Water Supply System for Bathrooms

another even though both layouts are patterned for the same building. The lines of pipe should, however, be as direct as possible and free from an excessive number of fittings to assure an ample flow of water at the fixture.

Installation of Pipe. After the plumber has decided on the design of the riser system and the holes have been cut in the wood framework, the actual pipe may be installed.

The first piece of riser, see Fig. 344, to be placed in the partition is the ¾″ cold hard-water line and is made up of a 3½″ nipple, *A*, which has been screwed into a ¾″ x ½″ tee and both pieces attached to a ¾″ piece of pipe, *B*, 9′ 6½″ from the end to the center of a ¾″ elbow. A piece of pipe, *C*, 10″ from end to center of a ¾″ ell is then screwed into the elbow to offset the pipe from the partition of the first-floor toilet room to the center of the partition of the second-floor bathroom. The piece *D* consists of a ¾″ x 4½″ nipple fitted with a ¾″ x 1½″ x ½″ tee to serve the second-floor water closet to which the pipe, *E*, a ½″ x 6″ nipple, and elbow have been attached. The riser may be extended to the lavatory with a piece of ½″ pipe, *F*, 30″ end to center of a ½″ elbow, a vertical piece, *G*, 13″ long and fitted with a ½″ elbow and a 2″ center to center offset made of a ½″ shoulder nipple and ½″ elbow to serve as a supply for the lavatory. The offset also acts as a swing joint so the connection between the lavatory and the riser can be made in a workmanlike manner.

The water supply to the closet is the next part of the riser to be fitted. This consists of a ½″ x 3½″ nipple, *J*, and ½″ elbow, which assembly is screwed into the ¾″ x ½″ x ½″ tee to pass around the soil pipe stack. The piece of ½″ pipe, *K*, 9″ long end to center of a ½″ elbow into which a ½″ nipple, *L*, 6″ long and fitted with a ½″ x ⅜″ elbow, which is to serve as a supply for the water closet, has been provided. The first-floor closet can now be furnished with water. These fittings consist of a piece of ½″ pipe, *M*, 11″ long, to which a ½″ elbow has been attached. The pipe is run horizontally between the joists to the east partition of the first-floor toilet room with a piece of ½″ pipe, *H*, 2′ 4″ end to center of a ½″ elbow, and then extended vertically with a piece of ½″ pipe, *I*, 13″ long end to center of a ½″ x ⅜″ elbow to the center of the water closet supply pipe.

The ¾″ hot soft riser is now made up of a vertical piece, *N*, 9′ 10½″ long from end to center of a ¾″ elbow. It is now extended horizontally to the second-floor partition with a ¾″ pipe, *O*, 10″ end to center of a ¾″ elbow. The vertical piece, *P*, is now made up with a ¾″ piece of pipe 8½″ end to center of a ¾″ elbow. The pipe is extended horizontally with a 12″ piece of ¾″ pipe, *Q*, equipped with a ¾″ x ½″ x ½″ tee. The side opening has been provided with ½″ close nipple and 45° elbow. The lavatory is supplied with a 16″ piece of ½″ pipe, Q', from this tee. The piece of ½″ pipe, R, 2′ 5½″ end to center of a ½″ elbow is now installed in the partition to extend the riser into the bathtub work panel.

A 2″ offset, S, made with a ½″ shoulder nipple and ½″ elbow, to serve the vertical extension to the bathtub fixture, is now made. The fixture, which has previously been installed in the end partition of the bathroom according to the measurements provided by the fixture jobber, is then connected with a piece of ½″ pipe, T, 1′ 10″ long end to end fitted with a ½″ globe valve and ½″ union.

The cold soft water riser, U, is made up of a piece of pipe 9′ 10½″ long fitted with a ¾″ x ½″ elbow and then extended horizontally to the partition with the ½″ pipe, V, 10″ end to center of ½″ elbow. The piece of pipe, W, consists of ½″ x 5″ nipple to which a ½″ elbow has been attached. The riser extends horizontally to the bathtub work panel with a ½″ pipe, X, 3′ 6″ long from end to center of a ½″ elbow. A 2″ offset, Y, consisting of ½″ shoulder nipple and ½″ elbow, is provided to serve the vertical extension to the bathtub fixture. The extension to the bathtub supply fixture, consisting of a ½″ globe valve and ½″ union cut into a piece of pipe, Z, 2′ 2″ long, is now installed.

All of the pipe must be well supported to the bathroom partitions to assure the plumber that no shifting of the supply outlets occurs through carelessness on the part of other workmen.

The basin supplies, A' and B', are placed in the partition. These consist of two pieces of ½″ pipe 2′ 11″ end to center of a ½″ elbow. A 2″ swing joint made with a ½″ shoulder nipple and ½″ x ⅜″ elbow is provided on each for the lavatory supply combination.

The kitchen sink supplies must also be extended into the basement. These consist of cold water soft, cold water hard, and hot water soft. See Fig. 343. The three pieces of ½″ pipe, $C'D'E'$, are cut 2′ 10″ from end to center of a ½″ elbow and fastened into the partition. A 4″ spread between pipes should be maintained.

Testing the Water Pipe. The fixture supply risers are now completed and after all the openings have been sealed with nipples and caps of sufficient length to extend them through the finished plastered wall, a rubber hose may be attached to them and the entire system filled with water under normal pressure as a means of test. Should any leaks occur, these naturally would have to be made watertight.

Insulation Against Frost. The supply pipes for the first-floor wash basin and kitchen sink are installed in outside walls and may become frozen in severe weather. To avoid this, the partition must be packed with hair felt or a good quality of antifreeze insulating material. Another way of preventing freezing is to allow a large opening in the plastered wall back of the apron of the fixture, thus permitting warm air from the basement to circulate. The opening cannot be seen after the fixture has been installed.

Fixture Backing or Reinforcements. Plumbing fixtures cannot be supported on plastered walls, hence backing or reinforcement is required for the wall type fixtures. This consists of a 6-inch board fastened between the studs of the partition behind the lath and plaster of the finished partition. The height of this member is indicated in the roughing-in measurement list furnished by the fixture jobber. Backing must also be provided for the shower curtain rod, which is installed sometime later.

Water Supply System (Basement). The basement water supply main, Fig. 345, for the building used as illustration of a plumbing job, is a three-pipe system as required by the specifications. It is constructed of galvanized steel pipe and galvanized malleable fittings and is suspended from the basement ceiling. The construction details of water distribution systems were presented in Chapters XVI to XXIII inclusive and it is advisable to refer to these chapters when technical details are not fully understood.

Before the mechanic can assemble the system, he must familiarize himself with general construction details so that he does not interfere with the work of other craftsmen. The entire construction process is one of cooperation and knowledge of each individual's problems.

The plumber must have a substantial amount of material on hand to complete the water system. The fittings are usually small, and neglect in having the proper one on hand when it is needed results in costly delay.

Cold Water (Hard). The first pipe which may be installed to give sequence to the construction of the untreated cold water, shown on Fig. 345, is a piece of ¾″ pipe, *A*, 10″ long from end to center of a ¾″ elbow. It has a stop and waste valve provided in it to serve as a main control valve for the entire distribution system.

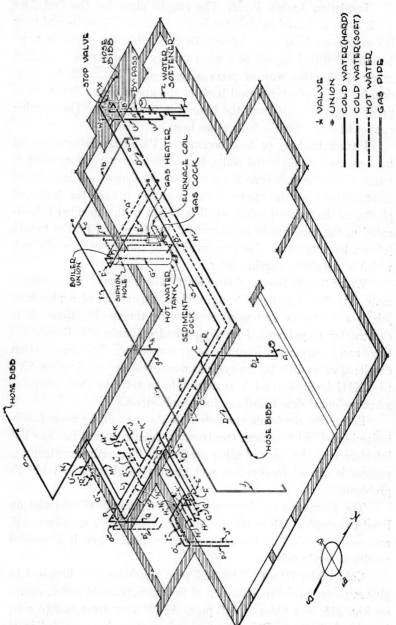

Fig. 345. Isometric Drawing Showing Layout of Water Supply Pipes in the Basement

A piece of ¾″ pipe, B, 6′ 5½″ long from end to center of a ¾″ tee is then run in a vertical position to begin the main suspended from the basement ceiling. The first branch from the horizontal main serves a hose bibb on the east side of the building and the connection for this run of pipe is made by screwing a ¾″ tee onto a ¾″ x 4½″ nipple, C. A 45° ell and shoulder nipple is then connected to the side opening of the tee and is extended to the outside of the building with a piece of ¾″ pipe, D, 8′ 11″ long and provided with a stop and waste valve approximately 1′ 6″ from the end of the pipe. The ¾″ pipe, E, equipped with a ¾″ tee and cut to a length of 7″ end to center, is installed to continue the main run of pipe. A 45° ell and shoulder nipple is screwed into the side opening of the tee, and a piece of ¾″ pipe, F, 6′ 1″ from end to the center of a ¾″ x ½″ elbow, is screwed into it. The run now changes to a vertical position (to allow the basement window to open) with a piece of ½″ pipe, G, 10½″ long from end to center of a ½″ elbow. The sink riser, which was previously extended into the basement, can now be connected with a piece of ½″ pipe, H, approximately 2′ long from center to center of two ½″ elbows. A ½″ stop and waste valve must be provided in this run of pipe and a ½″ union is used to complete the connection.

The main is now continued with a ¾″ pipe, I, 2′ 6″ long from end to center of a ¾″ tee which is to serve the bathroom risers. A ¾″ shoulder nipple and 45° elbow is screwed into the side opening of the tee. The branch is constructed with a piece of pipe, J, 3′ long end to center of a ¾″ elbow and provided with a stop and waste valve to control the bathroom water supply. The riser which was previously run to the basement can now be connected with a piece of ¾″ pipe, K, 1′ 3″ long end to center of a ¾″ elbow. A nipple and union must be used to make the vertical connection between the elbow and the bathroom riser.

The main run can now be continued to the next branch which supplies water to the lawn sprinkler on the west side of the building. A piece of ¾″ pipe, L, 3′ 10½″ long from end to center of a ¾ x ½ x ¾″ tee is fitted to the main run. A ¾″ 45° ell and shoulder nipple is screwed into the side opening of the tee and extended with a ¾″ pipe, M, 5′ 8″ end to center of a ¾″ elbow to a point on the east wall of the basement stairway. The run is then continued with

¾" pipe, *N*, 5' 9½" long from end to center of a ¾" elbow to the location of the hose connection. The branch is completed with ¾" pipe, *O*, 7' 11" long from end to end with a stop and waste valve provided just inside the west basement foundation wall.

The water supply to the laundry slop sink may now be completed. This connection consists of two pieces of ½" pipe, *P*, 1' 11" from end to center of a ½" elbow provided with a stop and waste valve about 6" from the elbow end, and a piece of ½" pipe, *Q*, 4' ½" long from end to center of ½" elbow. A nipple and cap may be provided at this point for testing purposes.

The cold water supply to the softener and hose bibb on the north side of the building can now be installed. This is done by screwing a piece of ¾" pipe, *R*, into the tee directly above the water meter. The pipe is 2' 9½" from end to center of a ¾" elbow turned on its side to effect a change of direction. The next piece of pipe to be run is ¾" in diameter, *S*, 19' 2" long from end to center of a ¾" tee fitted with a 7" nipple, *T*, and ¾" elbow to serve as a connection for the water softener. A piece of ¾" pipe, *U*, 2' 4" end to center of a ¾" tee is now put into place. This connection is to serve as a by-pass so that cold untreated water can be distributed throughout the soft water system. The softener can be connected to the main with a ¾" pipe, *V*, about 1' 8¼" long from end to center of a ¾" elbow, and two short nipples, using a ¾" union actually to complete the connection. The main is now extended to the lawn sprinkler on the north side of the building with two pieces of ¾" pipe, *W*, 1' 11½" from end to center of a ¾" elbow and, *X*, 2' 4" from end to end, with a stop and waste valve provided just inside the north foundation wall. The cold water (hard) is now completed except that the hose bibb must be connected to the system when the building is ready for occupancy.

Cold Water (Soft). The soft cold water system must be so organized that it will serve the fixtures specified by the architect. Building conditions must be considered to decide just where the main can be started and it is not likely that any two mechanics would run the pipe in the same way. For convenience and some sequence in this chapter, however, the main is started at the water softener and then extended to its various branch connections. The first connection is the by-pass, *A'*, between the treated and untreated water. This connection consists of two short ¾" nipples, a ¾" union, and ¾" stop and waste valve measuring 9" from end to center of a ¾" tee when it is finally made up. The water softener connection to the tee is completed with two short ¾" nipples and a ¾" ground joint union. The top opening of the tee is extended to the level of the cold water (hard) main with a piece of ¾" pipe, B', 2' 5" long from end to center of a ¾" elbow, where it changes direction with a ¾" pipe, C', 9" long from end to center of a ¾" elbow, in order that an even spread between the water lines can be maintained. The run of pipe now extends horizontally with a piece of ¾" pipe, D', 7' 5¼" long from end to center of a ¾" tee which serves as a branch to the hot water storage tank.

The branch to the storage tank consists of a ¾" short nipple and elbow screwed into the tee so the line may pass under the cold (hard) and hot (soft) water lines and then be extended to the location of the storage tank with a piece of ¾" pipe, E', 8' 5" long from end to center of a ¾" elbow. The storage tank is connected to the branch with two nipples and a stop and waste valve, F', made up to 10" from end to center of a ¾" elbow and a short piece of pipe and boiler union. The boiler tube, G', consists of a piece of ¾" pipe about 4' 6" long provided with a siphon hole. Refer to Chapter XXII, page 284, first paragraph.

The main run of pipe is continued with a piece of ¾" pipe, H, 13' 10½" long from end to center of a ¾" elbow, and then changes direction with a ¾" pipe, I', 12' 5½" long from end to center of a ¾" tee which serves as a branch connection for the sink and laundry tub. A ¾" 45° elbow and short nipple is screwed into the side opening of the tee and then extended with a piece of ¾" pipe, J', 3' 9" long from end to center of a ¾" x ½" x ½" tee.

The ½" run to the sink riser is continued with a piece of pipe, K, 11" long from end to center of a ½" elbow and then extended vertically with a piece of pipe, L', 10½" from end to center of a ½" elbow to permit opening of the basement window. The sink riser which was previously extended into the basement can now be connected with a piece of ½" pipe, M', approximately 2' long from center to center of two ½" elbows. A ½" stop and waste valve must be provided in this run of pipe and a ½" union and a short nipple are used to complete the connection.

The laundry tub faucet can now be connected to the water line. This is done by inserting a ½″ 45° elbow and short nipple into the ¾″ x ½″ x ½″ tee and then extending this connection with a piece of ½″ pipe, N′, 2′ 4½″ long from end to center of a ½″ elbow. A piece of ½″ pipe, O′, fitted with a stop and waste valve and 2′ 5½″ long from end to center, is screwed into the elbow and is then connected to the laundry tub faucet with a piece of ½″ pipe, P′, approximately 3′ 10″ long from end to end.

The main run is now continued with ¾″ pipe, Q′, 3′ 3″ long from end to center of a ¾″ x ¾″ x ½″ tee fitted with a short nipple and 45° elbow in its side opening. The bathroom riser is connected to the 45° elbow with a piece of pipe, R′, 3′ from end to center of a ¾″ elbow and provided with a ¾″ stop and waste valve so the bathroom water supply can be controlled. The riser may be connected to the branch with a short nipple and ¾″ ground joint union.

The wash basin of the first-floor toilet room now remains and must be connected to the main to complete the cold water (soft) installation. This is done with a piece of ½″ pipe, S′, 3′ 2″ long from end to center of a ½″ ell installed. The elbow must be provided with a short nipple and 45° elbow to pass above the other water lines. It is then extended to the center of the riser with a piece of ½″ pipe, T′, 5′ 10½″ long from end to center of a ½″ elbow, and then to the previously installed riser with a piece of pipe, U′, 1′ 8½″ long from end to center of a ½″ elbow provided with a ½″ stop and waste valve. The riser can be connected with a short nipple and ground joint union to complete the system.

Hot Water (Soft). The hot water system is organized in about the same manner as the other two lines. It is run in perfect alignment and centered between the cold water lines. The first piece of pipe, A″, suspended from the basement ceiling is approximately 7′ 3″ long from center to center of two ¾″ elbows. It is connected to the top opening of the range boiler with a ¾″ nipple and ¾″ boiler union. A short nipple and a ¾″ elbow are provided to attain the height of the cold water lines. The run is extended with a ¾″ piece of pipe, B″, 13′ 11½″ long from end to center of a ¾″ elbow, where it changes direction with a piece of ¾″ pipe, C″, 10′ 6½″ long from end to center of a ¾″ tee fitted with a short nipple and 45° ell to serve the branch for the laundry tub and kitchen sink.

The branch is started with a piece of ¾″ pipe, *D″*, 4′ 11½″ from end to center of a ¾″ x ½″ x ½″ tee fitted with a short ½″ nipple and a ½″ 45° elbow in the side opening. A ½″ x 5″ nipple, *E″*, on which a ½″ elbow has been attached, is now associated with the tee and is then extended vertically with a piece of ½″ pipe, *F″*, 10½″ from end to center of a ½″ elbow to permit the basement window to open. The sink riser can now be connected with a piece of ½″ pipe, *G″*, approximately 2′ long from center to center of two ½″ elbows. A ½″ stop and waste valve must be provided in this run of pipe and a short nipple and union must be used to complete the connection.

A piece of ½″ pipe, *H″*, 2′ 11½″ long from end to center of a ½″ elbow is run to the center of the laundry tub faucet and then extended to the wall with a piece of ½″ pipe, *I″*, 2′ long from end to center of a ½″ elbow and fitted with ½″ stop and waste valve. The faucet is connected to the elbow with a piece of ½″ pipe, *J″*, approximately 3′ 10″ long.

The main is now continued with a ¾″ piece of pipe, *K″*, 3′ 8½″ long from end to center of a ¾″ tee, the side opening of which has been fitted with a short nipple and 45° elbow. A piece of ¾″ pipe, *L″*, 3′ ½″ long and fitted with a stop and waste valve, is run to the center of the hot water bathroom riser. The riser is connected with a ¾″ pipe, *M″*, 7″ long from end to center of a ¾″ elbow and is then connected with a short nipple and ground joint union to complete the branch.

The run to the slop sink and wash basin riser can now be started with a piece of ¾″ pipe, *N″*, 4′ 5″ long from end to center of a ¾″ x ½″ x ½″ tee fitted with a short nipple and 45° elbow and then extended to the center of the slop sink faucet with two ½″ nipples and stop and waste valve, *O″*, assembled to make up 10″ from end to center of a ½″ elbow. The line now drops to the slop sink with a piece of ½″ pipe, *P″*, 4′ ½″ long from end to center of a ½″ elbow to complete the connection.

The first-floor toilet room wash basin can now be provided with hot water by screwing a piece of ½″ pipe, *Q″*, 4′ 1″ long from end to center of a ½″ elbow into the 45° elbow previously provided in the main and then run to the riser with a piece of pipe, *R″*, 1′ long from end to center of a ½″ elbow and fitted with a ½″ stop and waste valve. The branch is completed with a short nipple and a ½″ ground joint union.

Hot Water Storage Tank. The specifications call for a 40-gallon storage tank with a furnace coil and auxiliary gas heater as the heating medium. These connections must be fitted after the furnace has been placed. The top or flow opening of the furnace coil must be connected to the top opening of the storage tank and the bottom or return connections of the storage tank must be connected to the bottom of the furnace coil. The pipe fittings used for this connection are all ¾″ in diameter.

The gas heater must likewise be connected later. This is done by connecting the top or flow opening of the gas heater to a tee provided directly above the boiler union. The opening in the bottom of the tank is connected with the bottom opening of the gas heater and is provided with a sediment cock in order that the hot water system may be drained. The pipe and fittings used for this connection are also ¾″ in diameter. Storage tank connections of this variety were explained in Chapter XXII.

Water Pipe Test. After the entire system has been completed it should be tested with water and any leaks which occur must be corrected. It is advisable to use pressure in excess of that of the city water supply system to assure the plumber that precision workmanship prevails.

Gas Piping. The gas pipe installation in a modern residence is simple to construct as compared to the water supply and drainage system. It consists of a run of ¾″ pipe to the kitchen stove with ½″ branches to the gas heater and laundry stove. Unlike the water pipe it is usually strapped tightly to the underside of the first floor joist. Little attention is paid to grade except that it would be considered poor workmanship if the pipe contained traps that would allow condensation to accumulate. Where a condition of this kind is unavoidable, a drain in the form of a plug must be provided.

Installation of the Gas Pipe. The first pipe to be strapped to the ceiling joist is a piece of ¾″ black, *b*, 8′ 2″ long from center of a ¾″ elbow to center of a ¾″ x ½″ tee. The elbow is fitted with 1′ of ¾″ pipe, *a*, to which the gas meter can eventually be attached. A piece of ½″ pipe 1′ 1½″, *c*, from end to center is now run to the gas heater and then a ½″ drop, *d*. 7′ long from end to center of a

½″ elbow is installed. The gas heater is connected to the elbow with nipples and a gas cock and ground joint union.

The main is now extended with ¾″ pipe, *e*, 10½″ long from end to center of a ¾″ elbow, to a point where its direction changes. The branch to the kitchen stove is the next to be provided. This is done by screwing a piece of ¾″ pipe, *f*, 12′ 10½″ long from end to center of a ¾″ x ½″ x ¾″ tee. The branch is continued to a point below the first-floor partition with a ¾″ pipe, *g*, 1′ 6″ from end to center of a ¾″ elbow, and then rises in the partition with a piece of ¾″ pipe, *h*, 2′ 5¼″ from end to center of a ¾″ elbow. The elbow is extended through the plastered wall with a nipple which can eventually be removed when the gas stove may be connected.

The main is now extended to the east wall of the laundry with a piece of ½″ pipe, *i*, 12′ long from end to center of a ½″ elbow and the run is completed with a ½″ drop, *j*, 6′ ½″ from end to center of a ½″ elbow sealed with a nipple and cap for future use.

The gas pipe may now be tested with water under city pressure and if no leaks occur, the run of pipe is ready for use.

The gas service from the street is usually constructed by the gas company and is not under the jurisdiction of the plumber. The plumber's work begins at the house meter and consists of the piping within the building.

Finishing. The term *finishing* is one applied to the setting of the plumbing fixtures. This must be done after the rooms in which they are to be installed have been decorated and are ready for occupancy. The bathtub, because it is built in, is the only exception to this rule. It must be set just before the plasterer or tile-setter is ready to apply the first coat of material to the rough wall.

The quality of workmanship is usually judged by the appearance of the plumbing fixtures and many installations have been criticized by the layman, even though the roughing in may be perfect, because the manner in which the plumbing fixtures were installed was poor by comparison with the workmanship of the other crafts.

The methods employed in setting plumbing fixtures are discussed in Chapter XXVII and may be satisfactorily used for installing the fixtures specified in the building of this example.

Before the fixtures are set into place each one must be carefully inspected for imperfections.

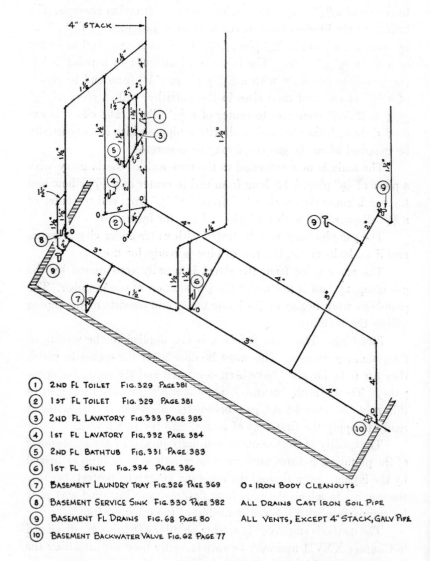

Fig. 346. Piping Layout Showing Drainage and Venting Systems. This Is the Maximum Requirement to Prevent Flooding of Basement in Event of Backflow from City Sewer During Heavy Storms and Wet Weather.

Should the fixtures become soiled during building construction it is up to the plumber to clean them thoroughly before turning the job over to the building owner. Care must be taken not to use cleaners which may be gritty and scratch the highly polished surface of the fixture.

Fig. 346 (representing maximum requirements) is designed to prevent flooding of the basement due to combined city sewer and storm drains which during heavy rains will overtax the sewers and back up, flooding resident basements. Also, first and second floor plumbing fixtures discharging into the basement sewer line will only exaggerate an existing bad condition. To prevent this, a back water valve (as shown in Fig. 62, Page 77) is installed back of a combination **Y** and ⅛″ bend that receives the discharge from all fixtures above the basement. The back water valve, or back flow valve, has very few working parts and seldom needs repairing.

Sometimes the basement will become flooded (if no back flow is installed) by a stoppage in the main street sewer or by stoppage in the house sewer between the house and the street sewer. In the latter case the flooding will only be caused by the discharge of fixtures on the upper floors.

Fig. 346 shows all drain lines from above the basement fixtures. Part of these drain lines are installed along the basement ceiling and are hung with substantial pipe hangers or heavy perforated pipe strap. **Note:** Only #20 gage or heavier straps should be used for hanging to the first floor joists.

The basement floor drains and fixtures are installed as before except that a smaller size 2″ pipe is used. This will be adequate to drain the basement and fixtures. However in some city codes, 3″ or 4″ drain pipe is required for all basement floor drains. **Note:** Always check the plumbing code in the location of the plumbing installation.

In this chapter a step by step plan is given which will be very helpful to a plumber or home owner who is doing all cutting and threading of water and gas piping and yarning and calking of drains and vents *on the job*. But where large housing developments are being installed and the prefabricating of all possible sections is practiced, the estimating of measurements from a set of plans will not always prove accurate.

The plan of the house shown in the blueprints (following page 374) is more detailed than the average, and generally speaking, the building contractor cannot place every brace, stop, or stud in such a manner that they will not interfere with pipes cut in the shop and brought to the job.

As previously stated, except for large scale building projects, piping measurements should be made on the job after the rough floors, partitions, studding, plates, and rafters are in place.

Fig. 346 represents actual requirements of a city located in the Southwest. All drains are cast iron soil pipe, and all vents either cast iron, galvanized, or copper pipe. All vents must rise at a 45° or 65° angle. Flat venting is prohibited except in extreme situations. Flat venting is a safe installation if properly sized and graded and composed of non-corrosive material.

Copper water piping is becoming more popular with builders and plumbers. The material cost is greater, but labor cost is much less due to increased speed of installation.

QUESTIONS PERTAINING TO PLUMBING

Chapter I—Municipal Sewage Disposal

1. How are public sewers classified?
2. What is a combination sewer?
3. Define a storm sewer.
4. What type of sewer carries only sanitary wastes?
5. What is the usual purpose of an intercepting sewer?
6. Of what materials are intercepting sewers constructed?
7. What procedure is followed in the construction of an intercepting sewer?
8. What is the function of a tributary sewer?
9. Of what materials is a tributary sewer constructed?
10. Why are public sewers provided with manholes?
11. How may a public sewer be aerated before it may be entered safely?
12. Why is it necessary to correct municipal sewage wastes?
13. What two methods are used in the treatment of sewage?
14. What do you understand by the term activated sludge?
15. How does a sprinkle filter treatment plant differ from an activated sludge system?
16. What is the function of a digester in the sprinkle filter disposal plant?

Chapter II—Private Sewage Disposal

1. Are cesspools considered a sanitary medium for the disposal of sewage?
2. Why is the outside privy an objectionable installation?
3. What factors must be considered in the construction of a septic system?
4. What two units compose a septic system?
5. What is the purpose of a septic tank?
6. How does decomposition of the organic materials contained in a septic tank occur?
7. What gases may be found in a septic tank?
8. What changes take place in the sewage after it enters the septic tank?
9. How are the nonsoluble solids removed from the septic tank?
10. Does a septic tank purify sewage?
11. Is it permissible to discharge large volumes of storm water into a septic tank?
12. Of what materials are septic tanks usually constructed?
13. Why must the outlet and inlet inverts of a septic tank remain open?
14. Name five essential qualities a septic tank must possess.
15. What is the purpose of a baffle plate?
16. What are the minimum permissible dimensions of a septic tank?
17. What is the purpose of the purification unit?
18. What does the purification unit of a septic system accomplish?
19. Give a definition of a dry well.
20. How does a dry well differ from a cesspool?

21. How does a filter trench function?
22. In what kind of soil may a distribution field be used?
23. Why must sewage be purified before it is discharged into an inland lake?
24. What is the purpose of a siphon compartment?

Chapter III—Materials Used for Sewer Pipe and Fittings

1. What materials are used in the construction of plumbing systems?
2. On what installations is vitrified clay pipe used?
3. Why must vitrified clay pipe be laid on substantial soil?
4. Under what circumstances may vitrified clay pipe be laid in unstable soil?
5. What do you understand by fittings of long radius?
6. On what installations is cast-iron pipe used?
7. Does cast-iron pipe offer any advantage over vitrified clay pipe?
8. Why is a coat of tar applied to cast-iron pipe?
9. For what purpose is galvanized steel pipe used in plumbing systems?
10. What acids affect galvanized pipe?
11. Why is wrought-iron pipe superior to steel pipe?
12. When does it become necessary to use acid-resistant pipe?
13. Why must acid-resistant pipe be well supported?
14. On what installation is brass pipe used?
15. What advantage is there in the use of lead pipe?
16. Why is copper pipe a good material for use in the plumbing system?

Chapter IV—Joints in Clay and Iron Sewer=Waste and Vent

1. Of what importance is the quality of the joints used on a plumbing system?
2. What essential qualities must a good joint possess?
3. What is the purpose of a swab?
4. What proportions of sand and cement are used in the mortar with which a cement joint is made?
5. What precaution must be taken when making a vitrified clay to cast-iron pipe joint?
6. What qualities must a cast-iron pipe which is joined to vitrified clay pipe possess?
7. Under what circumstances is a composition joint used?
8. What is a calk joint?
9. For what purpose is oakum used?
10. Explain how you would make a vertical calk joint.
11. When a fitting is cracked, what procedure is followed?
12. How does the process of making a horizontal calk joint differ from that of making a vertical joint?
13. Explain how you would cut off a piece of cast-iron pipe.
14. What materials are used in making an acid-resistant calk joint?
15. Explain the essential qualities of a screw thread joint.
16. Why must the pipe be reamed before making a screw thread joint?
17. What is a sweat joint?
18. What tools are used in producing a wiped joint?

Chapter V—The House Sewer

1. What portion of the drainage system is the house sewer?
2. What common difficulties occur in a house sewer constructed of vitrified clay pipe?
3. Explain the process involved in making the connection of the house sewer to a concrete public sewer.
4. Explain how you would insert a new **Y** in the public sewer to complete a house sewer terminal.
5. What do you understand by a slant?
6. What factors must be determined before a house sewer can be installed?
7. What is the permissible covering of a clay house sewer?
8. What do you understand by the term additional grade?
9. What is considered the smallest diameter of pipe practical for house sewer service?
10. Under what circumstances can one house sewer serve two buildings?
11. What is the recommended minimum pitch per foot for a house sewer?
12. What essential qualities should a house sewer possess?

Chapter VI—The House Drain

1. Define the term "house drain."
2. Of what materials should a house drain be constructed?
3. How are house drains classified, and what wastes does each type handle?
4. What constitutes a fixture unit?
5. Why is the unit system a practical method for sizing the house drain?
6. At what grade per foot should a house drain be extended?
7. What essential qualities should a house drain possess?
8. What is meant by a self-scouring flow?
9. Why is excessive grade of the house drain impractical?
10. What is the smallest diameter of house drain permissible?
11. How must change of direction of the house drain be made?
12. At what locations on the house drain is it necessary to place cleanouts?
13. Why must a cleanout be extended at least 2 inches above the finished floor?
14. What is the smallest diameter of cleanout that may be used on a house drain?
15. Why must the branches of a house drain be carefully planned?
16. What precaution must be taken when backfilling the trench in which the house drain has been installed?

Chapter VII—House Drain Appliances

1. What is the purpose of a house trap?
2. What gases are found in public sewers?
3. In what way is a house trap detrimental to a plumbing system?
4. Under what circumstances should a house trap be installed?
5. What is the purpose of a fresh air pipe?
6. What is the purpose of a back-flow valve?
7. Explain the construction of a balanced back-flow valve.
8. What precaution must be taken when installing a back-flow valve of the unbalanced variety?

9. Why should a gate valve be installed in conjunction with a back-flow valve?

10. Name a number of locations where a back-flow valve should be provided on a combination house drain.

11. What is an area drain?

12. Where should the trap of an area drain be located?

13. Explain how you would construct an area drain.

14. What advantage does a bar strainer offer over a strainer of the drilled variety?

15. What essential qualities should a floor drain possess?

16. Under what circumstances should a floor drain be protected with a back-flow valve?

17. What materials should be used in the construction of a floor drain?

18. Why should the trap of a floor drain be of deep seal?

19. What is the purpose of a yard catch basin?

20. What is the most important precaution to observe in the construction of a yard catch basin?

21. What essential qualities must a yard catch basin possess?

22. What is the function of a garage catch basin?

23. Is there any advantage in the use of a garage catch basin of the cast-iron variety?

24. On what does the efficiency of a garage catch basin depend?

25. Why must the water level of a garage catch basin be within 10 inches of the floor level?

26. What is the purpose of a local vent on a garage basin?

27. What is the purpose of a steam boiler blow-off basin?

28. Why must a steam boiler blow-off basin be provided with a local vent?

29. What would result if a steam boiler were connected directly to the house drain?

30. Why should the branch between the house drain and steam boiler blow-off basin be constructed of cast-iron pipe?

31. For what purpose is a drain-tile receptor used?

32. Of what materials should a drain-tile receptor be constructed?

33. What is the purpose of a flap trap?

34. What essential qualities should a drain-tile receptor possess?

35. When must a sewage ejector be installed?

36. Why must the outlet pipe of a sewage ejector be provided with a back flow valve?

37. What is the purpose of a strainer fitting?

38. In what way were the objectionable materials removed from the older type of sewage ejector installation?

39. How would you determine the size of the sump pit?

40. What essential qualities must a sewage ejector installation possess?

41. Why must a sump pit be provided with a fresh air pipe?

42. What objection is there to the use of an automatic water siphon?

43. When does it become necessary to install a grease catch basin?

44. What three methods are used to congeal grease in a grease basin?

45. Why must the inverts of a grease basin be allowed to remain open?

46. How would you determine the size of a grease basin?

Chapter VIII—Storm Drainage

1. Define a storm drain.
2. At what locations may a storm drain be terminated?
3. How are storm drains classified?
4. How must change in direction of the storm drain be made?
5. What elements affect the size of the storm drain?
6. What method is used in determining the size of the storm drain?
7. How must change of direction of the house drain be made?
8. Explain how you would construct a frost-proof curb terminal.
9. Define a roof leader.
10. Name the varieties of roof leaders.
11. Why is it impractical to use a roof leader for a soil pipe?
12. How is the base fitting of a roof leader constructed?
13. What method may be used to determine the size of a roof leader?
14. What type of roof terminal do you consider most practical for a conductor?
15. How may a roof leader be supported?

Chapter IX—Soil Pipe

1. Define the term soil pipe.
2. What is the smallest diameter permissible for a soil pipe?
3. Name the kinds of material which can be used for a soil pipe installation?
4. What advantage does a pipe clamp offer over a pipe rest?
5. What problems occur in sizing a soil pipe?
6. How many units are permissible on a 4-inch soil pipe?
7. What advantage is gained by starting the installation of a soil pipe from a test tee provided at the base fitting?
8. Why must a soil pipe be flashed where it passes through the roof?
9. Explain one method of flashing a soil pipe.
10. Why are roof flashings of the galvanized sheet metal variety impractical?
11. How can the terminal of a soil pipe be protected against frost?
12. What two methods may be used to join a soil pipe to the base fitting of the house drain?
13. How should a soil pipe be supported at its base?
14. Define a soil branch.
15. Why are cleanouts on the soil branch essential?
16. At what intervals should a horizontal soil branch constructed of cast-iron be suspended?
17. How should change of direction of the soil branch be made?
18. In what way may the soil pipe installation be protected against freezing?

Chapter X—Waste Pipe

1. How does a waste pipe differ from a soil pipe?
2. What do you understand by "materials in suspension" in the waste line?
3. How may the plumber add to the efficiency of a waste pipe?
4. In what manner should change of direction of the waste line be made?
5. What are the advantages of a lead waste pipe?

6. Why is a waste pipe extended with excessive grade impractical?

7. What factors were considered in organizing the unit system as a means of determining waste pipe size?

8. How are waste pipes classified?

9. What essential qualities must a sink waste possess?

10. Define a laboratory waste.

11. What is an indirect waste?

12. Name three terminals of an indirect waste.

13. Name three industrial enterprises whose wastes can be discharged into gutters which terminate at an approved receptor.

14. What is the purpose of a 45-degree fitting installed ahead of the drum trap on an underfloor work?

15. What essential qualities should an iron underfloor possess?

16. How would you proceed to make a lead bend?

Chapter XI—Traps Used on Plumbing Systems

1. Define a trap.

2. What do you understand by the term trap seal?

3. What is the depth of a common seal? A deep seal?

4. Is it the depth or the volume of a trap seal that offers most resistance against sewer gases?

5. Why is a mechanically sealed trap impractical?

6. Under what circumstances may a deep-seal trap be used?

7. Name three types of traps which can be used on plumbing installations.

8. Why is an **S** trap impractical?

9. Why are traps with metal partitions objectionable?

10. How much actual resistance does a trap of common seal offer against sewer gases?

11. Why should a trap be self-scouring?

12. On what fixtures are drum traps generally used?

13. What do you understand by the term anti-siphon trap?

14. When may a bell trap be used?

15. Name four traps which are obsolete.

Chapter XII—Ventilation

1. Why must a plumbing system be ventilated?

2. What three major difficulties are encountered in an unvented drainage system?

3. What gases are contained in the atmospheric blanket?

4. What do you understand by atmospheric pressure?

5. Define trap seal loss.

6. Explain trap siphonage and how it may be overcome.

7. Define back pressure, and state where it is likely to occur.

8. Explain trap seal loss by evaporation.

9. What is capillary attraction?

10. How would you account for trap seal loss in a well-ventilated plumbing system?

11. What is a minus pressure? A plus pressure?

12. Are fixture traps installed on flat-bottom fixtures more subject to trap seal loss than those which serve round-bottom fixtures? Explain.

13. What causes retarded flow in a drainage system, and how can it be overcome?

14. What method is used to size a vent pipe?

15. Define the main soil and waste vent.

16. Why must a main soil and waste vent be at least 4 inches in diameter where it passes through the roof?

17. Where a number of main soil vents are joined together, what precaution is necessary?

18. What is the main vent, and what purpose does it serve?

19. Define an individual vent. Why is this type of vent most practical?

20. What essential qualities should a main vent possess?

Chapter XIII—Soil, Waste, and Vent Pipe Principles

1. What two types of plumbing installations are described in the first section of this chapter?

2. When may the minimum class of plumbing be acceptable?

3. Why should a floor drain trap be supplied with water from another fixture?

4. Why should each trap be vented?

5. Why should a 4″ cleanout be installed?

6. Why should a basement cleanout plug be above floor level?

7. Why should there be at least one 4″ vent extending through the roof?

8. What could happen if two closets, one above the other, drained into one vertical stack?

Chapter XIV—Inspection and Test

1. Why must a plumbing system be tested?

2. Why is inspection of the plumbing system by outside authority important?

3. Why is a water test most practical?

4. How would you make a water test?

5. When would an air test be used?

6. When should a smoke test be used?

7. Explain how you would make a smoke test.

Chapter XV—Water Supply

1. Why is an adequate amount of pure water essential?

2. Give a description of the water cycle.

3. How can soil erosion cause water contamination?

4. What objectionable elements are eliminated in the process of water purification?

5. How are obnoxious gases removed from water?

6. In what manner does sand filtration remove bacteria?

7. What chemical compounds are used for coagulation?

8. How may water be disinfected?

9. What two methods are used in the distribution of public water supply?

10. Where are public water supply mains usually located?

11. Why are the ends of the distribution mains cross-connected?

12. When is the indirect method of water distribution used?

Chapter XVI—Materials Used for Water Distribution

1. What determines the type of pipe to be used for water distribution?

2. Name three types of fittings used on corporation pipe.

3. Name four more fittings used on special installations.

4. Of what materials is transite pressure pipe composed?

5. Why are cast-iron screw fittings used for temporary water and steam lines?

6. Why is it considered poor practice to use malleable iron fittings on brass pipe installations?

7. What are the advantages of brass pipe and fittings?

8. Threading is eliminated in what type of pipe material?

Chapter XVII—Joints on Water Supply Systems

1. Why is it essential to make a good joint on a water-supply system?

2. What elements are responsible for failure of a joint?

3. What procedure is followed in securing a good screw joint?

4. How is a horizontal calk joint made? A vertical calk joint?

5. How is a swedge joint made?

6. Name three types of wrenches used by plumbers.

Chapter XVIII—The House Water Supply

1. Define the house service.

2. What essential factors must the plumber observe when constructing a house service?

3. How may the water service be protected from soils of an acid nature?

4. What problems are involved in determining the size of the water service?

5. What do you understand by the term maximum demand?

6. In what way does simultaneous use of plumbing fixtures affect the size of a distribution system?

7. Define friction.

8. Why should the house service be run as direct and free of short offsets as possible?

9. What is the purpose of the corporation stop?

10. Under what circumstances should the curb stop be used?

11. Why is it inadvisable to use the meter stop as a control valve?

12. What is the purpose of a water meter?

Chapter XIX—Cold Water Distribution Systems

1. Define domestic cold water.

2. How are distribution systems classified?

3. Why must basement supply mains be properly graded and aligned?

4. Where should gate valves be used?

5. Where should globe valves be used?

6. What advantage does a globe valve offer over a gate valve?
7. What is the purpose of a drip?
8. Where should drips be installed?

Chapter XX—Pumps and Lifts

1. What is the purpose of a pump in connection with plumbing?
2. Name two kinds of pumps most commonly used.
3. On what principle does a piston pump operate?
4. How close to the source of water supply must a piston pump be located?
5. What is the theoretical lift of a pump?
6. What is the actual lift of a pump, and what causes the variation between actual and theoretical lifts?
7. What distinguishes the double-action pump from the single-action pump?
8. What is the purpose of an air chamber on a piston pump?
9. What is a centrifugal pump?
10. What advantage does a centrifugal pump offer over a piston pump?

Chapter XXI—Cold Water Distribution Systems

1. What two methods are used for the distribution of water in tall buildings?
2. What advantages does a pneumatic system offer over the overhead supply system?
3. What units comprise the pump assembly of the pneumatic air pressure system?
4. On what scientific principle is the pneumatic system operated?
5. What is the purpose of the air compressor?
6. What do you understand by the term pressure range?
7. How much pressure would be required to elevate water 2.31 feet?
8. How does the overhead system differ from the pneumatic system of water supply?
9. What are the disadvantages of an overhead system?
10. How is the level of water in the tank of an overhead system controlled?
11. Where is the logical place to terminate the overflow from the tank?

Chapter XXII—Domestic Hot=Water Supply

1. Define domestic hot water.
2. Why have domestic hot-water systems troubled the plumber?
3. What do you understand by the molecular theory?
4. Does the boiling point of water increase or decrease under pressure greater than atmospheric?
5. What is a storage tank?
6. Why are storage tanks of cylindrical design?
7. What factors determine the size of a storage tank?
8. What do you understand by the term working load?
9. Name five heating devices and the types of buildings which may be served by them.

10. Why is the cold water delivered into the storage tank through a tube terminated close to the bottom of the tank?

11. What factors should be taken into consideration when connecting a heating device to a storage tank?

12. Why is the control of temperature of more importance than the control of pressure in a storage tank installation?

Chapter XXIII—Hot=Water Distribution Systems

1. What two methods are used for the distribution of heated water?

2. On what types of buildings is the upfeed system used?

3. What is the purpose of the circulating return on the upfeed system?

4. What causes the water to circulate within the system?

5. How does the length of one riser affect the other as far as circulation is concerned?

6. Why should the circulating return riser be connected to the feed riser below the highest draw cock?

7. What type of valves should be used on this design of water distribution system? Why?

8. On what types of buildings is the overhead system generally used?

9. Why is the overhead system more efficient than the upfeed system of water supply?

10. Why must the highest point of the overhead system be provided with an air-relief valve?

Chapter XXIV—Private Water Correction

1. In what way do the mineral elements, calcium and magnesium, affect water?

2. What do you understand by temporarily hard water?

3. What chemical change occurs in a water softener?

4. Name the types of water softeners in use today.

5. What chemical compound is used to reactivate a water softener, and what chemical change occurs?

6. What is the purpose of a pressure filter, and on what installation is it used?

7. What material is used as the filtering agent?

8. Explain the working principle of a pressure filter.

Chapter XXV—Cross=Connections

1. Define a cross-connection.

2. Why are cross-connections in a plumbing system objectionable?

3. What two varieties of cross-connections are most common, and what scientific principles are involved in each?

4. In what way can an undersized water supply riser result in a cross-connection?

5. Name two plumbing fixtures that may produce cross-connection where the principle of the siphon is involved. Where gravity is involved.

6. On what principle does a siphon breaker function?

7. What do you understand by the term submerged inlet? On what fixtures is this condition likely to occur?

8. How may cross-connection resulting from a submerged inlet be corrected?

9. Why are cross-connections between steam and water supply dangerous?

Chapter XXVI—Fire Line Installation

1. What benefits are derived from the installation of a fire line?

2. What types of fire lines are in use today?

3. What advantage does one offer over the other?

4. On what types of buildings are wet standpipes generally used?

5. Why is a wet standpipe with a Siamese connection considered most practical?

6. What is the purpose of the Siamese connection?

7. How may cross-connection occur on a fire line installation?

8. Why should all fire lines be painted a color differing from that of the domestic water system?

9. How are buildings located in places where fire-fighting apparatus is unavailable protected against fire?

10. Why must all valves and fittings be of standard design?

Chapter XXVII—Plumbing Fixtures

1. Define a plumbing fixture.

2. Of what materials are plumbing fixtures constructed?

3. What two types of water closets are considered most sanitary?

4. How does a siphon washdown closet operate?

5. How does a siphon jet closet eliminate organic materials?

6. On what types of buildings would you recommend a siphon jet water closet? Why?

7. Why is a washout closet unsanitary?

8. Explain how you would set a modern closet on a tile floor.

9. Explain how a flush valve operates.

10. What constitutes the mechanism of a low closet tank?

11. What is the purpose of a refill tube?

12. At what height is a wash basin generally set?

13. Explain how you would set a wash basin.

14. What type of trap generally is used on a bathtub?

15. What is the purpose of sheet-lead safing under a shower bath?

16. Why should a mixing valve be used as a shower bath water supply device?

17. Explain how you would set a stall urinal.

18. Why should automatic flushing devices be used on a urinal?

19. Why is a trough urinal unsanitary?

20. What precaution must be observed in locating a laundry tub?

FILM SOURCES

Following is a list of films which can be obtained for showing at no cost. It is suggested that at least thirty days notice be given the organization lending the films.

Whenever possible, the instructor should preview the film and prepare 10–20 questions for use as a quiz at its conclusion.

The film sources are listed below with the chapter of the book where they can best be used. The number appearing in parentheses after the film title (under the heading WHERE OBTAINABLE) refers to the number appearing before the following film sources:

(1) The National Association of Plumbing Contractors
1016 20th Street, N. W.
Washington 6, D. C.

(2) Modern Talking Picture Service
140 E. Ontario Street
Chicago, Illinois

(3) Revere Copper & Brass, Inc.
230 Park Avenue
New York 17, New York

(4) Sales Manager
Red Jacket Mfg. Co.
Davenport, Iowa

CHAPTER	FILM TITLE	WHERE OBTAINABLE
I	A. Municipal Sewage Treatment Process	(1)
	B. Sewers, Guardians of Community Health	(1)
II	A. Modern Pipe for Modern Living	(1)
III	A. Permanent Investment	(1)
	B. Pipe Schemes	(1)

Note: Films listed for A & B, Chapter 3, may also be used for Chapters 4, 5, & 6.

VII	A. Toward a Uniform Plumbing Code	(1)
	B. Flow of Water in Plumbing Drains	(1)

INDEX

Fire lines   325
Sizing Water   235
Vent Size   171
Sizing waste pipe 113, 114, 66